Wild Plants
of the Sierra Nevada

T0289513

Wild Plants of the *Sierra Nevada*

Ray S. Vizgirdas and Edna M. Rey-Vizgirdas

Illustrations by Edna M. Rey-Vizgirdas

University of Nevada Press
RENO & LAS VEGAS

University of Nevada Press, Reno, Nevada 89557 U S A

www.unpress.nevada.edu

Copyright © 2006 by University of Nevada Press All
rights reserved

Manufactured in the United States of America

Design by Cameron Poulter

Library of Congress Cataloging-in-Publication Data

Vizgirdas, Ray S., 1960–

Wild plants of the Sierra Nevada / Ray S. Vizgirdas and
Edna M. Rey-Vizgirdas.

p. cm.

Includes bibliographical references and index.

I S B N 0-87417-535-6 (hardcover : alk. paper)

1. Botany—Sierra Nevada (Calif. and Nev.) 2. Plants—Sierra Nevada
(Calif. and Nev.)—Identification. I. Rey-Vizgirdas, Edna. II. Title.

Q K 142.7.V59 2005

581.9794′4—dc22

2005004867

The paper used in this book meets the requirements of American
National Standard for Information Sciences—Permanence of Paper
for Printed Library Materials, A N S I Z.48-1984.
Binding materials were selected for strength and durability.

This book has been reproduced as a digital reprint.

ISBN 978-0-87417-789-3 (paperback : alk. paper)

For Tomas,
from Mom and Dad

We are losing our ancestral knowledge because the technicians only believe in modern science and cannot read the sky.

—Andean peasant expression

Contents

Illustrations

I wrote this book because, after thirty years of exploring the Sierra Nevada and its forests as a student, teacher, professional naturalist, and field biologist, I believed that I should impart some of my own enjoyment and appreciation of these mountains to all those—young and old, amateur and professional—who want to learn more about the richly diverse wild plants of this important mountain range and the exceptionally wide variety of plant habitats enfolded within its expanse. My objective is to provide a comprehensive inventory of all the wild plants endemic to the range—more than 400 plant taxa, over half of which are considered rare. I also provide information that will help readers identify individual plant species, as well as information on the traditional and modern uses of Sierra plants—for example, as food or beverage, medicine, tools, or containers.

Additionally, I want to guide readers to additional information about Sierra plants, and to this purpose I provide an extensive list of suggested readings and other sources of information about the plants, their habitats, and their uses. To encourage readers to practice conservation and to respect the flora they are observing, I identify endangered plants when they are described in the text and I list them in one of the appendices.

My interest in the outdoors started when I was young and demanded to be told the names of all the plants and animals I saw within and around the camp-sites during family vacations. My curiosity only grew with time as I acquired the skills and knowledge that enabled me to live comfortably in the mountain environment. This ability eventually grew to the point where I was able to explore the Sierra Nevada (and northern Rockies) for weeks and months at a time, sustained by my knowledge of plant (and animal) uses. I also learned to respect these mountains and their fragile resources, and I always had a conservation ethic foremost in my mind as I was learning and practicing my outdoor skills. I understood early on that to wantonly destroy plants, animals, or other resources is unwarranted, as no animal or plant that lives in the mountains destroys the resources it depends on. These lessons I continue to apply to my present life.

The introduction offers a brief overview of the Sierra Nevada's biological, geological, and cultural significance; introduces the reader to the science of ethnobotany; and gives some insight into the nutritional value of wild plants. It also takes you on a trek across the Sierra Nevada to examine the many and extremely diverse plant habitats and communities found in this range. The heart of the book, "Major Plant Groups," surveys the major vascular plant groups, including ferns and their allies, gymnosperms, and flowering plants, both dicots and monocots. Within each group, the families (and genera and species therein) are arranged in alphabetical order. Information in these groupings includes a brief description of the plants, noting any special characteristics and any interesting natural history information, as well as their uses—be it traditional uses by the Native American people who once made the Sierra Nevada their home, or more contemporary uses, or both. The book also includes a number of "quick keys" to help readers differentiate between similar species.

Appendix 1 of the book has been created to make the book more useful for people in the field. It is a table that identifies the major plant communities of the range (from west side to east side) and lists the plants commonly associated with them. Once you've chosen a plant about which you want to learn more, simply refer back to "Major Plant Groups" to read more about it. Appendix 2 provides a list of rare and endangered plants found in the Sierra Nevada. Readers should be aware of these plants and make efforts to protect and conserve them during excursions to the mountains. The indices in appendix 3 are designed to allow the reader to locate information about Sierra Nevada plants by cross-referencing their common and scientific names. There is also a glossary defining many of the common terms used in botany, ethnobotany, and medicine.

The book includes eighty-two line drawings of all the major plant species, intended to help readers identify plant specimens. They were done by my wife, Edna Rey-Vizgirdas.

Albert Einstein once said that he never had an original idea. Rather, his ideas were built upon the ideas of others. In much the same way, this book is not an original idea but rather an assemblage of information that has been passed down to us from people of many previous generations, including the Native Americans who inhabited these mountains for millennia and the European and Asian pioneers who settled there later. All that I have done is gather what information I could on plants of the Sierra Nevada and make it available here. Therefore, I first acknowledge those that came before me, who in a sense had to learn by trial and error which plants were useful and which were not. Additionally, I would like to thank the University of Nevada Press for the opportunity to make this book a reality. There were many reviewers who provided thoughtful and constructive criticisms that greatly improved its content and presentation. To all of you, a very sincere thank-you. However, I accept responsibility for any errors.

Edna Rey-Vizgirdas

My interest in the environment began early in childhood. When we first visited the Sierra on a camping trip over thirty-five years ago, I was enthralled. The view of the majestic Minarets took my breath away, and I promised myself I would someday move to the mountains, which I did as soon as I graduated from high school. I went on to live in June Lake, Lake Tahoe, and Sequoia National Park working as a park ranger, researcher, and ski instructor. Spending time outdoors continues to provide me with inspiration, enjoyment, and fulfillment. For me, it is an ongoing practice of meditation—being aware of my surroundings in every moment, with each step and each breath, whether I'm in a forest, sun-drenched meadow, or snowstorm—and realizing that life has unlimited possibilities. It is in this spirit that I dedicate this book to all those who find hope and beauty and peace in the wilderness.

Introduction

*The greatest service which can be rendered to any country
is to add a useful plant to its culture.*

—Thomas Jefferson (1790), on introducing upland rice to America

THE SIERRA NEVADA

The Sierra Nevada is one of the most magnificent mountain ranges in the world. This snowy and sawtooth range is high in the East, low in the West, and was born of convulsions beneath the earth. Since its birth, the range has been finely molded by wind, water, fire, ice, and time.

In Spanish, *Sierra Nevada* means "snowy range." It was in 1776 that Padre Pedro Font on the second Juan Bautista de Anza expedition gave that name to the mountains that were seen in the distance to the east of San Francisco. Another commonly used nickname is the Range of Light, which was first used by John Muir in 1894 in his book *The Mountains of California*:

> Looking eastward from the summit of Pacheco Pass one shining morning, a landscape was displayed that after all my wanderings still appears as the most beautiful I have ever beheld. At my feet lay the Great Central Valley of California, level and flowery, like a lake of pure sunshine, forty or fifty miles wide, five hundred miles long, one rich furred garden of yellow Compositae. And from the eastern boundary of this vast golden flower-bed rose the mighty Sierra, miles in height, and so gloriously colored and so radiant, it seemed not clothed with light but wholly composed of it, like the wall of some celestial city. . . . Then it seemed to me that the Sierra should be called, not the Nevada or Snowy Range, but the Range of Light. And after ten years of wandering and wondering in the heart of it, rejoicing in its glorious floods of light, the white beams of the morning streaming through the passes, the noonday radiance on the crystal rocks, the flush of the alpenglow, and the irised spray of countless waterfalls, it still seems above all others the Range of Light. (1)

Today, millions of different people come to these mountains. There are skiers who seek the speed of snow-covered slopes, backpackers and hikers who kick up dust as they advance to some distant lake or peak, and naturalist types who take the time to count the number of petals and stamens of wildflowers in meadows and who watch for familiar and new species of birds, butterflies, and mammals. There are also wilderness connoisseurs who search for nothing more than the peace, quiet, and solitude these mountains have to offer. However, to the Native people of the past, these mountains meant something more.

The Sierra Nevada is the largest continuous mountain range in the United States and lies almost entirely within the state of California, extending into Nevada only on the eastern side near Lake Tahoe (page 2). The range is more than four hundred miles long and between sixty and eighty miles wide, and it extends in a southeasterly direction from about Lake Almanor in the North to approx-

THOREAU VERSUS MUIR IN THE HIGH SIERRA
In examining the lives of Henry David Thoreau and John Muir, there are those who believe that Thoreau might not have made it in the high county of the Sierra Nevada. Consider the following: he could not have built a cabin for twenty-eight dollars, grown vegetables, taken odd jobs, and gone for ramblings and contemplative walks in the woods. To survive in the High Sierra requires considerable effort.

John Muir, on the other hand, was the quintessence of adaptation to the range. For food, he took a loaf of bread and sometimes tea. To sleep, he used no blankets but kept the campfires burning throughout the night. He drank simply from the streams or sucked on snow, and for shelter he did what most of the denizens of the mountains did: he crawled under the lower branches of a pine. Without anything to slow him down, he went where he pleased, and survival was no problem.

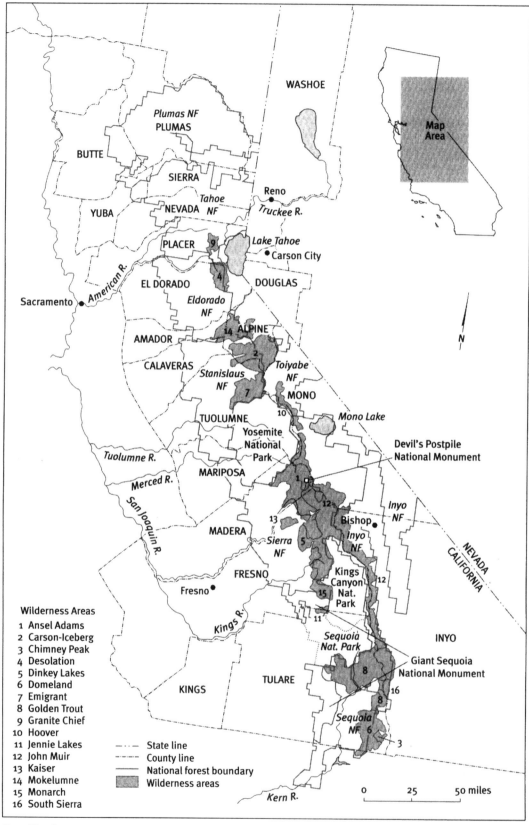

WASHOE

Plumas NF
PLUMAS

BUTTE

SIERRA

YUBA   NEVADA   Tahoe
                          NF

                    Reno
                    Truckee R.

PLACER   9   Lake Tahoe
                       Carson City
              4

EL DORADO   DOUGLAS

              Eldorado
              NF

Sacramento   American R.

AMADOR   14   ALPINE

CALAVERAS   2
            Stanislaus   Toiyabe
            NF   NF
                 7
                          MONO
TUOLUMNE   10

                    Mono Lake

Yosemite
National
Park   Devil's Postpile
                            National Monument

Tuolumne R.

Merced R.   MARIPOSA

                          1   Inyo
San Joaquin R.   12   NF

            MADERA   Bishop
            13   Inyo
                    Sierra   NF
                    NF   5

FRESNO   Kings
              Canyon
              Nat.   12
              Park

Fresno   15   NEVADA
                               CALIFORNIA
              11

         Kings R.   Sequoia   INYO
                     Nat. Park
                              Giant Sequoia
                     TULARE   National Monument

Wilderness Areas   8
 1 Ansel Adams
 2 Carson-Iceberg   KINGS   16
 3 Chimney Peak   8
 4 Desolation
 5 Dinkey Lakes   Sequoia
 6 Domeland   NF   6
 7 Emigrant   3
 8 Golden Trout
 9 Granite Chief
10 Hoover
11 Jennie Lakes   —··— State line
12 John Muir   —·—   County line
13 Kaiser   ——   National forest boundary
14 Mokelumne   ▓▓   Wilderness areas
15 Monarch
16 South Sierra   Kern R.

                    0   25   50 miles

N

Map
Area

Sierra Nevada

imately Tehachapi Pass, located southeast of Bakersfield. To the early settlers and travelers, the Sierra Nevada was an impediment to communication and progress and was of little interest until the discovery of gold in the foothills in 1848. Today, there are three national parks (Yosemite, Sequoia, and Kings Canyon), one national monument (Devil's Postpile), nine national forests, and numerous state parks (Whitney 1979; Barbour et al. 1993).

At its foundation, the Sierra Nevada is an enormous deposit of granitic rock whose exposed slopes are readily visible at the crest of the range. The gradual west slope rising from the expansive Central Valley to the Sierra crest is dissected by deep, west-trending river canyons. At the eastern edge of the uplift, the high peaks dominate the uppermost elevations, forming rolling highlands in the North—with elevations mostly less than 9,000 feet—and expansive, highly dissected mountains in the broad southern alpine zones, where Mount Whitney (the highest peak in the contiguous forty-eight states) rises to 14,495 feet. The range ends abruptly at the eastern escarpment, dropping with a shallow gradient in the North but in the South plunging more than 10,000 feet from the Sierran crest to the floor of the Great Basin. In all, there are about eleven peaks that top 14,000 feet and an additional five hundred peaks above 12,000 feet. Some sixty glaciers can also be found.

With so much topographic relief, the Sierra Nevada possesses a high diversity of plant species, and many species are endemic (restricted) to the range. In fact, because the range represents about 20 percent of the California land base, it also contains more than 50 percent of the state's flora. Of this, approximately 405 plant taxa are endemic to the Sierra Nevada, and 218 taxa are considered rare by conservation organizations, state and federal agencies, or both (Hickman 1993).

Plants as Food and Medicine
A Word about Ethnobotany

As this book explores the uses of Sierra Nevada plants, we provide a brief overview to the science of ethnobotany. In the 1800s, there were a number of scientists actively studying all forms of the plant world that the primitive or aboriginal people used for medicine, food, and textile fabrics. As early as 1874, this was referred to as "aboriginal botany," and this term persisted for a number of years. Then in 1895, a professor of biology, John W. Harshberger, initiated the fashion of using the prefix *ethno* to indicate the study of local people's natural history (Powers 1874; Harshberger 1896). Therefore, ethnobotany is the study of "plants used by primitive and aboriginal people" (Harshberger 1896, 146). Other definitions for ethnobotany include "the study of the interrelationships of primitive man and plants" and "the study of plants in their relations to human culture." In a real sense, ethnobotany is a science born in the United States. The following list shows how the meaning of ethnobotany has changed from the mid-1800s to the present.

Interpretations of Ethnobotany

DATE	THE INTERPRETATION OF ETHNOBOTANY
1873	*Aboriginal Botany*—the study of all forms of vegetation that aboriginal people used for commodities such as medicine, food, textiles, and ornaments (Powers 1874)
1895	*Ethnobotany*—the use of plants by aboriginal peoples (Harshberger 1896)
1916	Not just a record of plant use, but the traditional impressions of the total environment as revealed through custom and ritual (Castetter 1944)

1932	Not only tribal economic botany but also the whole range of traditional knowledge of plants and plant life (Gilmore 1932)
1941	The study of relations that exist between humans and their ambient vegetation (Castetter 1944)
1941	The study of the interrelations between "primitive" humans and plants (Castetter 1944)
1981	The study of direct relationships between humans and plants (Ford 1978)
1990	The study of useful plants prior to commercialization and eventual domestication (Wickens 1990)
1993	The recording and evaluation of environmental knowledge that different cultures have accumulated throughout millennia (FEB 1993)
1994	All studies (concerning plants) that describe local people's interaction with the actual environment (G. J. Martin 1995)

Adapted from Cotton 1996

Before the term *ethnobotany* arose in 1895, the study of traditional botanical knowledge centered on the applications and economic potential of plants used by Native peoples. However, during the first part of the 1900s, anthropological and ecological aspects became more important, and later, as ethnobotanical studies intensified in the 1980s, a whole spectrum of interpretations appeared as many other disciplines become involved in "ethnobotany," ending with a reinterpretation in 1995 (adapted from Cotton 1996).

An ethnobotanist, in turn, explores how plants are used as food, shelter, medicine, and clothing; in hunting; and in religious ceremonies. The roots of ethnobotany lie within botany, and botany originated, in part, from an interest in finding plants to help fight illness. Until very recently, medicine and botany were always studied together, and many of the drugs we have today were derived from plants. Because ethnobotany is a multidisciplinary science, to become an ethnobotanist, a person's interest in botany and ecology is supplemented by such disciplines as linguistics, anthropology, chemistry, sociology, and pharmacology. The following list highlights some of the various fields of study in ethnobotany today.

Some Areas of Modern Ethnobotanical Studies

FIELD	MAIN AREAS OF STUDY
Ethnoecology	Traditional knowledge of plant phenology, adaptations, and interactions with other organisms. Natural and environmental impact of traditional vegetation management.
Traditional Agriculture	Traditional knowledge of crop varieties and agricultural resources. Natural and environmental impact of crop selection and crop management.
Cognitive Ethnobotany	Traditional perceptions of the natural world (through the analysis of symbolism in ritual and myth) and their ecological consequences. Organization of knowledge systems (through ethnotaxonomic study).
Material Culture	Traditional knowledge and use of plants and plant products in art and technology.
Traditional Phytochemistry	Traditional knowledge and use of plants for plant chemicals (for example, pest control and traditional medicine).
Paleoethnobotany	Past interactions of human populations and plants based on the interpretation of archaeobotanical remains.

Ethnobotany is a diverse field of study that examines all aspects of the relation-ships between plants and traditional peoples. As such, it is interdisciplinary and draws from a wide range of subject areas (adapted from Cotton 1996).

For most of the one to two million years that humans have existed, they lived in much closer association with the natural world than now. In fact, day-to-day survival required an entirely different set of skills and knowledge than we have today, which included detailed knowledge of plants, animals, location of water, and other natural resources. Then, over the years (and many generations) and through trial and error, humans maintained a repertoire of edible, useful, and medicinal plants. It is because of these experiments that we have information on the uses of many plants. Unfortunately, today most people are unaware of these riches and have become dependent on only a few "domesticated" species of plants. Those who seek out wild plants find a continual source of stimulation and enjoyment, particularly as it relates to the survival of one's spirit.

Today, we still interact with many plants, but in more indirect ways. However, there are still cultures that maintain strong connections with their local plants, and there are still people involved in recording this knowledge for the benefit of those who have preserved the information as well as for the benefit of all human-ity. The plant kingdom continues to provide treatments for illnesses in the world. For example, a number of species have been useful in the treatment of certain types of cancer: taxol is the active compound from *Taxus brevifolia* used in the treatment of ovarian cancer, and colchicine from crocus is used in chemotherapy. Other plants such as periwinkle (*Catharanthus roseus*) are being used for Hodg-kin's disease and leukemia, milk thistle (*Silybum marianum*) for liver disease, and foxglove (*Digitalis*) for heart disease (Lewis and Elvin-Lewis 1977; Duke 1992a). In some cases, the curative substance is available only from the plant, and the demand for the substance can be satisfied only by collecting it in the wild or by growing the plants as crops.

FARMING
Farming has often been blamed for materialism in human society because it gave people something worth hoarding and squabbling about—food. It also allowed the start of commerce, with some people living by providing goods and services, and others by growing food.

Nutrition and Seasonality of Plants

Wild plants are a good source of vitamins and minerals. In fact, much of the medicinal value originally associated with many wild plants was due simply to their high vitamin and mineral contents. Wild plants can also provide proteins, carbohydrates, fats, vitamins, and minerals that are needed for good mental and physical condition (C. E. Smith 1973; Elias and Dykeman 1982). Certain amino acids, however, can be obtained only from animal products (for example, meats, milk, cheese, and eggs). The following summarizes some of the important vita-mins and minerals that are needed for good health.

SOME IMPORTANT VITAMINS AND MINERALS FOR HEALTH
Vitamins

These are organic compounds that are necessary in small quantities to prevent dis-ease and help regulate the body's biochemical processes. Prolonged excessive doses of vitamins A, D, and K can have toxic effects. In addition to the vitamins listed below, biotin, choline, folic acid, and pantothenic acid are also essential nutrients.

Niacin   (Nicotinic Acid) A vitamin of the B complex, niacin occurs in both plant and animal tissues in various forms. In the body, niacin from plants is changed to niacinamide for use. Niacin takes part in enzyme reactions in-volved in the production of body energy and tissue respiration. Pellagra is a niacin-deficiency disease. Niacin is water soluble and is not sensitive to heat, acids, or alkali.

Vitamin A   Vitamin A is not found in plants but rather is manufactured by animals from pigments called carotenes that are common in plants. Vitamin A is essential for night vision and promotes healthy skin and mucous membranes. It is also important for bones and teeth, proper digestion, and the production of red and white blood cells. It is fat soluble and sensitive to oxygen.

Vitamin B_1   (Thiamine) Found in both plant and animal tissues, thiamine is important for the body's production of energy through the breakdown of carbohydrates. It appears to be important for normal functioning of the nervous system and is involved in the action of the heart. Vitamin B_1 is water soluble and sensitive to heat. Most plants contain trace amounts.

Vitamin B_2   (Riboflavin) Riboflavin usually occurs in the same foods as vitamin B_1. It is essential for cell growth and enzymatic reactions by which the body metabolizes proteins, fats, and carbohydrates. Vitamin B_2 is water soluble and sensitive to light.

Vitamin B_6   (Pyridoxine) B_6 is still a relatively little-known vitamin. It participates in many enzymatic reactions and is particularly important for brain and nervous system function. Vitamin B_6 is water soluble and sensitive to oxygen and ultraviolet light.

Vitamin B_{12}   (Cyanocobalamin) Little or no B_{12} is found in plants. Strict vegetarians sometimes suffer from pernicious anemia, a disease associated with B_{12} deficiency. Vitamin B_{12} is necessary for proper functioning of cells, especially in the nervous system, bone marrow, and gastrointestinal tract. It is involved in the metabolism of fats, proteins, and carbohydrates. Vitamin B_{12} is water soluble and sensitive to light, acids, and alkalis.

Vitamin C   (Ascorbic Acid) Vitamin C occurs in almost all plants to some degree. Because our bodies cannot make or store vitamin C, a continuous supply must be present in the food we eat. Body cells require vitamin C for proper functioning, as does the formation of healthy collagen (the basic protein of connective tissue), bones, teeth, cartilage, skin, and blood vessels. Vitamin C also promotes the body's effective use of other nutrients such as iron, B vitamins, vitamins A and E, calcium, and certain amino acids. By promoting the formation of healthy connective tissue, vitamin C helps to heal wounds and burns. Stress, fever, and infection tend to increase the body's need for vitamin C. A deficiency of vitamin C is called scurvy. Vitamin C is water soluble and is sensitive to air, heat, light, alkalis, and copper ware.

Vitamin D   Vitamin D does not occur in plants. However, some plants contain compounds called sterols, which when irradiated with ultraviolet light make Vitamin D. Vitamin D is necessary for healthy bones and teeth, for proper assimilation of calcium and phosphorus, and in preventing rickets. It is a fat-soluble vitamin that is not sensitive to heat, light, or oxygen.

Vitamin E   (Tocopherol) Vitamin E is found in both plant and animal tissues. It is an antioxidant, acting to protect red blood cells, vitamin A, and unsaturated fatty acids from oxidation damage. It also helps maintain healthy membrane tissue. In laboratory experiments, it was found to be necessary for fertility in rats. Vitamin E is fat soluble and is sensitive to oxygen, alkali, and ultraviolet light.

Vitamin K   Though vitamin K occurs primarily in plants, it is also synthesized by bacteria found in the small intestine. It is necessary for the liver's synthesis of the blood-clotting enzyme prothrombin. Vitamin K is fat soluble and is sensitive to light, oxygen, strong acids, and alcoholic alkalis.

Minerals

These are chemical elements necessary for proper functioning of the body. Most are obtained from the foods we eat. There are two groups of minerals: *macrominerals* and *microminerals*. Macrominerals are found in relatively large amounts in the body, whereas microminerals are found in smaller amounts. Following is a list of minerals known to be necessary in human nutrition. There are other minerals, but their functions are not clearly understood.

MACROMINERALS

Calcium	Calcium is the most abundant mineral in the body. It occurs in plants, dairy products, and seafood. Calcium is necessary for healthy bones and teeth, clotting of blood, the functioning of nerve tissue and muscles (including the heart), enzymatic processes, and controlling movement of fluids through cell walls.
Chlorine	As a gas chlorine is poisonous, but in the form of chloride compounds, it is an essential mineral. It acts with sodium to maintain the balance between fluids inside and outside the cells. Gastric juices in the stomach contain hydrochloric acid, the production of which requires chloride. Table salt (NaCl) is our main source.
Magnesium	Found in both plant and animal tissues, magnesium is essential as an enzyme activator and is probably involved in the formation and maintenance of body protein.
Phosphorus	Occurring in plant and animal tissues, phosphorus takes part in the production of energy for the body and is second only to calcium as a constituent of bones and teeth. Phosphorus is necessary for metabolic functions relating to the brain and nerves, as well as for muscle action and enzyme formation.
Potassium	Potassium is abundant in plant and animal tissues. It promotes certain enzyme reactions in the body and acts with sodium to maintain normal pH levels and balance between fluids inside and outside the cells.
Sodium	A common mineral in plants and animals, sodium regulates the volume of body fluids, and when balanced with potassium, it helps maintain cell-fluid equilibrium. It is also necessary for nerve and muscle functioning. The ideal amount can be obtained through a diet of vegetables such as dandelion greens, spinach, mustard greens, watercress, and carrots.
Sulfur	Sulfur supply comes from sulfur-containing amino acids and from the B vitamins thiamine and biotin. Its main sources are dairy products, meats, legumes, nuts, and grains. Sulfur is involved in bone growth, blood clotting, and muscle metabolism. It also helps to counteract toxic substances in the body by combining with them to form harmless compounds.

MICROMINERALS

Copper	Copper is found in plant and animal tissues and is essential (with iron) for formation of hemoglobin in red blood cells. Copper is also important for protein and enzyme formation, as well as for the nervous and reproductive systems, bones, hair, and pigmentation.
Iodine	Iodine's only dependable sources are seafood and seaweeds. Other plants will contain iodine if grown in iodine-rich soils. It is necessary for normal physical and mental growth and development, as well as for lactation and reproduction. An iodine deficiency is called goiter.
Iron	Iron occurs in plant and animal tissues. The body retains iron very well, and only trace amounts are needed in the diet. Iron is essential to form the oxygen-carrying hemoglobin in red blood cells and is involved in muscle function.
Manganese	Plants are the best source for manganese. Trace amounts are necessary for healthy bones and for enzyme reactions involved in energy production.
Zinc	Zinc is found primarily in animals but also occurs in plants growing in good soil. It is important for various enzyme reactions, the reproductive system, and for the manufacture of body protein.
	Adapted from Lust 1987

There are nine categories of plant foods discussed in this book. They include root vegetables; green vegetables; fleshy fruits; seeds, nuts, and grains; inner bark; and flowers. Plant food categories in the following list show the approximate number of species providing food within each category in the Sierra Nevada. Many plants have multiple food uses.

PLANT FOOD CATEGORY	APPROXIMATE NUMBER OF SPECIES
Roots (roots, bulbs, corms, rhizomes, and tubers)	247
Greens (stems, leaves, shoots, and buds)	676
Fleshy Fruits	387
Seeds, Nuts, and Grains	349
Inner Bark, Cambium, and Sap	246
Flowers	137
Sweetening Agents	69
Beverages (tea and juice)	92
Miscellaneous (flavoring, casually edible, and gum)	147
TOTAL	2,350

Root vegetables (that is, tubers, corms, bulbs, rhizomes, and true roots) include plants such as wild onions, blue camas, spring beauty, lilies, bitterroot, and balsamroot. Roots are the storage organs high in carbohydrates. The greatest amount of energy from roots is available at the end of the growing season. These carbohydrates come in a variety of forms and flavors and are not always readily digestible by humans. One type of carbohydrate found in some roots is inulin, which becomes sweet after cooking due to its conversion to fructose. If the skin of a plant's root is consumed, it can provide minerals and a small amount of vitamins.

Green vegetables include leaves, stems, shoots, and buds. Examples are fireweed (shoot and stem), lamb's-quarters, nettles, and mustards (leaf). Many green vegetables are most palatable and digestible when they are young. Green vegetables are high in moisture and often contain carotene, vitamin C, folic acid, and various minerals (for example, iron, calcium, and magnesium).

Fleshy fruits include serviceberry, gooseberries, currants, huckleberries, wild plums, cherries, and rose hips. Fleshy fruits are a good source of ascorbic acid and contain high amounts of other nutrients such as calcium, vitamin A, and folic acid.

Seeds, nuts, and grains are good sources of protein, fat, carbohydrates, vitamins, and minerals. Oils can also be rendered from these foods. Nuts are especially good sources of B vitamins, amino acids, and iron (Doebley 1984).

CARBOHYDRATES, FATS, AND PROTEINS
The body's primary source of energy is glucose, a carbohydrate. Fats produce more energy for each gram consumed than carbohydrates. Fats and carbohydrates are the best sources of energy. Protein is the least-preferred source of energy because it has to be extensively metabolized by the body to make glucose.

The cambium or inner bark of coniferous and deciduous trees and shrubs is another category of plant foods. The inner bark may be scraped off trees in the spring. Many species have a high sap content. For example, maple sap is high in carbohydrate- or sugar-energy value for an inner bark food (Gottesfeld 1992).

The final category of plant foods are the flowers. Rose petals, fireweed flowers, and mariposa lily buds are high in moisture. Flowers are low in proteins and fats, but some are rich in vitamin A (carotene) or vitamin C. There is little published information on the mineral content of flowers.

The nutritional value of plants changes with the seasons. During spring and summer, many plants are tender and rich in vitamins. Roots and tubers are high in carbohydrates and other nutrients. But as summer progresses, roots become less desirable because the stored energy is shifted to the aboveground parts. Fall is a time of nuts and berries, which provide a good source of protein. Roots again begin to store carbohydrates. Winter, however, can be bleak. The aboveground edibles may be limited to berries that have persisted into winter, bark and pine needles for teas, and inner bark. Teas can be restorative and do provide some food value. Teas can be upgraded into stews by adding insect larvae, birds, or mammals to make them more nutritious and sustaining.

The nutritional value of plants also depends on preparation methods. For example, cooking greens in two changes of water makes them more palatable but can reduce the nutritional value. Generally, the preferred order of preparation for plants foods is: raw, quick-cook or steamed, baked, then boiled. Frying is the least desirable cooking method because it destroys many useful vitamins and minerals.

Active Principles of Medicinal Plants

The medicinal value of plants is due to the presence in the plant tissue of a chemical substance—an active principle—producing a physiological effect. Many of the active principles are highly complex, and their precise chemical nature is still unknown. However, some have been isolated, purified, and even synthesized or simulated. Active principles commonly fall into one of six categories: alkaloids, glycosides, essential oils, gums and resins, fatty oils, and antibiotic substances (Hocking 1949; Lewis and Elvin-Lewis 1977; Duke 1992b; Chatfield 1997).

Alkaloids are a diverse group of alkaline compounds with marked physiological activity and always contain nitrogen. Alkaloids include morphine, cocaine, nicotine, quinine, plus more than five thousand others. The plant families richest in alkaloids include the Solanaceae, Fabaceae, Rubiaceae, Liliaceae, and above all Apocynaceae.

Glycosides are compounds that, when hydrolyzed, produce a component of one or several sugars (that is, glycone) and a nonsugar component (aglycone). The various types of glycosides are classified by their aglycone part. Among the most important glycosides in modern medicine are the cardiac glycosides that are found in the dogbane (Apocynaceae), milkweed (Asclepiadaceae), lily (Liliaceae), buttercup (Ranunculaceae), and figwort (Scrophulariaceae) families. Digitalis is a widely prescribed drug of plant origin and owes its activity to cardiac glycosides. Additionally, cyanogenic glycoside yields hydrocyanic acid and is reported from some two thousand species of flowering plants.

Essential oils usually have various chemical constituents, oftentimes terpene derivatives or aromatic compounds. They rarely consist of a single constituent but often contain alcohols, ketones, aldehydes, phenols, ethers, esters, and other compounds, as well as sometimes nitrogen and sulfur. Many are highly germicidal, a property owing to their ability to penetrate into protoplasm. However, they are usually too insoluble in water to be important in medicine as antiseptics. They are valuable as carminatives in cough drops, mouthwashes, gargles, sprays, and healing ointments.

Gums are polymers of various rarer sugars, and *resins* are oxidation products of essential oils. Both are used as purgatives and in ointments. *Fatty oils* (lipids) are used in emulsions and as purgatives. *Antibiotic substances* are various complex organic compounds, usually from molds, actinomycetes, and bacteria that are capable in small amounts of inhibiting life processes of microorganisms.

DOCTRINE OF SIGNATURES
In ancient times, it was thought that if a plant part was shaped like, or in some other way resembled, a human organ or disease characteristic, then that plant was useful for that particular organ or ailment. This concept has no scientific basis, though sometimes uses conceived centuries ago have persisted and may even have been corroborated by scientific evidence of their efficacy.

A couple of examples of this doctrine would be that kidney beans were good for the kidneys, and the leaves of Arrowhead (*Sagittaria* spp.) would be useful for wounds caused by arrows.

THE NATURAL HISTORY AND ECOLOGY OF SIERRA NEVADA PLANTS

Sierra Nevada Forests and Plant Communities

To better facilitate the learning of useful plants, it is helpful to know something about plant communities and plant distribution. For example, if you found yourself in a forest dominated by ponderosa pine (*Pinus ponderosa*), then you will also know that there are often a great many other plant species associated with this forest type. Each species has its specific distribution based on a variety of requirements. Appendix 1 provides a list of the habitats within the Sierra Nevada and associated useful plant species.

Life-Zone Concept

One of the earliest attempts to explain the distribution of plants (and animals) was introduced by C. Hart Merriam, chief of the U.S. Biological Survey. He attempted to correlate climatic conditions, more specifically temperature, as the most important factor in fixing the limits beyond which a particular species of plant or animal cannot go. Merriam's life-zone system, or life zones as they are commonly referred to, found great favor with biologists in the early years of the nineteenth century and was widely used. It is not uncommon to still see some field guides today using the terminology developed by Merriam. However, the limitations to this scheme were quite evident.

Climate of the Sierra Nevada

The present-day climate of the Sierra Nevada is dominated by a Mediterranean pattern of a cool, wet winter followed by a long dry period in summer. Temperature and precipitation help in determining to a great extent the kinds of plants and animals that live at different elevations within the Sierra Nevada. In general, temperature decreases about one degree Fahrenheit for each three hundred–foot rise in elevation, and as it cools the air drops its moisture as rain or snow. Annual precipitation tends to increase by approximately one inch for every one hundred–foot rise in elevation, reaching a maximum at about fifty-five hundred feet on the central western slope of the range. Most precipitation falls in the form of snow at the higher elevations and may persist in areas well into the summer months. The elevation at which maximum precipitation is higher in the southern portion of the Sierra Nevada is at about six thousand feet, whereas in the northern Sierra it is at about four thousand feet. What is generally seen with an increase of one thousand feet in elevation are changes similar to if one was moving three hundred miles northward. Because more precipitation occurs on the western slope of the range, the eastern slope is drier and is often referred to as being in the *rain shadow* of the Sierra Nevada (Whitney 1979).

Plant Communities

Patterns of plant distribution are controlled not only by changes in elevation but also by factors such as the availability of moisture during the growing season and the amount of snow accumulation during winter. Plant communities of warm south-facing slopes can differ markedly from cooler north-facing slopes that lie just across a valley. Soil type also affects plant distribution. For example, soils that have developed on glacial till, as an example, often harbor different species than soils formed from decomposed granite (Whitney 1979; Barbour et al. 1993).

In the late 1940s, two botanists, P. A. Munz and D. D. Keck, were developing a new flora for California and desired a more precise classification to describe the distribution of plants than that provided by Merriam. In 1949 they published "California Plant Communities," which divided the state into five biotic provinces that were further divided into eleven vegetation types. Within the eleven vegetation types, they recognized twenty-four plant communities, including valley grassland, foothill woodlands, chaparral, montane coniferous forests, alpine, sagebrush, and pinyon-juniper woodland. More recently, *The Jepson Manual: Higher Plants of California* took a slightly different and more convenient approach to dividing the state. The area covered by this book is primarily the Sierra Nevada and east of the Sierra Nevada (in part) (Hickman 1993).

Trek across the Sierra

As is already evident, mountains have a major influence on climate. In general, they force moisture-laden air from the prevailing winds upward along their steep slopes. As the air rises, it cools, and the moisture condenses into rain or snow. Thus, the western Sierra Nevada slopes are cool and wet, whereas the eastern slopes are relatively dry. Due to changes in elevation and the rain-shadow effect, mountains provide habitat for considerably more kinds of plants (and animals) than the surrounding lowlands. Although there are many more plant communities and species to be encountered on the western slope of the Sierra Nevada, this book does provide information for as many species as possible on both sides of the range.

For a closer look at the plant communities within the range, we will climb up and over—first up the western slope through the foothill woodlands with its gray pine and oaks and through the thickets of the chaparral. Once we breach this nearly impenetrable rampart, we will be in the midmountain forests of ponderosa (*Pinus ponderosa*) and sugar pine (*P. lambertiana*) and groves of giant sequoias (*Sequoiadendron giganteum*), then higher still through the red firs (*Abies magnifica*) and lodgepole pines (*P. contorta*) until we reach the subalpine world of gnarled trees. Proceeding even higher yet, we will go beyond the timberline and into the alpine world. Once we reach the crest of the High Sierra, we will start down the steep east side into a land totally different from the one where we began. The following discussion summarizes the works of M. G. Barbour and J. Major (1988), S. Whitney (1979), M. G. Barbour et al. (1993), and V. R. Johnston (1994).

The Foothills

Our trek begins on the west side of the Sierra where we first encounter the foothills. The foothills form a belt about ten to thirty miles wide and five hundred to five thousand feet in elevation above the Central Valley floor. The zone is defined by a distinctive vegetation mosaic comprising grassland, woodland, and chaparral (at its upper limits). The four major plant communities recognized include valley grassland, foothill woodland, riparian woodland, and chaparral.

Valley grassland covers the gentle rolling hills and extends upward in elevation as an understory beneath the foothill woodland. Like the chaparral and foothill woodland, valley grassland is a widespread and characteristic element in the foothill landscape but is now dominated by alien plant species. At one time, the grassland was dominated by perennial bunchgrasses such as purple needlegrass (*Stipa pulchra*) and foothill bluegrass (*Poa scabrella*). The Sierra grasslands are most beautiful in the spring when grasses are brilliant green and decorated with great swaths of native wildflowers. Appendix 1 provides an additional list of useful plants to be found in the valley grassland.

Foothill woodland occurs on fairly deep, well-developed soils and is dominated by blue oak (*Quercus douglasii*) and gray pines (*P. sabiniana*). These two species are more or less constant associates throughout their geographic ranges.

The foothill-woodland community occupies a variety of habitats between three hundred and five thousand feet, depending on latitude and other factors. A number of different woodland phases and types have been described and include the valley oak phase, blue oak phase, live oak phase, and north slope phase. Each has its own species components. Appendix 1 provides an additional list of useful plants to be found in the foothill woodland.

Snaking through the dry landscape, foothill streams with their characteristic willows (*Salix* spp.), cottonwoods (*Populus* spp.), and other moisture-loving species are the *riparian areas*. Because these species lack the adaptations to heat and drought, their existence is dependent on the assurance of water during the hot summer. As one moves upslope through the foothills, the stream channels become increasingly narrow and steep sided. The steeper canyon walls tend to drain water rapidly, and the riparian woodland is reduced to a thin margin of vegetation along the stream bank, often losing its identity as a community. Appendix 1 provides an additional list of useful plants to be found in the riparian habitat.

On steep slopes or other areas underlain by thin, rocky, or otherwise inhospitable soils, the foothill woodland is replaced by *chaparral*. This is a dense, fire-adapted shrub community dominated by chamise (*Adenostoma fasciculatum*) and several species of manzanita (*Arctostaphylos* spp.). In most areas, chaparral usually forms extensive cover on steep south-facing slopes but occurs on other aspects as well. In mature chaparral, grasses and wildflowers are rare, as are seedlings of the dominant shrubs. However, these grasses and wildflowers are quite abundant in recently burned areas. In contrast to mature chaparral, young chaparral stands exhibit greater species diversity. Appendix 1 provides an additional list of useful plants to be found in the chaparral.

COMMON YELLOW PINES
The two common yellow pines are ponderosa and Jeffrey. They are so closely related that at one time they were considered to be varieties of the same species. However, the differing chemistry of their resins and the size and structure of their cones separate them. In Jeffrey pine, the cones are larger and less prickly than the ponderosa. On the ponderosa, the needles are greener against the Jeffrey's grayer ones. Finally, the Jeffrey pine gives out the fragrant aroma of vanilla or butterscotch from cracks in its bark on warm days.

SLOPE AND ASPECT
In mountains, aspect can have nearly as great an influence on habitat as elevation. Aspect is the direction a slope faces. This determines the amount of sunlight, frost, wind, and snowpack to which it is exposed, and in turn it affects the types of plants that grow there. North- and south-facing slopes may be dominated by completely different plant species. Aspect helps create a mosaic of habitat types, enhancing biological diversity of an area. Finally, aspect been shown to be important in determining land use. North-facing slopes can be about six degrees Fahrenheit colder than south-facing slopes, so humans have opted to settle south-facing slopes first, building villages, clearing cropland, and planting orchards.

The Montane Forest Belt

Above the elevations of two thousand feet in the northern Sierra Nevada and five thousand feet in the southern Sierra Nevada, foothill woodland gives way to montane coniferous forest. This broad vegetation belt spans three life zones: lower montane, upper montane, and subalpine. Within these zones are four distinct coniferous communities—mixed coniferous forest, red fir forest, lodgepole pine forest, and subalpine forest—as well as montane chaparral and meadows. In the northern Sierra Nevada, the coniferous belt extends to about eight thousand feet, whereas in the southern Sierra Nevada, the trees occur as high as eleven thousand feet. Although these four communities are more or less arrayed by elevation, their shared boundaries are seldom distinct. As a general rule, the communities tend to range higher on south-facing slopes and ridge tops and lower on northern exposures and in canyon bottoms.

The *mixed coniferous forest* ranges from two thousand to six thousand feet in the northern Sierra and between four thousand and eight thousand feet in the southern Sierra. Three subtypes (ponderosa pine forest, white fir forest, and Jeffrey pine forest) have been identified within this forest.

The first subtype, the *ponderosa pine forest*, occurs from two thousand to six thousand feet in the northern Sierra and four thousand to seven thousand feet in the southern Sierra. It is replaced in the North above five thousand feet and above six thousand in the South by Jeffrey pine (*P. jeffreyi*). This species is able to tolerate lower air temperatures and deeper snows than the ponderosa pine. Additionally, Jeffrey pine also replaces ponderosa pine on the east side of the Sierra from Carson Pass south.

The second subtype is the *white fir forest*. It is dominant on cool, mesic sites from four thousand to seven thousand feet on the west slope, and common associates include sugar pine, incense cedar (*Calocedrus decurrens*), and Douglas-fir (*Pseudotsuga menziesii*). On drier sites, it is found with ponderosa or Jeffrey pine. The giant sequoia is also a member of the community and is found growing on moist, unglaciated flats in the central and southern Sierra between forty-five hundred and eighty-four hundred feet. Giant sequoias are only found on the west side of the Sierra Nevada in about seventy-five scattered groves from Placer County to southern Tulare County.

The third subtype is the *Jeffrey pine forest*. This species is the cold-weather version of ponderosa pine, replacing it on the east side of the Sierra and above five thousand to seven thousand feet on the west slope. Common associates on the west slope include incense cedar, black oak (*Quercus kelloggii*), white fir (*Abies concolor*), and sugar pine on drier, rockier slopes. Appendix 1 provides an additional list of useful plants to be found in the mixed coniferous forest.

The *red fir forest* occupies deep, well-drained soils between five thousand and eight thousand feet in the northern Sierra and between seven thousand and nine thousand feet in the South. This forest is primarily dominated by red fir (*Abies magnifica*), and mature stands often contain no other conifer because of the dense canopy that shades out competitors. What understory vegetation that does exist in this forest is largely confined to forest openings; otherwise, the forest is a realm of deep shade. The red fir forest is also called the "snow forest" of the Sierra because it occupies the zone with the greatest reported snowfall. In general, soil moisture is the primary factor governing the upper and lower elevational limits of the red fir forest. Few species of plants are able to tolerate dense shade and deep litter in mature red fir forests. Appendix 1 provides an additional list of useful plants to be found in the red fir forest.

Above the red fir forest is the *lodgepole pine forest*. This forest is typical of the glacial basins encountered in the lower subalpine and forms rather open stands at elevations between six thousand and eight thousand feet in the North and eight thousand and eleven thousand feet in the southern Sierra. In general, lodgepole pine is the dominant conifer of the glacially scoured ridges, valleys, and basins. Understory is sparse or absent in this forest. However, what does occur as understory is usually confined to sunny openings or moist soils near lakes and streams. Lodgepole forests have little in terms of understory, and as such these forests are sometimes called biological deserts (V. R. Johnston 1994). Appendix 1 provides an additional list of useful plants to be found in the lodgepole pine forest.

The *subalpine forest* consists of scattered stands and individuals trees (dwarf or shrubby) in the rocky slopes above eight thousand to ten thousand feet in the northern Sierra and ninety-five hundred to twelve thousand feet in the South. The community is poorly represented north of Lake Tahoe and occurs only in isolated stands on a few of the higher peaks. Subalpine forest is dominated by whitebark pine (*P. albicaulis*), mountain hemlock (*Tsuga mertensiana*), and in the southern Sierra Nevada by foxtail pine (*P. balfouriana*). Subalpine soils are rather rudimentary, consisting largely of disintegrated granite with little humus and minimal horizon development (lithosols). Appendix 1 provides an additional list of useful plants to be found in the subalpine forest.

The *mountain meadows* that occur in the Sierra Nevada range in size from a few acres to large open flats covering many square miles. Although these meadows range from dry to wet types, as a group they occur only where moisture is abundant in the upper few inches of soil at least during part of the growing season. It is this factor that accounts for the persistence of meadows that would otherwise be covered by forest (Whitney 1979).

Meadows are numerous and extensive in the subalpine zone, but they also occur downslope in the red fir and mixed conifer forests as well. Their lower limit in the northern Sierra is about four thousand feet, whereas in the southern Sierra it is about six thousand feet. This elevational limit more or less coincides with reliable winter snow cover. Farther downslope, the precipitation (snowfall) is irregular and melts quickly (Whitney 1979). Appendix 1 provides an additional list of useful plants to be found in the mountain meadows.

The Alpine Environment

The alpine environment extends from the tree line to the summit of the high peaks, usually above 9,900 feet near Lake Tahoe, 10,500 feet in the Yosemite region, and 11,000 feet in the southern Sierra. The region is devoid of trees and is dominated by large expanses of bare or sparsely vegetated rock. What vegetation does exist is usually sparse and consists entirely of dwarf perennial herbs growing close to the ground where the winds are gentler and rock faces are warmed by the sun. The two broad categories of plant associations found in the alpine zone are the *alpine meadows* (dominated by sod-forming sedges and grasses) and *alpine rock communities* (dominated by widely spaced bunchgrasses and cushion plants). Appendix 1 provides an additional list of useful plants to be found in the alpine habitats.

The East Side of the Sierra

Now, our trek takes us onto the desert face of the range. The winter storms that move up the west face of the Sierra Nevada are nearly wrung dry by the time they reach the crest, and so the east flank receives significantly less precipitation. Below seven thousand feet on the east slope, the vegetation is dominated by two communities—pinyon-juniper woodland and sagebrush shrubland (together called the pinyon-sagebrush zone). The east slope forests and meadows are comparable to those on the west slope but are less extensive and represented by fewer species.

The *pinyon-juniper woodland* community ranges from four thousand to fifty-five hundred feet in the northern Sierra and from five thousand to eight thousand feet in the southern. Sagebrush usually forms the understory in this woodland that is dominated by single-leaf pinyon pine (*Pinus monophylla*). The woodland occurs sparsely but more or less continuously from Alpine County south to Kern County. Interestingly, throughout most of the Great Basin, Utah juniper (*Juniperus osteosperma*) is a constant associate with single-leaf pinyon, but for some unknown reason it is missing from the woodland on the east slope of the Sierra (Whitney 1979).

The treeless community of the *sagebrush shrubland* occurs on coarse, dry, well-drained soils below six thousand feet (seven thousand feet in the South). rubs are widely spaced, with grasses and forbs forming a sparse but characteristic understory between the larger plants. The overwhelming dominant species in most stands is sagebrush (*Artemisia tridentata*). Appendix 1 provides an additional list of useful plants to be found in the pinyon pine and sagebrush habitats of the eastern Sierra Nevada.

Scientific and Common Names

In the discussion of useful plants, plant families are arranged alphabetically within each of the four major groups of higher plants (ferns and their allies,

The mountains are fountains of men as well as of rivers, of glaciers, of fertile soil. The great poets, philosophers, prophets, able men whose thought and deeds have moved the world, have come down from the mountains—mountain-dwellers who have grown strong there with the forest trees in Nature's workshops.
—John Muir, *John of the Mountains*

gymnosperms, and flowering-plant dicots and monocots). Although this arrangement may seem awkward to professional botanists, it has been adopted with the realization that this arrangement of families (and genera within families, species within genera) will be more easily consulted by readers in nonbotanical fields who may have occasion to use this book. The following list identifies the number of species that occur within each of the four basic categories of vascular plants.

FOUR BASIC CATEGORIES OF USEFUL MOUNTAIN PLANTS IN THE SIERRA NEVADA

CATEGORY (PLANT GROUP)	APPROXIMATE NUMBER OF SPECIES
Ferns and Fern Allies	>47
Gymnosperms	>23
Flowering Plants—Dicots	>1,126
Flowering Plants—Monocots	>175
TOTAL	

The scientific and common names are given for each species. A brief description of the plant stressing key features is provided. Plant nomenclature (scientific and common names) used in this book follows that of *The Jepson Manual: Higher Plants of California* (Hickman 1993). Common names for plants can be misleading and do not always distinguish among the species. Additionally, a species known by a common name in one region may have another common name elsewhere, leading to further confusion. However, common names have been retained because they are generally of more interest, and more likely to be known, by the public. You are encouraged to learn to identify plants by both their scientific and their common names.

Finally, the dichotomous keys (called "Quick Keys") found in this book are provided for those who are familiar with their use. These keys are associated with only those groups of plants that the beginning or experienced botanist might consider to be "safe" and easy to recognize. The various keys are adapted from several sources, including N. F. Weedon (1996), Hickman (1993), and P. A. Munz and D. D. Keck (1973). No attempts have been made to develop concise keys for all the species that may occur in the Sierra Nevada.

Rare and Protected Plants

The Native Americans and other aboriginal peoples were dependent upon nature for all of their needs and had an extensive knowledge of which plants (and animals) were edible or useful. Because of this dependence, they shared a strong conservation ethic based on the sanctity of life. Today, with increasing human population and our demands upon natural resources, many species are becoming rare due to habitat destruction, competition with nonnative species, or other means. As previously mentioned, the Sierra Nevada is rich in vascular plant diversity, with more than 3,500 native species of plants (Hickman 1993). Of this total, about 400 plant species occur only in the Sierra Nevada, including 3 trees, 20 shrub species, and several hundred herbaceous plants. Appendix 2 lists some of the threatened, endangered, and sensitive species of plants that are known to occur in the range.

Mountain ecosystems have evolved in the absence of human activities and have no ready response to some kinds of disturbance. Though appearing rugged, mountain environments are fragile, are highly susceptible to disturbance, and at

WILLIS LINN JEPSON
Willis Linn Jepson (1867–1946) is often referred to as the father of California botany and was one of the first people to receive a Ph.D. in botany from the University of California, in 1898. While he was a professor of botany at the University of California, he is best known for writing *The Jepson Manual: Higher Plants of California*. Mount Jepson (13,390 feet), located in the Palisades on the border of Inyo and Fresno Counties, is named in his honor. The 1993 revision of the Jepson manual is currently the standard reference for the identification of wild vascular plants in California.

Of recent years there has come into man's life a new joy. This joy is the acquaintanceship with plants. Nature has long been ready to reveal her secrets, but only to those prepared to hear and see. Gradually a new understanding has arisen between Nature and mankind, and as a result we obtain from such a revelation a joy undreamed of a few years ago.... Plants no longer are lifeless things labeled and grouped under ponderous Latin titles; they are highly developed organisms, which ... walk, swim, run, fly, jump, skip, hop, roll, tumble, set traps, and catch fish; decorate themselves that they may attract attention; powder their faces; imitate birds, animals, serpents, stones; play hide and seek; blossom underground; protect their children, and send them forth into the world prepared to care for themselves.
—Royal A. Dixon

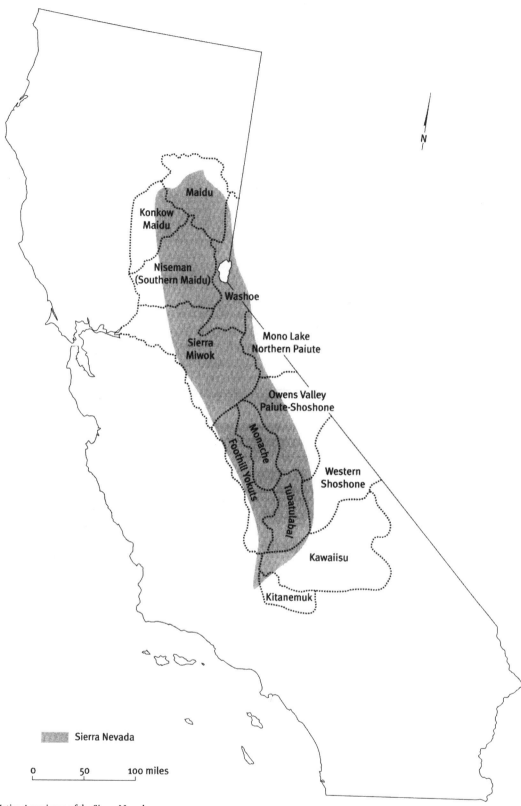

N

Maidu

Konkow
Maidu

Niseman
(Southern Maidu)

Washoe

Sierra
Miwok

Mono Lake
Northern Paiute

Owens Valley
Paiute-Shoshone

Monache

Foothill Yokuts

Tubatulabal

Western
Shoshone

Kawaiisu

Kitanemuk

Sierra Nevada

0   50   100 miles

Native Americans of the Sierra Nevada

times have a low ability to rebound and heal themselves after damage. The degree to which this is true is variable, but the vulnerability of mountain environments to disturbance is well documented (Zwinger and Willard 1972; Price 1981). The flora and fauna of mountain ecosystems are composed of species that are well adapted to cope with environmental extremes, low productivity, and fluctuations within the system. Because of climatic extremes, a brief growing season, lack of nutrients at higher elevations, low biological activity and productivity, their islandlike character, steepness of slopes, and the basic conservatism of the dominant life forms all make for the rate of restoration to original conditions after disturbance in mountain environments to be rather slow (Price 1981).

Therefore, the plants you encounter will vary in their ability to withstand harvesting. For example, collecting berries may not directly kill a plant but may affect its ability to survive. Additionally, because digging up the roots of a plant will destroy it, you must select your specimens carefully. And remember, many species of wildlife rely on plants for survival, whereas most hikers and nature enthusiasts collect plants for pleasure, not necessity.

It is strongly recommended that you obtain a list of threatened, endangered, or sensitive plant species before you start collecting. Currently, about 218 species are considered rare or threatened by the California Native Plant Society, California Department of Fish and Game, and the U.S. Fish and Wildlife Service. By avoiding rare species and using common sense, you should be able to enjoy wild plants without appreciably affecting either their population or their surroundings. You should also check in with the local land-management agency (that is, the Bureau of Land Management or the Forest Service) for its policy on collecting native plants.

California Native Groups

To most California Natives, it is said that their ancestors were created tens of thousands of years ago in the place where their group lived. That is, they did not migrate here, but rather they were created here.

Most California Natives thought of themselves as the "People" and often did not have a specific "name" for themselves. The names depicted on the map were given to the California Natives by nonnatives or are the names currently used by the California Natives. The areas outlined are approximate to provide the reader of this book with a reference to their general location (see page 16).

Guidelines for Gathering

There will undoubtedly be people who may be curious in sampling the wild plant foods. If you are one of those people, please keep in mind that there are no general rules for distinguishing an edible plant from an unsavory or poisonous species, and one must identify a plant correctly before attempting to use it. There are some books that suggest if you do not know a plant, you can eat a small quantity and wait to see if it has any adverse effects. This is a potentially serious mistake (Kingsbury 1964; Kinghorn 1979; Vizgirdas 1999a). For instance, if the unknown plant happens to be death camas (*Zigadenus*), not only would it cause much discomfort (such as a burning sensation in the mouth), but it could even kill you. Additionally, anyone who intentionally plans to search out and consume edible plants should exercise extreme caution. Correct identification of plants is necessary to avoid similar species or parts that may be unpalatable or poisonous. One of the best ways to learn about plants is to consult a knowledgeable botanist or qualified individual.

THREE IMPORTANT RULES BEFORE CONSIDERING ANY PLANT AS A SOURCE OF FOOD:
1. If you cannot positively identify a plant, do not consider it as a food source.
2. There is no such thing as a safe "plant edibility test."
3. Just because you see an animal eating a particular plant does not mean that we are able to eat it. We have seen squirrels eat the deadly amanita mushrooms and bears eat death camas (*Zigadenus*).

ECOLOGICAL REASONS NOT TO PICK WILDFLOWERS
1. Flowers are more than beautiful structures that appeal to humans: they exist so the plant can reproduce itself. Many of the most spectacular blossoms are specially designed to attract certain pollinating animals. The number of flowers pollinated combined with their arrangement on the stem can make a difference between reproductive success or failure for the entire year.
2. Removing wildflowers from annuals (plants that bloom for only one year and then die) means the seeds the plant would have made will not be there for next year's wildflower season.
3. Many species of wildflowers have already suffered great reductions in numbers over the past one hundred years because of increasing alterations of their habitat.
4. It is often difficult to distinguish between common and rare or endangered species of wildflowers. Species that are in danger of extinction may look abundant to the casual observer who is possibly looking at one of only a few remaining areas where the plant is found.

If you do decide to harvest plants, it is important to harvest them with wisdom and respect. The uncontrolled harvesting of mountain plants could severely damage delicate plant communities. In addition, it is illegal to injure or uproot a living plant in some areas covered by this book (for example, national parks and monuments and state parks). If a plant is rare or endangered, look for other plants. If you are not in an emergency survival situation, you should be even more frugal and thoughtful.

Also, be mindful of your own safety when dealing with edible and useful plants. In California and elsewhere, state and federal agencies often spray chemicals to control noxious weeds, especially in areas where logging, mining, and grazing activities occur or in developed campgrounds. Although such chemicals may be considered "safe," there are no guarantees. You should avoid collecting in areas affected by pollutants such as along roads or in drainages affected by mining activities.

In *Stalking the Wild Asparagus,* Euell Gibbons (1962) described the taste for wild edible plants as an acquired one, ranging from awful to barely palatable. Although we have sampled many wild plants that fall into those two categories, we have also found wonderful delicacies that make supermarket food seem pale by comparison. If you have a positive outlook in your endeavor, it may someday help you if you are ever in a survival situation. Otherwise, it is a wonderful excuse to explore the western mountains.

A final thought:

> Where you find a people who believe that man and nature are indivisible, and that
> survival and health are contingent upon an understanding of nature and her processes,
> these societies will be very different from ours, as will their towns, cities and landscapes.

—Ian McHarg, *Design with Nature*

Ferns and Fern Allies

On the mountains there is freedom!

The world is perfect everywhere, save where man comes with his torment.

—Johann Christoph Friedrich von Schiller, *The Bride of Messina*

Fern and fern allies include the clubmosses, horsetails, and ferns. They are herbaceous plants that reproduce by spores, which develop inside structures called sporangia.

ADDER'S-TONGUE FAMILY (Ophioglossaceae)

The herbaceous plants have fleshy rhizomes with numerous fibrous, often fleshy roots. The leaves (fronds) consist of two parts, a sterile simple or compound, sessile or stalked blade, and a stalked spore-bearing spike or panicle, sterile or fertile parts borne on an erect common stalk. There are three genera and seventy-plus species of this family. Two genera are found in the United States, but only moonwort (*Botrychium*) may be found in the Sierra Nevada. None of the members of this family are of economic importance.

Moonwort (*Botrychium*)

Moonwort
(*Botrychium* spp.)

Description: Worldwide, there are about forty species of moonwort, of which eight occur in California and five in the Sierra Nevada. This is a difficult genus to study and requires fully developed specimens for identification. The spore sacs, technically termed the sporangia, are borne in grapelike clusters on a naked stalk, not on leaves as in the true ferns. The genus name is from the Greek *botrys*, meaning a cluster, referring to the grapelike arrangement of the sporangia.

1. Scalloped Moonwort (*B. crenulatum*)—This is a rare species found in marshes and meadows throughout the Sierra Nevada between 4,000 and 8,000 feet.

2. Common Moonwort (*B. lunaria*)—Look for common moonwort in open, dry places, meadows, slopes, and banks below 8,000 feet in the central Sierra Nevada.

3. Mingan Moonwort (*B. miganense*)—This is a rare species occurring in forests along streams below 6,000 feet in the southern Sierra Nevada.

4. Leathery Grape Fern (*B. multifidum*)—This species grows in moist, open habitats between 3,000 and 10,000 feet throughout the Sierra Nevada.

5. Yosemite Moonwort (*B. simplex*)—This species is found in open places, grassy meadows, and damp places from 5,000 to 11,000 feet throughout the Sierra Nevada.

Interesting Facts: Early records appear to indicate that juice extracts from common moonwort were used to stop bleeding and vomiting, and also for the treatment of bruises. They may have been used to concoct balsams for healing internal wounds (Grillos 1966).

Additionally, a root poultice or lotion made from *B. virginianum* (rattlesnake fern) was used for snakebites, bruises, cuts, or sores. In folk medicine, the root tea was used as an emetic (Foster and Duke 1990).

Fern Family (Polypodiaceae)

Many botanists consider that a number of families (for example, Dennstaediaceae, Dryopteridaceae, Polypodiaceae, and Pteridaceae) are represented in this assemblage. For convenience, they are treated here as a single family. Generally speaking, fern fronds can be used as makeshift place mats and interspersed between layers of food in cooking pits.

A number of survival handbooks (for example, Risk 1983) suggest that most ferns in their young stage are edible. However, in most cases, they are referring to bracken fern (*Pteridium*). It is best to avoid ferns unless you are able to positively identify them. Many plants such as poison hemlock (*Conium maculatum*) have delicately cut leaves, which to the untrained eye could easily be mistaken for a fern. Also, there are deadly ferns (Wherry 1942; Kingsbury 1964).

Cliffbrake (*Pellaea*)

Description: These are small tufted ferns growing in crevices of rocks. The rhizomes are short, thick, creeping, densely brown, scaly, and covered with old stipes. The leaves are singly pinnate or bipinnate. The stipes are dark reddish brown. The sori are covered by the recurved margin of the leaf segments. The generic name comes from the Greek *pellos* (dusky) and refers to the appearance of the leafstalks. The following six species can be found in the Sierra Nevada.

IDENTIFYING FERNS
In order to identify the ferns, first note the sori. Sori vary in their shape and position on the leaf surface and are the first step in the identification process.
1. Sori following course of veins or completely covering them—*Pentagramma, Pityrogramma*
2. Sori oblong, linear to lunate, or horseshoe shaped—*Asplenium, Woodwardia, Athyrium*
3. Sori round to oval—*Polypodium, Dryopteris, Thelypteris*
4. Sori borne at or near the margins of the leaves—*Pteridium, Adiantum, Pellaea, Cheilanthes, Aspidotis, Notholaena*

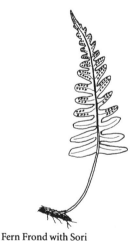

Fern Frond with Sori

Quick Key to the Cliffbrakes

Blades once pinnate	
Leaflets entire and thick	Bridge's Cliffbrake (*P. bridgesii*)—This species occurs in dry habitats, especially around rocks between 5,000 to 11,000 feet.
Leaflets thin and mitten shaped	Brewer's Cliffbrake (*P. breweri*)—This species is found on exposed rocky slopes between 7,000 and 11,000 feet.
Blades two to four times pinnate	
Leaflets with a short abrupt tip	
Fronds two to three times pinnate, the secondary divisions three-parted	Bird's-foot Cliffbrake (*P. mucronata*)—This species can be found in exposed or partially shady, rocky places mainly below 6,000 feet.
Fronds twice pinnate, leaflets simple	
Leaflets five to eleven per pinnae	Sierra Cliffbrake (*P. brachyptera*)—Look for this species on exposed, rocky habitats between 3,500 and 8,000 feet.
Leaflets eleven to fifteen per pinnae	Coffee Fern (*P. andromedifolia*)—This species is found in the foothill elevations throughout the Sierra Nevada.
Leaflets rounded at the end	Desert Cliffbrake (*P. compacta*)—Look for this species in dry, rocky habitats between 4,500 and 8,500 feet.

Interesting Facts: A refreshing tea was said to be made by the Luiseno Indians from *P. mucronata*. They also used the tea medicinally as a decoction to stop hemorrhages, to reduce fevers, and as an emetic. The decoction was used as a wash for skin problems. Other Native Americans used the brown fibers from the rhizome to make basketry patterns (Strike 1994).

Caution: The younger leaves and stems are occasionally eaten by sheep and other grazing animals. However, they are poisonous and frequently cause death when eaten.

Deer Fern (*Blechnum spicant*)

Description: The sterile fronds are eight to twenty inches long, prostrate, and evergreen. The fertile fronds are fourteen to thirty inches tall and erect. The genus name comes from the classical Greek, *blechnon,* meaning fern. Deer fern has a sporadic circumpolar distribution. In North America, it is distributed from coastal Alaska to California. It is found below 4,000 feet in the northern Sierra Nevada.

Interesting Facts: Deer fern fronds were eaten as food in emergencies or to alleviate thirst on long journeys (Coville 1897). A decoction of the root was taken or the root held in the mouth for diarrhea.

Fragile Fern (*Cystopteris fragilis*)

Description: This six- to sixteen-inch-tall plant is loosely tufted from a short, creeping rhizome. The leaves are thin and delicate in texture. Stipes are brown below, yellowish above, and smooth. The indusia are small, attached at one side, and arching back to form a hood. The genus name comes from the Greek *kystis,* meaning bladder, and *pteris,* for wing, referring to the shape of the indusium. This is a widely distributed fern and is found in the crevices of cliffs and ledges, in soil under rocks, shrubs, or trees, between 5,000 and 9,000 feet.

Interesting Facts: This plant was used as a dermatological aid by the Navajo. Here, a cold compound infusion of the plant was made and used as a lotion for injuries (Elmore 1944; Zigmond 1981).

Giant Chain Fern (*Woodwardia fimbriata*)

Description: This species occurs in wet seeps, mainly below 5,000 feet throughout the Sierra Nevada. The genus name honors Thomas Jenkinson Woodward (1745–1820), an English botanist.

Interesting Facts: The Luiseno tribe of southern California steeped chain fern roots in water and utilized the resulting liquid to alleviate the pain of wounds and bruises (Balls 1970; Callegari and Durand 1977).

Goldback Fern (*Pentagramma triangularis*)

Description: This perennial fern has a rhizome with slender scales that have a thickened, blackish midstrip and thinner, brown, narrow margins; petioles are chestnut brown to purplish brown and notably longer than the blades. The blades are glabrous above and have a pale to bright yellow waxy powder on the underside. The few pinnae or segments are opposite, and the lowest pair is the largest. The leaf margins are very narrowly revolute but not covering the sporangia. Goldback fern is common in moist habitats below 4,500 feet throughout the western Sierra Nevada.

Interesting Facts: The Miwok chewed the plant for toothaches (Barrett and Gifford 1933). To other Native Americans, the plant was used to mitigate the afterpains of childbirth (Strike 1994).

Indian's Dream (*Aspidotis densa*)

Description: The leaves of this species are long and of two forms. Both are glabrous and densely tufted and originate from a short, much branched rhizome with persistent, glossy-brown petiole bases. This species is found in exposed, rocky habitats between 5,000 and 9,000 feet in the southern and central Sierra Nevada.

Interesting Facts: Related species are said to prevent baldness (Strike 1994).

Ladyfern (*Athyrium*)

Description: The sori are oblong to horseshoe shaped. The indusium is thin and fragile, soon disappearing or sometimes lacking. The name is from the Greek *a*, meaning without, and *thurium*, referring to a long, oblong shield. This species is widely distributed in North and South America.

Quick Key to the Ladyferns	
Indusium is crescent or horseshoe shaped	Ladyfern (*A. filix-femina*)—This species is found in cool habitats between 4,000 and 8,000 feet throughout the Sierra Nevada.
There is no indusium present	Alpine Ladyfern (*A. alpestre*)—Look for this species in moist habitats from 6,000 to 11,000 feet throughout the Sierra Nevada.

Interesting Facts: Ladyfern fiddleheads were eaten in the early spring when they were two to six inches tall. They were boiled, baked, or eaten raw with grease. Ladyfern has certain chemical properties for medicinal use. The underground stems, pulverized to a powder, have been used to drive worms out of the intestinal system, although this use has not been medically recognized (Grillos 1966). A tea made from the root was used as a diuretic, and the powdered root was used externally for sores.

Lipfern (*Cheilanthes*)

Description: These are small, densely tufted ferns from thick rhizomes that grow in the crevices of rocks or on soil under rocks. The genus name is from the Greek words *cheilos* (margin) and *anthos* (flower), because the sori are clustered on the edge of the frond. Three species are known to occur in the Sierra Nevada.

Quick Key to the Lipferns	
Blades twice pinnate	Lace Lipfern (*C. gracillima*)—Lace fern is found in dry, rocky places below 9,000 feet from the central Sierra Nevada northward.
Blades three to four times pinnate Upper surface of leaflets glabrous	Coville's Lipfern (*C. covillei*)—This species is found in rocky places below 8,000 feet throughout the Sierra Nevada.
Upper surface of leaflets with some scales	Lipfern (*C. intertexta*)—This species is found in dry, rocky habitats below 6,500 feet.

Interesting Facts: Coville's lipfern was used medically by Hupa women during childbirth. The Kawaiisu drank a tea made from the stems and leaves as a general tonic (Strike 1994).

Maidenhair Fern (*Adiantum*)

Description: There are about two hundred species of *Adiantum* that occur world-wide. The sporangia in sori are borne on the veins under the frond surface. The genus name is from the Greek word *adianthos,* meaning unwetted and referring to the impermeable leaves of some species shedding water.

Quick Key to the Maidenhair Ferns	
Blade about as wide as long and divided into two equal parts	Five Finger Fern (*A. aleuticum*)—This is a common species in moist, shady habitats throughout the Sierra Nevada below 10,000 feet.
Blade much longer than wide	Common Maidenhair (*A. capillus-veneris*)—This species is infrequently found growing on moist canyon walls below 4,000 feet in the southern Sierra Nevada.

Interesting Facts: These black-stemmed ferns can be used in basketry. The herbage is reported to be bitter and causes an increased secretion of mucus. The leaves were used as a tea or syrup to treat colds, coughs, and hoarseness. Rhizomes were used as a stimulant, to soothe the mucous membranes of the throat, and to loosen phlegm. Some of these medicinal uses have been recorded since classical times (Romero 1954; Strike 1994).

Rock-brake (*Cryptogramma acrostichoides*)

Description: Rock-brakes are small ferns of rocky places, mostly at high elevations. These plants grow in dense tufts up to sixteen inches tall. The sterile leaves are dark green and tripinnate, and the stipes are yellow or straw-colored to the base. Fertile leaves are larger and longer. This species occurs on cliffs, ledges, or talus slopes from 6,000 to 11,000 feet. The genus name is from the Greek *cryptos,* meaning hidden, and *gramme,* meaning line, and refers to the sori being covered by the infolded margins of the pinnules

Interesting Facts: A related species (*C. sitchensis*) was used as an eyewash. Here they made an infusion of the washed, strained fronds. The infusion was also taken for gallstones (Strike 1994).

Shield Fern (*Polystichum*)

Description: These ferns have evergreen leaves clustered on a short, vertical rhizome. Leaves have scaly petioles and blades that are one to two times pinnately divided into numerous toothed or lobed segments. The name is from the Greek *poly* (many) and *stichos* (row), because the sori of some species develop in several rows. The membranous indusium arises from the center of the sori.

1.   Sword Fern (*P. imbricans*)—Look for this species in rocky habitats below 8,000 feet throughout the Sierra Nevada.
2.   Kruckeberg's Sword Fern (*P. kruckebergii*)—This species occurs on rocky slopes and cliffs between 7,000 and 10,000 feet throughout the Sierra Nevada.
3.   Lemmon's Sword Fern (*P. lemmonii*)—This species is found in rocky habitats below 7,000 feet.
4.   Holly Fern (*P. lonchitis*)—This rare species occurs in cool, rocky habitats between 6,500 and 8,500 feet in the northern Sierra Nevada.

5. Sword Fern (*P. munitum*)—Sword fern can be found in open and shaded moist habitats below 7,000 feet throughout the Sierra Nevada.
6. Eaton's Shield Fern (*P. scopulinum*)—This species is found on dry cliffs and rocky places from 5,000 to 10,500 feet.

Interesting Facts: In general, the leaves of these ferns can be used as a protective layer in pit cooking, flooring, or bedding. Although the edibility of these species is unknown, the large rhizomes of *P. imbricans* can be roasted over a fire or steamed in a pit oven, then peeled and eaten (Frye 1934; Willard 1992). The cooked rhizomes are also said to be a cure for diarrhea. Sword fern (*P. munitum*) is frequently used in floral arrangements.

Sierra Water Fern (*Lastrea oregana*) (= *Thelypteris nevadensis*)
Description: This is a delicate, pale green, deciduous fern emerging in a small compact clump from a slender rhizome. The leaves are few, erect to arching, two to four feet tall, and quite narrow (up to three inches wide). Leaves are twice-pinnate (twice-divided leaflets) with thirty-five to forty opposite or off-set leaflet pairs. The upper leaflets are larger and close together; the lower are greatly reduced in size and spaced quite far apart. Spore clusters (sori) are covered by an inconspicuous, horseshoe-shaped cover (indusium) on the leaflet undersides. This species occurs along wet stream banks below 5,000 feet in the central Sierra Nevada.

Interesting Facts: A related species (*T. kunthii*) was used as an orthopedic and psychological aid (Weedon 1996). The leaves were used for treating insanity and weakness of the limbs.

Spleenwort (*Asplenium*)
Description: This is a genus with about 650 to 750 species of terrestrial and epiphytic ferns with divided leaves. These are ferns that grow in moist crevices of cliffs and ledges or in soil, usually in cool, shaded canyons. The fronds are pinnately compound, and the sori are elongate and occur on the outer edge of the veins.

1. Northern Spleenwort (*A. septentrionale*)—This species is found in the northern Sierra Nevada near Sawtooth Pass and Lassen National Park.
2. Green Spleenwort (*A. trichomanes-ramosum*)—The only known location of this species in the Sierra Nevada is on a north-facing cliff on South Butte in Sierra County at about 7,500 feet.

Interesting Facts: The genus name comes from the Greek *a* (meaning not) and *splen* (meaning spleen), alluding to the plant's supposed medicinal properties. Several species are grown as ornamentals.

Western Brackenfern (*Pteridium aquilinum*)
Description: This is a medium- to large-sized plant with decompound, broadly triangular leaves up to seven feet long including the stipe. The stipes are green or yellowish, and on the surface there are fine white hairs. The sori are marginal and continuous, and they are partially covered by the recurved leaf margins. This is a widely distributed species found in open woods, rock slides, or slopes in damp or dry places, up into the high mountains (below 10,000 feet). The generic name comes from the Greek *pteris* (wing) and is applied to ferns because of their feathery leaves.

Interesting Facts: The young fronds, or fiddleheads, of brackenfern can be collected, boiled, and dried in the sun. The dried product can then be used as a winter food. Old fronds may be poisonous in large amounts (see "Caution" below). The starchy rhizome (underground stem) is edible after roasting or boiling but is usually tough. The leaves can be used as one of the protective plant layers for pit cooking. Some Native Americans would consume only fiddleheads so that their scent would not scare off deer. A root tea was used for stomach cramps and diarrhea, and poulticed roots were used for burns and sores. Ashes of the plants have been used as an ingredient to make glass and soap.

Caution: Although this plant has traditionally been accepted and harvested as a suitable edible, there is evidence indicating that eating sufficient quantities over a period of time may be dangerous to your health (Foster and Duke 1990). The plant is known to contain several poisonous compounds, including a cyanide-producing glycoside (prunasin); an enzyme, thiaminase, that reduces the body's thiamine reserves; and at least two potent carcinogens, quercetin and kaempferol. Another unidentified toxin is believed to be a naturally occurring, radiation-mimicking substance, also apparently mutagenic and carcinogenic. Bracken has caused many livestock deaths. The risks to humans of eating bracken fiddleheads and rhizomes have not been fully established, but their safety is questionable. Schofield (1989) reports that it is currently suspected of causing stomach cancer in Japan.

Western Polypody (*Polypodium hesperium*)

Description: This is a shade-loving fern. The fronds of this fern are pinnatifid to nearly pinnately compound and attached to creeping rhizomes by a distinct articulation. The sori are round in outline, and there is no indusium. This species occurs in the Sierra Nevada, growing on rock ledges and crevices between 5,000 and 8,500 feet. The genus name is Greek, meaning many and foot, alluding to the numerous protuberances on the rhizomes.

Interesting Facts: The rhizomes were chewed, or an infusion of the rhizomes was made for colds and sore throats. The rhizomes were also used as medicine for sore gums. The rhizome has a pleasant, sweet taste, almost like licorice. The roots of related species (*P. californicum* and *P. glycyrrhiza*) were also eaten by California Natives (Grillos 1966). Other Native Americans used bruised roots to heal sores and to relieve rheumatic pain. An extract of the root of *P. californicum* was used to treat internal injuries and kidney ailments.

Woodfern (*Dryopteris arguta*)

Description: This species occurs in cool habitats below 6,000 feet throughout the Sierra Nevada. The name is from the Greek *drys*, meaning oak, and *pteris*, meaning fern.

Interesting Facts: The edibility of the various species of woodfern is unknown, but some species are reported to be edible, poisonous, or of medicinal value. Several species contain phloroglucinol derivatives ("filicin"), which paralyze intestinal parasites. A root tea from a related species, *D. cristata* (crested woodfern), was traditionally used to induce sweating, clear chest congestion, and expel intestinal worms (Foster and Duke 1990). In addition, an oleoresin was extracted from the roots of *D. filix-mas* (male fern) to expel worms.

STEAM-PIT COOKING
The steam-pit cooking method has been used by many different peoples around the world for thousands of years. This method of cooking locks in the natural juices and flavor of the food during the cooking process. There are many variations of the pit, but basically all you need to do is simply dig a hole in the ground about two feet deep and two feet across and line the bottom and sides with flat rocks. One rule of thumb is to make the hole no larger than three times the size of the total amount of food. Construct a fire in the pit until the rocks are red-hot. After about an hour, carefully remove the coals and place about six to eight inches of green grass, fern fronds, or other nonpoisonous vegetative matter on top of the hot rocks. Then place your food on top of this layer of plants and cover with an additional six to eight inches of grass. On top of the grass, place a thin layer of bark slabs or brush to prevent dirt from sifting through to the food. Finally, cover with a mound of dirt and leave it alone for a few hours. This will give you time to go do other things.

Woodsia (*Woodsia*)

Description: These small ferns commonly grow in rocky places. The underground stem is densely tufted and clothed with broad, thin scales. The leaves are clustered, numerous, small, linear to lanceolate-ovate, and once- or twice-pinnate. The sori are round and seated on the back of the free veins, and the indusium is under the sori with star-shaped divisions. The genus name honors Joseph Woods, an English botanist. Two species may be encountered in the Sierra Nevada.

Quick Key to the Woodsias	
Blades without hairs	Oregon Woodsia (*W. oregana*)—This species occurs in dry rocky places from 4,000 to 11,000 feet.
Blades finely glandular to distinctly pubescent	Rocky Mountain Woodsia (*W. scopulina*)—Similar to Oregon woodsia, but the blades are finely glandular to distinctly pubescent. This species occurs on exposed rocks from 4,000 to 11,000 feet.

Interesting Facts: Rocky Mountain woodsia was used as a sign of water when the Natives traveled through the mountains farther to the north (Turner and Kuhnlein 1991). Some species are cultivated as ornamentals.

HORSETAIL FAMILY (Equisetaceae)

In this family, there is only one genus with about fifteen species worldwide. The plants have hollow, joined stems, and the leaves are whorled at the nodes, minute, and toothlike. None of the species are of economic importance.

Horsetail (*Equisetum*)

Description: In general, these are rhizomatous herbs with hollow, grooved, regularly jointed stems that are impregnated with silica. The leaves are reduced in size, appearing as a series of teeth around a joint. Spores are produced in conelike structures atop the stems. They are found in moist soil along streams and rivers, marshes, and other damp habitats. The name comes from the Latin *equus,* for horse, and *seta,* for bristles. Three species may be encountered in the Sierra Nevada.

Horsetail (*Equisetum* spp.)

Quick Key to the Horsetails	
Stems of two kinds, sterile ones profusely branched	Common Horsetail (*E. arvense*)—Common horsetail occurs in damp or wet soils along streams or dryer sites following a disturbance.
Stems of one kind, erect and rarely branched	
Cones abruptly rounded or acute apically	Smooth Scouring Rush (*E. laevigatum*)—This species occurs in wet or dry streamside soils and in marshes. It can be seen from June to July.
Cones ending in an abrupt slender tip	Common Scouring Rush (*E. hyemale*)—This species is found along streams, ditches, and ponds, often growing with *E. arvense.* It can be seen from June to July.

Interesting Facts: Although all species are useful and identical in application, common horsetail is the most popular. The tough outer tissue can be peeled away, and the sweet inner pulp of all species can be eaten in small amounts. In large quantities, defined as greater than 20 percent of body weight by some authorities, they can be toxic. Certain chemicals in this plant are said to destroy specific B vitamins such as thiamine. The enzyme thiaminase is apparently responsible for the poisoning. Cooking destroys this enzyme and renders the plants safe for consumption. The tuberous growth on the roots (actually rhizomes) can be eaten raw in the early spring or boiled later in the season.

Horsetails have an unusual chemistry. Some species contain alkaloids (including nicotine) and various minerals, whereas other species have been known to concentrate gold in their tissues, although not in sufficient amounts to warrant extraction. In the fall, the stems become impregnated with silicon dioxide and can be used to scour pots and pans or as a type of sandpaper for wood. Many Native Americans used horsetails to polish arrow shafts. The silica-rich stems are reputed to be superior to the finest grades of steel wool. Additionally, the high silica content of this herb is said to be effective in strengthening bones and connective tissue (Tilford 1997).

Caution: The waters within which these plants grow may be contaminated.

QUILLWORT FAMILY (Isoetaceae)

These are small submerged or partially immersed plants with a cormlike stem crowned by numerous subulate or nearly filiform leaves. There are two genera and seventy-five species worldwide. None are of economic importance.

Quillwort (*Isoetes*)

Description: These are small usually erect grasslike plants that are submerged or partially immersed in streams, ponds, and lakes. The stem, called a corm, is short, fleshy, two to three lobed, and has many roots developing from the base. The sporangia develop on the upper surface of the expanded leaf base. They are solitary, orbicular to ovoid, and usually covered by a thin membranous tissue called velum. The common name is derived from the plants' resemblance to a bunch of quills. The approximately 150 species of quillwort worldwide can be distinguished only by microscopic examination of their spores. The following two species may be found in the Sierra Nevada.

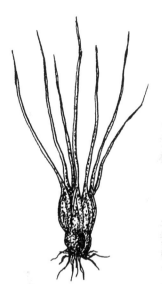

Quillwort (*Isoetes* spp.)

Quick Key to the Quillworts	
Plants are submerged in water	Quillwort (*I. occidentalis*)—Look for quillwort in shallow mountain lakes and ponds, from 5,000 to 11,500 feet.
Plants growing in wet places, but not underwater	Nuttall's Quillwort (*I. nuttallii*)—Nuttall's quillwort is found in damp places, along banks of streams and rivers, below 9,000 feet.

Interesting Facts: None of the species is known to be of economic importance. It is reported they have been occasionally used as food by people in Europe (Pfeiffer 1922). We found no record of anything injurious about the plants. The corm serves as a storehouse of food for the plant, as do the subterranean winter organs of many perennials; thus, some food value is quite probable (Frye 1934). Others indicate that quillworts are rich in starch and oils and could be edible raw or cooked (Weedon 1996). The corms are edible but not palatable.

SPIKEMOSS FAMILY (Selaginellaceae)

This is a family of mosslike plants that includes many important natural ground covers on rocky or gravelly soils. The sporangia are borne in terminal spikes of sporophylls, the larger ones containing three or four megaspores, the smaller ones numerous minute microspores. Members are sometimes called "little clubmoss" because of their resemblance to the clubmoss family (Lycopodiaceae).

Spikemoss (*Selaginella*)

Description: In general, spikemosses are low, mosslike, leafy, evergreen terrestrial plants, and their leaves are scalelike, less than one-eighth of an inch long. Two species are known to occur in the Sierra Nevada.

1. Hansen's Selaginella (*S. hanseni*)—This species forms mats on rocky outcrops usually below 6,000 feet.
2. Alpine Selaginella (*S. watsoni*)—This species is found in rocky habitats in the northern Sierra Nevada between 7,500 and 13,000 feet.

Interesting Facts: The spores of spikemoss were once used like those of clubmoss (*Lycopodium*) in making pills not adhere to one another. Many campers still gather the plants with mosses and other plants to make a soft bed when sleeping on the ground, a rather extravagant practice with such a small plant (Kephart 1909).

Although there are no edible or medicinal uses recorded for Sierra Nevada species, related species, such as *S. concinna* and *S. obtusa* found on the Reunion Islands in the Indian Ocean, were used medicinally as astringents, blood purifiers, and carminatives in cases of dysentery (Frye 1934). In Mexico, a variety of *S. rupestris* was used as a home remedy, and in the East Indies, *S. convoluta* was considered to be an aphrodisiac.

WATERCLOVER FAMILY (Marsileaceae)

These are aquatic or semiaquatic perennial herbs with creeping, hairy rhizomes that root in the mud. The leaves are four-foliate and cloverlike. The sporocarps are borne at the base of the stipe. The family comprises three genera and seventy species distributed worldwide. Two genera are native to the United States. They are of limited importance as ornamentals.

Hairy Pepperwort (*Marsilea vestita*)

Description: This species is found in shallow lakes and ponds or on their muddy borders below 6,500 feet throughout the Sierra Nevada. It has also been found along slow-moving streams. Arising from the rhizome are the distinctive leaves arranged like those of a four-leaf clover. The leaflets are pale green and hairy. Spores are produced within an elliptical, nutlike body borne on short stalks.

Interesting Facts: Although there are no documented uses for this species in the Sierra Nevada, related species are supposedly edible when dried and ground into flour. For example, the sporocarps of nardoo (*M. drummondii*) found in central Australia can be crushed into a powder between stones, made into a dough by adding water, and baked into cakes (Weber 1987).

Hairy Pepperwort
(*Marsilea vestita*)

WATER FERN FAMILY (Azollaceae; also known as Salviniaceae)

These are small floating plants on water. Two species are known from California, but only the following species may be encountered in the Sierra Nevada at the lower elevations.

Mosquito Fern (*Azolla filiculoides*)

Description: This is a free-floating or stranded-on-mud aquatic fern that is triangular or polygonal in shape. It floats on the surface of the ponds, ditches, and other slow or sluggish waters. From a distance, it looks like a green-reddish carpet floating over the surface. It is sometimes found at elevations below 5,000 feet.

Interesting Facts: Living within the cavities of this aquatic fern is a blue-green alga (*Anabaena azolla*). This is a nitrogen-fixing alga that excretes nitrogenous compounds into the cavity from which *Azolla* can absorb them, making nitrogen available for both species. In addition to fixing nitrogen by utilizing light energy, the blue-green alga associated with *Azolla* also releases hydrogen from water. This is the first reported known photosynthetic system for producing hydrogen from water that is stable in air and requires only water as a hydrogen source. In nature, the fixed nitrogen combines with the hydrogen to form ammonia that fertilizes the host *Azolla* plant. However, in the laboratory, the fern-alga relationship can be diverted from producing ammonia to producing only hydrogen gas. Although these are small-scale experiments to date, there may be some promise in producing hydrogen on a larger scale via this biological method (Ashton and Walmsley 1976). Hydrogen yields more energy (on a weight basis) than any other non-nuclear fuel, which is an important finding in light of dwindling petroleum supplies.

Azolla has other important contributions to many developing countries. In areas such as Southeast Asia, Vietnam, and Singapore, it is used as a green manure. In these areas where rice is grown for human consumption, *Azolla* is deliberately cultivated. It is reported that in these rice fields, the yield of rice is 50 percent higher than normal. Pigs, cattle, and ducks are also fed *Azolla* in many parts of the world. In other areas, the plant is also useful in controlling mosquitoes and other weeds by blocking the water surface.

Gymnosperms

Nearness to nature . . . keeps the spirit sensitive to impressions
not commonly felt, and in touch with the unseen powers.
—Ohiyesa, *The Soul of the Indian*

Gymnosperms are a group of plants that do not have flowers and fruits. The seeds are exposed, usually on a conelike structure. Hence, the majority of the gymnosperms are cone-bearing, most commonly known as conifers: spruce, fir, pines, and junipers.

The only gymnosperm that does not resemble other gymnosperms is an ancient plant called ephedra or Mormon tea. Ephedra is a desert shrub with broomlike branches that appear leafless. Leaves are minute and scalelike, and opposite or whorled at the nodes. At flowering time in the spring, small cones form at the nodes.

BALD-CYPRESS FAMILY (Taxodiaceae)

This family is sometimes combined with the cypress family (Cupressaceae). The two noted species that occur in this family in California include the redwood (*Sequoia sempervirens*) and giant sequoia (*Sequoiadendron giganteum*). The latter is discussed below.

Big Tree, Sequoia (*Sequoiadendron giganteum*)

Description: This is the most massive tree in the world, growing 255 to 275 feet tall and 10 to 20 feet in diameter. Mature trees have long, clear, stout stems and rounded, spreading crowns, whereas the ¼-inch-long leaves are sharp pointed, awl-like, and bluish green. Cones are barrel shaped, 1½ to 3 inches long, serotinous, and brown, with scales that are umbrella-like and wrinkled. The thick, cinnamon-red bark is fibrous and deeply furrowed, and on mature trees it can be up to 2 feet thick. Giant sequoia is found in the western Sierra at elevations between 2,700 and 8,800 feet in areas that were not glaciated.

Interesting Facts: There are about 75 scattered groves of this species, constituting approximately 35,600 acres. Groves range in size from approximately 2,400 acres with 20,000 sequoia to small groves with only 6 living trees from Placer County to southern Tulare County (Barbour and Major 1988).

Young sequoia has favorable wood properties. It is decay-resistant and is used in dimensional lumber, veneer, and plywood. Old growth has low tensile strength and brittleness, making it unsuitable for most structural purposes.

The tree was named in honor of Sequoiah, the son of a Cherokee Indian who helped devise an alphabet for the Cherokee tribe that ensured they became literate.

The steam from this plant's sap in boiling water was used to stop excessive bleeding from wounds. The Monache and Yokuts chewed the bitter sap to relieve cold symptoms (Strike 1994). Bark slabs and planks were used to construct dwellings and sweat houses (Romero 1954).

SNAGS
Standing dead trees are referred to as "snags" or wildlife trees. These trees are valuable to the local ecosystem as nesting and denning sites for many animals. Woodpeckers excavate nesting holes, which may be appropriated by other birds and small mammals like flying squirrels. In addition to providing a home for animals, snags are sources of food for birds, as the dead trees also have beetles and other insects. Hawks and flycatchers use the snags as perches while hunting, too. New logging practices today include leaving some standing trees.

CYPRESS FAMILY (Cupressaceae)

There are 19 genera and 130 species in this family. Five of the genera are native to the United States. The family is economically important as timber trees and ornamentals.

The heartwood of many species of Cupressaceae is resistant to termite damage and fungal decay, and therefore it is widely used in contact with soil (for example, for fence posts). The architect Frank Lloyd Wright preferred *Taxodium* as a siding for wooden residences, whereas *Sequoia* is preferred for lawn furniture throughout the vast suburbs of California. The premier coffin wood of China, *Cunninghamia lanceolata,* is another member of the family, and *Chamaecyparis* wood is in similar demand in Japan. Many genera are incorrectly called cedars because their heartwood is as aromatic as that of the true cedars, *Cedrus* (Pinaceae). Wooden pencils are made from incense-cedar (*Calocedrus decurrens*) and eastern redcedar (*Juniperus virginiana*), which is also used for lining "cedar" chests. Wood from species of redcedar (*Thuja*) is used for roofing shingles and for house siding.

Quick Key to the Cypress Family

Leaves are less than one-quarter inch long and triangular in shape	
The fruit is fleshy and berrylike, and the bark is fibrous	Junipers (*Juniperus*)
The fruit is woody and conelike, and the bark is scaly	Cypress (*Cupressus*)
Leaves are longer than one-quarter inch and oblong to lanceolate in shape	Incense-cedar (*Calocedrus*)

Baker Cypress (*Cupressus baker*)

Description: A number of different trees around the world go by the name cypress. However, only those in the genus *Cupressus* are true cypresses. They look very much like junipers (*Juniperus*), but the tiny, twig-hugging leaves are not in flat sprays. This species is rare and known in only two locations in Plumas County. There is also a second species found at the lower elevations in Kern and Tulare Counties, *C. arizonica,* and it too is rare.

Interesting Facts: A decoction of the stems was taken for rheumatism and colds. Berries were ground into meal and used with bread dough. The berries were also eaten raw or roasted.

Incense-cedar (*Calocedrus decurrens*)

Description: This is a tall tree that grows 75 to 125 feet tall. It has evergreen, aromatic herbage and red-brown shedding bark. The branches form graceful, flat sprays. The leaves are scalelike and closely appressed to the branches. The cone is small, about one inch long, and appears almost birdlike, with six scales. Incense-cedar is fairly common on the higher mountain slopes, particularly where there is sufficient moisture. In most places in the Sierra Nevada, it is found between 6,000 and 8,000 feet.

Interesting Facts: This species was formerly assigned to the genus *Libocedrus*. This plant was used as a drug, for food, and for fiber. As a food, the leaflets were added as flavoring while acorn meal was leached. As a drug, the branches and twigs were used in sweat baths. The Paiutes made an infusion of the leaves, and the steam was inhaled for colds, whereas other Natives made a decoction of the leaves for stomach troubles (Strike 1994).

The wood of incense-cedar can be used in fire-by-friction fire-starting methods. The bark and wood can be used in weaving baskets. The roots were used as warps and weft in baskets, too.

Junipers (*Juniperus*)

Description: These are evergreen trees and shrubs, with opposite or whorled, scalelike or linear leaves. The male cones are small, and the female cones are larger and "berrylike." There are one to three seeds within the berries, which often have a waxy bloom on them. Three species can be found in the Sierra Nevada.

U-MA-CHA
The typical shelter of some Native Americans in the Yosemite area was the conical *u-ma-cha*. Ten- to twelve-foot poles were set in the ground around an area twelve feet in diameter with the tops of the poles inclined together. Over this framework, slabs of incense-cedar bark were placed. This structure was easily built, fairly waterproof, and readily kept warm. The entrance was placed on the south side, and there was an opening at the top to allow smoke to escape from a fire that was often kindled in the middle of the dwelling. The *u-ma-cha* easily housed a family of six and all their possessions.

Quick Key to the Junipers

Leaves awl-like; plants prostrate shrubs	Dwarf Juniper (*J. communis*)—This species occurs in stony or wooded slopes between 6,000 and 11,000 feet in the northern Sierra Nevada.
Leaves mostly scalelike; plants more or less trees	
Mature berrylike cones are bluish black under a waxy white coating	Mountain Juniper (*J. occidentalis*)—This is a low tree, fifteen to sixty feet tall with a well-defined trunk growing in dry habitats between 3,000 and 10,500 feet throughout the Sierra Nevada.
Mature berrylike cones are reddish brown under a waxy white coating	California Juniper (*J. californica*)—This shrub has no distinct trunk below the branches. California juniper is common on dry, rocky slopes at elevations below 5,000 feet in the southern Sierra Nevada.

Interesting Facts: Junipers offer countless products. The berries and twigs can be used to make tea, season game, smoke fish, repel moths, soothe rheumatic pains, and kill infectious germs. The fleshy cones are edible raw but taste better if dried, ground, and used as a flour or flour extender or made into cakes. Cooking the flour with other foods can make it more palatable. The berries can also be roasted and ground and used as a coffee substitute. The Swedes make an extract from the berries, which they generally eat with bread, in much the same way we use butter. In addition, the berries of *J. communis* have been used to give gin its characteristic flavor.

ASH CAKES
Make a dough by mixing flour with water. Pat the dough into a patty. The thicker it is the more doughy it will be, whereas the thinner it is the crispier it will be. Throw the patty into the ashes (hot coals), let it cook a bit, and you have an ash cake. The ash acts as leavening.

The boughs can be steeped in hot water for five to ten minutes to make a beverage. Cooking them in an uncovered pot is recommended to allow the volatile oils to escape. A leaf or berry infusion was used to relieve urinary problems.

The shredded bark is excellent tinder for primitive fire-starting techniques and can be used as bedding and padding. It is said that some Native American tribes ate the inner bark in times of famine. The inner bark was also used for clothes and mattresses and could be worked with the hands until soft enough to use for baby diapers or sanitary pads.

Juniper-oil extract can be rubbed on skin as an insect repellent and to relieve pain in muscles and joints. The bark, roots, twigs, and cones furnish red dyes. The white film covering the berries contains a type of yeast that can be used to make a primitive sourdough starter (Webster 1980).

EPHEDRA FAMILY (Ephedraceae)

These are shrubs or small trees with opposite or whorled, jointed branches. Leaves are foliaceous or scalelike. There is one genus with about forty species that is found in warm, arid regions of the Old World and New World. An Asiatic species is the source of ephedrine.

Ephedra (*Ephedra viridis*)

Description: This is an erect shrub with broomlike, leafless, yellow-green branches. The leaves are scalelike at the joints and are seldom noticed. Leaf scales are opposite, about one-eighth-inch long, and deciduous. Cones are about three-quarters of an inch long and occur at two or more at the nodes. Ephedra is common on dry, rocky slopes from 4,500 to 7,000 feet on the eastern side of the Sierra Nevada.

Interesting Facts: The name "ephedra" is from ancient times, as the plant has been in human history for thousands of years. Long ago, the Chinese realized the medicinal properties of various species of ephedra for treating respiratory ailments, and we now know that ephedra taken orally does stimulate the body in a manner similar to injected adrenaline.

The active alkaloids ephedrine and pseudoephedrine are synthetically produced under the name of ephedrine and are two of the leading over-the-counter and prescription treatments for allergies, congestion, and asthma. These alkaloids are present only in the Asian species. Our common species, *E. viridis*, has little or no ephedrine in it (Moore 1989). Western U.S. residents have for many years brewed what some consider a soothing tea from stems. *Viridis* is Latin for "green," referring to the green stems.

Note: Pseudoephedrine is now synthesized commercially and is an ingredient in commercial asthma and cold remedies. Pseudoephedrine is also a precursor in the production of the dangerous illegal drug methamphetamine ("speed"). A tea with stimulant properties is made by steeping dried stems. It has been used medicinally to treat a variety of ailments including syphilis, diabetes, and pneumonia. A Chinese species is the source of ma huang, a tea so potent that it has caused deaths from overstimulation of the heart.

PINE FAMILY (Pinaceae)

There are 10 genera and 250 species in this family worldwide. Six genera are native to the United States. The pine family is composed of trees with needlelike leaves. Except for larches (*Larix*), most species are evergreen. Many are economically

---

Quick Key to the Pine Family

Leaves needlelike, two to many per cluster, each clustered at the base by a papery sheath, cones maturing in two or three seasons	Pines (*Pinus*)
Leaves linear, often flattened, occurring singly along the branchlets	
Cones hanging on branchlets, the entire cone falling as one unit	
Branchlets rough because of scattered peglike stubs that remain after the leaves have fallen	Hemlock (*Tsuga*)
Branchlets smooth, without peglike structures; cones with bracts longer than the scales	Douglas-fir (*Pseudotsuga*)
Cones upright on the branches, the cone scales falling away after the seed is shed, leaving a single spikelike central axis	True Firs (*Abies*)

---

important timber trees. Male cones are small, soft, and deciduous after shedding pollen. Females cones (the pinecone) are more robust, with woody, spirally arranged scales bearing the seeds on their upper surface. Both winged and non-winged seeds have evolved depending on the reproductive strategy of the species.

Douglas-fir (*Pseudotsuga menziesii*)

Description: This is a tall tree, growing thirty-five to sixty feet, with drooping branches. The leaves are needlelike, blue-green, and spirally arranged on the branches, but they appear to be in a flat spray because the needles are turned at the petiole base. Needles are three-quarters to one and one-half inches long and pointed at the tip. Cones are cylindrical, four to six inches long, with three-fingered bracts overlapping the scales. These bracts are characteristic of the genus.

Interesting Facts: Douglas-fir is an important timber species. The wood is resinous with close, even, well-marked grains and is of medium weight, strength, stiffness, and toughness. It is very durable and when well seasoned does not warp. It is used in piles, ties, floors, and millwork and to make a variety of items such as spear handles, spoons, fire tongs, and fishing hooks.

A tea can be made from the needles of Douglas-fir (Davidson 1919). Similar to pines, the pitch can be used as glue. It can be used for sealing implements and caulking water containers. Medicinally, the sap provided a salve for wounds and skin irritations. The pliable roots have been used in weaving.

Douglas-fir
(*Pseudotsuga menzesii*)

Mountain Hemlock (*Tsuga mertensiana*)

Description: Mountain hemlock grows up to one hundred feet tall in some areas. The crown is narrowly pyramidal when young, becoming irregular with age. Needles are nearly circular in cross section, round tipped, and up to an inch long. Bark is reddish brown with narrow ridges, becoming heavily furrowed on old trees. Immature male pollen cones are purple, while female ones are green to purple. Both sexes are often borne on the same branch. The seed cone is cylindrical, light to mid-brown, up to two and one-half inches long. It opens and releases seeds in the fall and then falls off. This species is found growing on north-facing slopes between 6,000 and 11,000 feet.

Interesting Facts: In general, *Tsuga* furnishes cheap coarse lumber for framing, sheathing, lathes, rafters, and other types of rough construction. Some other uses include pulp, ties, boxes, and planks. The wood is coarse grained and splintery but is very strong, stiff, tough, and easily worked. It is also a source of alpha cellulose for making cellophane, rayon yarns, and plastics.

A tea can be made from the needles or inner bark of either species by steeping them in hot water. The inner bark can also be eaten raw, boiled like noodles, cooked with berries, or dried and used as a flour substitute, or it can be blended with sap and pressed into cakes. Indians of southeastern Alaska made a coarse bread from the inner bark (Bank 1951). Hemlocks are high in tannins and can be used as a tanning agent, pigment, and cleansing solution. A red dye was derived from the bark to color basket materials. The wood is durable and easy to carve into implements such as spoons, combs, spear shafts, and fishing hooks. The branches make good bedding material. Hemlock pitch is used topically as a poultice, liniment, and salve. The young branch tips were chewed by aboriginal peoples as a hunger suppressant (Turner and Kuhnlein 1991). Finally, the powdered inner bark is a good ingredient in foot powder, as it reportedly eliminates foot odor.

Pines (*Pinus*)

Description: Pines may be divided into two major subgroups—"soft" pines and "hard" pines. In soft pines, the needles are usually in bundles of five, and the cones have no prickles. The wood of the soft pines is straight grained, comparatively free from resin, and easy to work. It is used for rough carpentry, cabinetwork, patterns, toys, crates, and boxes. In contrast, hard pines have two to three needles per bundle, and the cones have prickles. The strong, resinous wood of hard pines is used in buildings, bridges, ships, and other types of heavy construction. Because of its durability, the wood of hard pines is valuable for floors, stairs, planks, and beams.

Interesting Facts: All pines have edible seeds. However, they are an erratic food source, yielding an abundant crop in some years and a sparse crop in others. To

Barbour, B. Pavlik, F. Drysdale, and S. Lindstrom, *California's Changing Landscape: Diversity and Conservation of California Vegetation*

Quick Key to the Pines

Needles occurring in bundles of three or less
  Needles three per bundle

Needles appearing more grayish in color; plants found in lower elevations on west side of the range (usually below 4,500 feet)	Gray Pine (*P. sabiniana*)—Gray pine grows on dry slopes and ridges below 5,000 feet on the west side of the Sierra Nevada.

  Needles green
  Cones three to six inches long — Ponderosa Pine (*P. ponderosa*)—Ponderosa pine is found from 5,000 to 9,000 feet.

  Cones six to twelve inches long with in-curved prickles — Jeffrey Pine (*P. jeffreyi*)—Jeffrey pine is a quite common or dominant species on the eastern side of the Sierra Nevada.

Needles one or two per bundle
  Needles occurring singly, one to one and a half inches long — One-leaf Pinyon Pine (*P. monophylla*)—Common on desert mountain slopes between 3,500 and 5,500 feet.

  Needles in bundles of two, one and a half to two and a half inches long — Lodgepole Pine (*P. contorta*)—Lodgepole pine is common in dry to moist areas, in some places forming pure stands.

Needles occurring in bundles of four or five
  Needles usually two inches or less; cones are rounded in shape
    Needles one and a half to two inches long, cones remaining closed — Whitebark Pine (*P. albicaulis*)—This species occurs in rocky habitats near timberline between 7,000 and 12,000 feet.

    Needles three-quarters to one and a half inches long, cones opening at maturity — Foxtail Pine (*P. balfouriana*)—Look for this species in rocky slopes between 9,000 and 11,500 feet in the southern Sierra Nevada.

  Needles usually between one and a half to four inches long, cones longer than wide
    Needles one and a half to three inches long; cones with a short or no stalk — Limber Pine (*P. flexilis*)—Limber pine occurs on dry mountain slopes, from 7,500 to 11,000 feet.

    Needles two to four inches long; cones about four to eight inches long

      Cones four to eight inches long; plants growing at elevations between 6,000 and 9,000 feet — White Pine (*P. monticola*)—This species occurs in various habitats between 5,000 and 10,000 feet throughout the Sierra Nevada.

      Cones ten to sixteen inches long; plants mainly below 7,000 feet — Sugar Pine (*P. lambertiana*)—Sugar pine is common in the upper montane zone, from 6,000 to 10,000 feet.

35
Major Plant Groups

One does not usually think of conifer trees and forests as sources of food. To Native Californians of the past, however, these forests were great orchards. While oak acorns symbolically provided bread for subsistence, pine nuts provided the cake.
  —M.

collect the cones, long poles were used to knock them from the branches. One of the best ways to gather seeds is to heat the green cones until they open. The seeds are best when harvested in fall or early winter when cones normally release their seeds. The nutritious seeds can then be shelled and eaten or ground or roasted and made into flour. Seeds may contain as much as 15 percent protein, 50–62 percent fat, and 18 percent carbohydrates, with approximately three thousand calories per pound (Harrington 1967; Farris 1980; Vizgirdas 2003a, 2003b).

The inner bark is also edible in an emergency. Though a tedious process, the tender mucilaginous layer between the bark and wood was scraped or peeled off. It was then cooked or ground into meal. The use of the inner bark by Native Americans was so extensive that early explorers reported large areas of trees stripped of bark (Strike 1994).

The firm and unexpanded pollen cones can be boiled and eaten. They have a surprisingly sweet and nonpitchy taste. The edible pollen is usually mixed with flour and used as a soup thickener.

The needles of most pines can be steeped in hot water to make a satisfying tea and are a good source of vitamin C. It also takes some practice to steep the right amount of leaves, as too much may be too strong. Additionally, the pine cleaning fluid can be extracted from boiling the needles and skimming off the oil-like substance from the surface. It may take a lot of pine needles to get a small cupful.

Pine sap can be collected in quantity from cuts, burns, and broken branches. The collected sap is then heated and formed into balls for future use. Be careful not to expose the sap to flames, as it is very flammable.

All species of pine have been used medicinally for many centuries. Chewing the sap was said to be soothing for a sore throat. The sap can be dried, powdered, and applied to the throat with a swab. It was also heated and used as a dressing to draw out embedded splinters or to bring boils to a head. Smeared on a hot cloth, the sap was used much like a mustard plaster in treating pneumonia, sciatic pains, and general muscular soreness. Pine oil is widely used in massage oil for muscular stiffness, sciatica, and rheumatism and in vapor rubs for bronchial congestion. All pines are rich in resin and camphoraceous volatile oils, including pinene, that are strongly antiseptic and stimulating. The needles and resin produce a brown dye.

Pine roots were valued as twining material for baskets. The roots, about as thick as a pencil, can be several feet long. Roots were collected from opposite sides of a tree each year to prevent destroying the tree. After cleaning, roots were buried in a heated pit. Fire built over the pit for two days turned the roots a light tan color, and they were then ready to be split into smaller strips.

THE PANDORA MOTH
Known to the Paiutes as "peaggies," the larvae of the pandora moth (*Coloradia pandora*) were a choice food. To capture them, they dug shallow trenches around infested trees and set smudge fires around the tree bases. The smoke forced the larvae to drop to the earth where they were collected by the Indians who cooked and dried them and then mixed them with vegetables in stews. The pandora moth is usually associated with pines such as the ponderosa, Jeffrey, and lodgepole. In general, outbreaks of this insect occur in areas with soils loose enough for the larvae to bury themselves prior to pupation—chiefly, pumice soils and decomposed granite. Other Native Americans in the Klamath region preferred the pupae, which were dug from the ground and roasted.

Pine seeds have outstanding food value, and are especially high in fat content. Modern diets tend to focus on protein and to play down the fat, but fat is highly important to people exposed to low temperatures and lacking warm clothing, as was the case among aboriginal Californians. The fifty percent fat content may have been a more important survival factor than the twenty-five percent protein content.
—M. Barbour, B. Pavlik, F. Drysdale, and S. Lindstrom, *California's Changing Landscape: Diversity and Conservation of California Vegetation*

Table 1 Nutrients and Calories of Pine Seeds, Acorns, and Modern Wheat

Plant Material	Protein (%)	Fat (%)	Carbohydrate (%)	Energy Content (calories/3.5 oz)
Sugar pine seed	21.4	53.6	17.5	594
Coulter pine seed	25.4	51	14.4	574
Pinyon pine seed	8.1	23	56.3	450
Black oak acorn	3.8	19.8	64.8	443
Corn flour	7.8	2.6	76.8	361
Camas bulb	0.7	0.2	27.1	110
Wild onion	2.2	0.4	20.8	96
Modern wheat	13.3	2	71	352

Adapted from Farris, G. 1980. "A Re-assessment of the Nutritional Value of *Pinus monophylla*." *Journal of California and Great Basin Anthropology* 2(1):132–36.

True Firs (*Abies*)

Description: Firs are tall evergreen trees with the bark of immature stems having "resin blisters." They have whorled branches and flattened, linear, needlelike leaves. Male cones are small and pendant from the lower side of the branchlets near midcrown. At the top of the crown, the female (seed) cones are borne erect at the tips of the previous season's growth. When these cones have ripened in the fall of the first season, they disintegrate, leaving an upright, persistent, bare central axis. The genus name comes from the Latin *abire*, meaning to rise, alluding to the great height some species attain.

Quick Key to the True Firs

Leaves flattened, cones narrow	White Fir (*A. concolor*)—This species is rather common below 7,000 feet but may be found up to 10,000 feet throughout the Sierra Nevada.
Leaves four-sided, cones broad	Red Fir (*A. magnifica*)—This is a dominant species in the mid-elevation western slopes between 5,000 and 9,000 feet.

White Fir (*Abies concolor*)

Interesting Facts: Native Americans used the needles of white fir for tea. Medicinally, an infusion of the foliage was taken and used as a bath by the Acoma and Laguna Indians for rheumatism. Some Native peoples in Nevada used a decoction of white fir–bark resin and needles to help pulmonary troubles. A simple poultice or warm pitch of resinous sap from the bark or large branches could be applied to sores or boils. The Tewa Indians used the sap from the main stem and larger branches for cuts (Wilford, Harrington, and Freire-Marreco 1916). A decoction of the resin can be taken for venereal diseases. The resin has been known to be used by early New Mexico natives to fill decayed teeth. Extracts from the bark have shown antitumor activity. One of the active materials might be a complex tannin.

Historically, white fir was considered undesirable for timber. Now that the availability of premium timber species has declined, white fir is being recognized as a highly productive and valuable tree species and is widely used in the wood products industry. White fir is a general, all-purpose, construction-grade wood used extensively for solid-construction framing and plywood and to a lesser extent for pulpwood. It is also used for poles and pilings because of its straight grain and low taper but requires large amounts of preservatives because the heartwood decays rapidly. It is poorly suited for firewood because of its low specific gravity and heat production (80 percent as much heat by volume as Douglas-fir produces), but it is used for firewood anyway.

YEW FAMILY (Taxaceae)

There are five genera and twenty species distributed in the Northern Hemisphere. Two of the genera are native to the United States. The family is of economic importance as a source of timber and ornamentals and pharmaceuticals (for instance, taxol).

Quick Key to the Yew Family

Fruit a seed partially enclosed in a scarlet or coral-red fleshy cup, maturing in one season	Pacific Yew (*Taxus*)
Fruit olivelike, consisting of a seed totally enclosed in a purple-striped green skin, maturing in two seasons	California Nutmeg (*Torreya*)

California Nutmeg (*Torreya californica*)

Description: The genus has seven species of trees and shrubs, and two are native to the United States. California nutmeg is an erect, single-stemmed, small- to medium-size tree, usually fifteen to ninety feet tall and up to twenty inches in diameter. The crowns of young trees are broadly conical, whereas mature trees have long, clear stems and domed crowns. The leaves are linear, stiff, spine tipped, and arranged in flat sprays, and when bruised the leaves emit a foul odor. The arils are olive shaped and green with purplish streaks. California nutmeg grows in coniferous forests, riparian zones, and hot, dry chaparrals at elevations ranging from 100 feet to 7,000 feet and is very shade tolerant (or sun intolerant).

Interesting Facts: The species is often called "stinking cedar," because of its strong aroma. It has also been called nutmeg cedar because of the shape of the fruit and its smell. Unlike Pacific yew (*Taxus*), California nutmeg is not harvested as a source of taxol because it produces the substance in extremely small quantities. The common name of nutmeg is a rather poor choice and can lead a person to think that the spice nutmeg can be obtained from it. In reality, the commercial nutmeg comes from the nutmeg tree (*Myristica fragrans*), which is not found in the Sierra Nevada.

The white inner nut is edible after roasting (Romero 1954). The seed oil has potential use in cooking, being similar in quality to olive and pine-nut oils. The seeds of a related Asian species (*T. nucifera*) are harvested in Japan for rendering high-quality cooking oil.

The plant was prized by many Native Americans for its wood in making bows. Early settlers used the wood for posts, pilings, and bridges due to its decay-resistant qualities. The tree roots were also used in basketry.

BOWS AND ARROWS
The construction of bows and arrows was a demanding task. Incense-cedar and California nutmeg furnished the wood for the bow. Incense-cedar had to be treated for several days with deer marrow to prevent brittleness when dry. The finished bow was three to four feet long, sinew backed, and often had recurved ends. Glue used for applying the sinew to the back of the bow was made by boiling deer bones and combining the product with pine pitch. The arrows often had two parts. The detachable foreshaft remained in the wound, preventing it from closing and hastening the animal's death from loss of blood. Arrow shafts were made from syringa or wild rose by removing the bark, stripping and trimming the pieces to an even thickness, and then straightening them with stone tools. When the shaft was ready, obsidian points and feathers were attached.

Pacific Yew (*Taxus brevifolia*)

Description: This is an evergreen shrub with spreading branches and drooping twigs. The very thin bark is composed of outer purplish scales covering the newly formed reddish to red-purple inner bark. The genus name is the classical Latin name. This coniferous tree has flat, linear leaves that are borne spirally but twisted so as to be arranged in two ranks. Male and female cones are borne on separate plants. Male cones are round and consist of a tight bundle of stamens subtended by roundish bracts. The female cone consists of a single ovule that develops a fleshy red aril.

Interesting Facts: In general, yews are toxic (Kingsbury 1964, 1965). However, the twigs of some eastern species were used for tea. Additionally, the red fruits borne on the female plant are said to be edible in small amounts and have a cherry Jello-like taste, but the seeds are considered to be very poisonous (Willard 1992).

The bark of Pacific yew contains taxol, an anticarcinogen that slows tumor growth (Duke 1992a; Foster and Duke 1990). It is presently being tested against breast, ovarian, and kidney cancers. This species has become an important source of income in the Northwest, but because of its slow growth, there is growing concern that the yew may become endangered due to overharvesting. There is only a small amount of taxol in the yew's bark, and it takes ten tons of bark to make two pounds of medicine. Fortunately, taxol is now produced synthetically.

The leaves are poisonous if eaten. However, some Native Canadians used the leaves in smoking mixtures, which have been described as very potent (Turner and Kuhnlein 1991).

The yew's hard wood is ideal for carving and takes on a polished look. Many useful implements can be made, including archery bows, clubs, paddles, digging sticks, eating utensils, snowshoe frames, and awls.

Warning: Because yews are known to be toxic, the use of any part as food is not recommended. 39

Flowering Plants
Dicots

A man is rich in proportion to the number of things which
he can afford to let alone.
—Henry David Thoreau, *Walden*

Dicotyledonous plants (dicots) can be distinguished by several key characteristics. In dicots, the embryos have two cotyledons (seed leaves), and the leaves are usually net veined. The flower parts are in fours or fives, rarely in threes, and the plants are herbaceous or woody. Most important, the possession of cambium cells distinguishes dicotyledons from the monocotyledons.

AMARANTH FAMILY (Amaranthaceae)
The amaranth family contains more than sixty genera and nine hundred species distributed worldwide. Many are weedy species of little economic importance. Some species of amaranthus are cultivated for their red pigmentation. The family name originates from the Greek *amarantos,* which means unfading, possibly alluding to the "everlasting" quality of the papery perianth parts.

Pigweed, Amaranth (*Amaranthus*)
Description: Approximately sixteen widespread species of amaranthus occur in the Sierra Nevada. In general, they are herbaceous annuals with small greenish flowers and alternate entire or wavy-margined leaves. Pigweeds occur in many different habitats and often hybridize, making identification difficult.

Interesting Facts: Used by Native Americans, the dried small black seeds of amaranth have been found in many archaeological remains. Seeds of all species can be eaten whole as a cereal or ground into meal and made into cakes. The seeds are best collected in summer when the plants are fully mature. To free the seeds from their husks, rub the seed clusters between your hands. You can then winnow the seeds if there is a breeze, or if the air is calm slowly pour the seeds out of your hands and blow the chaff away. The seeds contain approximately fifteen grams of protein per one hundred grams, more than is found in rice and corn, and equal to, if not surpassing, that found in wheat. When ground into a flour, amaranth has a distinctive flavor that is a bit strong used alone. We find it is better when mixed with other flours for breads and pancakes.

The highly nutritious amaranth contains more fiber and calcium than any other "cereal" in addition to a wide spectrum of vitamins and minerals, including vitamins A and C, calcium, magnesium, and iron. Amaranth is rich in the amino acid lysine, a product scarce in true cereal grains, thereby providing a more balanced source of protein.

The edible young shoots and leaves have a pleasant taste if eaten as a potherb soon after collection. Because the plants can accumulate nitrates, it is wise not to consume large quantities where nitrate fertilizers are used. Livestock losses have occurred as a result of excessive amaranth consumption. The leaves of amaranth contain oxalic acid, which binds with calcium, restricting its absorption by the

LYSINE
Lysine is one of numerous amino acids that is needed for growth and tissue repair. It is considered to be one of the nine "essential" amino acids because it comes from outside sources such as foods or supplements. Like all amino acids, lysine functions as a building block for proteins and has a key role in the production of various enzymes, hormones, and disease-fighting antibodies. Although many foods supply lysine, the richest sources include red meats, fish, and dairy products. Vegetables, on the other hand, are generally a poor source of lysine, with the exception of legumes (beans, peas, and lentils).

body. As long as your diet contains plenty of calcium from other sources, eating amaranth and other vegetables that contain oxalic acid (such as spinach and wood sorrel) should not cause any health problems.

Amaranth also has astringent properties and can be used for treating diarrhea, excessive menstrual flow, hemorrhaging, and hoarseness. Amaranth is also helpful in treating mouth and throat inflammations and sores and in quelling dysentery. Steeping dried leaves in boiling water (the more leaves steeped, the stronger the tea) was considered a valuable remedy.

In the Midwest, several species of amaranth are being grown as agricultural crops for use in cereals and bread. It is photosynthetically efficient and produces a high yield of both greens and seeds. Amaranth was an important food in the past and may become an important one for the future (National Research Council 1985).

BARBERRY FAMILY (Berberidaceae)

There are 9 genera and 590 species in this family, distributed throughout the Northern Hemisphere and South America. Two genera are native to the United States. This is a diverse family of perennial herbs and shrubs. The flowers have six or more stamens that split open by two hinged valves to splatter pollen over insects as they crawl by. Several species are cultivated as ornamentals.

Barberry, Oregon-grape (*Berberis*)

Description: The genus *Berberis* has about six hundred species, with six species occurring in California. In California these are rhizomatous, viney, or upright shrubs with pinnately compound, evergreen leaves. The leaflets have spiny margins, and the yellow flowers are in three whorls that are interpreted as bracts, sepals, and petals. The fruits are blue to purple in color and have a waxy covering.

Oregon-grape (*Berberis* spp.)

In some books, barberries are those plants with simple leaves and spiny stems, whereas Oregon-grapes have compound leaves and spiny blades. As such, they have been treated as separate genera, *Berberis* and *Mahonia*, respectively.

BERBERINE
Berberine-containing plants are used medicinally in virtually all traditional medical systems and have a history of usage in Ayurvedic and Chinese medicine dating back at least 3,000 years. Berberine has demonstrated significant antimicrobial activity against bacteria, fungi, protozoans, viruses, helminths, and chlamydia.

Quick Key to the Barberries	
Leaves generally have eleven to twenty-three leaflets	Barberry (*B. nervosa*)—This rarely encountered species grows in rocky areas in the central part of the range. It flowers from April to May.
Leaves generally have five to eleven leaflets	Oregon-grape (*B. aquifolium*)—This is a creeping or erect shrub that may grow up to ten feet tall. It occurs below 7,000 feet in forests.

Interesting Facts: The blue berries are edible raw or can be dried for winter use or added to soups to improve flavor. We like them best when picked right off the plant. They also make good jellies.

Medicinally, the bark may be boiled and the infusion used to wash sores on the skin and in the mouth. The plants contain berberine, a bitter alkaloid that gives roots their distinctive yellow color and usefulness as a digestive tonic. Berberine stimulates the involuntary muscles and possesses antipyretic, laxative, and antibacterial qualities. A tea from the berries of *B. vulgaris* (common barberry) can be drunk to stimulate an appetite. The liquid obtained from the root by chewing was placed on injuries and wounds.

A yellow dye can be obtained by boiling bark and roots. The whitish film on the berries is a yeast that can be used in making a primitive sourdough starter (Webster 1980).

Caution: The roots should be considered toxic, and the spines on the leaves may inject fungal spores into the skin.

BEECH FAMILY (Fagaceae)

Members of this family are trees and shrubs, either deciduous or evergreen. The flowers are typically unisexual, with the staminate flowers being catkinlike and the pistillate in an involucre. The fruit is a nut fused to a cuplike organ (cupule) that is composed of fused bracts. There are approximately eight genera and nine hundred species in this family. The family is a source of lumber, edible fruits, cork, and many ornamental shade trees.

Quick Key to the Beech Family

Nuts enclosed in a spiny or prickly burr	Chinquapin (*Castonopsis* = *Chrysolepis*)
Nuts only partially enclosed by an open scaly cup ("acorn")	
Male flowers borne in many-flowered, narrow, upright catkins	Tanbark Oak (*Lithocarpus densiflora*)
Male flowers borne in many-flowered, narrow, hanging catkins	Oaks (*Quercus*)

Chinquapin (*Castonopsis* = *Chrysolepis*)

Description: These are evergreen shrubs with leathery two-ranked or spirally arranged leaves and erect catkins of small flowers. The fruit consists of a spiny case (cupule) surrounding three nuts.

Quick Key to the Chinquapins

The tips of the leaves are rounded and broad; bark is smooth and thin; shrub	Bush Chinquapin (*C. sempervirens*)—This species occurs on dry, rocky slopes and ridges between 2,500 and 11,000 feet throughout the Sierra Nevada. It flowers from June to August.
The tips of the leaves are pointed and tapered; bark is furrowed and thick; tree or shrub	Golden Chinquapin (*C. chrysophylla*)—This species is infrequently encountered on gravelly or rocky ridges and slopes below 6,000 feet. It flowers from June to September.

Interesting Facts: The seeds can be eaten raw or stored for winter use. They can also be roasted and pounded into flour. The Paiute made a tea from the leaves.

Oaks (*Quercus*)

Description: There are about six hundred species of oaks, with about sixty native to the United States and nineteen native to California. They are described as either trees, shrubs, or shrubs that can become trees. The leaves are either winter deciduous, drought deciduous, or evergreen. The margins can be lobed, entire, serrated, or toothed. Fruits are acorns—nuts enclosed by a scaly cup that takes one to two years to mature.

Leaves are deciduous

Leaf lobes have one to four coarse, bristle-tipped teeth and pointed lobes	California Black Oak (*Q. kelloggii*)—This winter deciduous tree with deeply lobed leaves is common between 1,000 and 8,000 feet. It may produce 200 to 300 pounds of acorns a year. These acorns were prized by the Native Americans, as they were good tasting and stored well.	43 *Major Plant* *Groups*
Leaf lobes lack the bristle-tipped teeth and have rounded lobes	Brewer's Oak (*Q. garryana*)—Look for this species on dry slopes throughout the Sierra Nevada.	

Leaves are evergreen

Leaves have shiny lower surfaces — Interior Live Oak (*Q. wislizenii*)—This evergreen tree is common on slopes and in valleys below 5,000 feet. It flowers from March to May. It may produce 200 pounds of acorns a year. The acorns take two years to mature.

Leaves have dull lower surfaces

Leaves have olive green, gray-green, or yellow upper surfaces — Huckleberry Oak (*Q. vacciniifolia*)—This species is found on dry ridges and rocky places between 3,000 and 10,000 feet from Fresno County northward.

Leaves have dark green or green upper surfaces

Leaf margins have teeth or spines and have rounded tips — Scrub Oak (*Q. berberdifolia*)—This is a common species on dry slopes from Tehama County southward. It flowers from April to May.

Leaf margins may have teeth or spines, and they have acute tips — Canyon Live Oak (*Q. chrysolepis*)—This evergreen tree is common on slightly moist slopes and in canyons below 6,500 feet. Catkins appear in April and May. The tree may live more than 300 years and produce 150 to 200 pounds of acorns a year. The acorns take eighteen months to mature.

Interesting Facts: With the exception of the highest mountains and the desert areas in the southwestern part of the state, oaks are an abundant tree in California. These plants were particularly important to many Native Californians as a primary food source. In some places, the acorns constituted as much as 75 percent of their daily diet (Merriam 1918).

The acorns of oaks were a very important food staple for many Native peoples. At least half of the Native American tribes in the United States ate acorns, but they were a staple for those in California. There are definite differences in the flavor of acorns, and you may also develop preferences as the Natives did. In fact, even people today when collecting acorns may pass up nearby oak groves of less favored species and travel further for stands offering a preferred acorn.

The acorns of most species ripen in the fall. In most cases, long poles can be used to knock acorns from the trees. Once collected, the acorns can be cracked with a stone or teeth and then spread out in the sun to dry. Native peoples often shelled the acorns before carrying them home so as to lighten the load.

Once acorns were brought back to camp, they were stored in granaries (for instance, on platforms, on bare rock or the ground, in woven baskets, and so on) in such a way as to prevent molding. In more arid climates, basketlike granaries might be less tightly constructed. Where insects were a problem, the leaves of such plants as California bay (*Umbellularia californica*) or wormwood (*Artemisia ludoviciana*) were used in the granaries to deter insects.

Once the acorns are shelled, the kernels should be rubbed by hand on a basket tray to remove the thin, papery membrane covering it. These are then winnowed, allowing the wind to blow away the lightweight skin while the kernels fall back into the basket, ready to be pounded into flour.

The dry acorns can be pounded using a stone pestle (a large flat stone) and a bottomless basket. Another method to produce meal is to use a pestle in conjunction with a mortar (depressions in another rock). Here, the depression is filled with acorn and then pounded. It is important to try to limit the pounding of the rock, as it will then, in the form of grit, become part of the flour. This may be a tedious and time-consuming process, but the food value of the acorns obviously makes it worthwhile. The pulverized meal is winnowed often during the pounding process, with the larger pieces being returned to the mortar for more work.

The pulverized acorn meal contains tannins and glucosides that need to be leached out of the meal in order to make it edible. One of the usual primitive methods in completing this task is to scoop a depression in sandy soil, line the depression with large nonpoisonous leaves such as maple, put the meal in the pit, and pour water through the meal until the inedible substances are washed away. Another method would be to use loosely woven baskets, perhaps partially buried in sand, to leach the meal.

Sometimes, the acorns can be leached before pounding. Some Natives buried acorns in swampy places for up to a year (until they turned black), then roasted and ate them without further leaching. Other groups left the shelled acorns to mold in a basket and then buried them in clean sand in a riverbed until they turned black and were ready to eat. The boiled whole or broken kernels can also be boiled in water for forty-five minutes or so, changing the water several times.

After leaching and tasting the acorn meal to be sure it is sweet, the meal is then ready for use. A porridge can be made by putting some meal into a watertight basket with some water. Then, rocks that have been heated in a fire and cleaned by dipping them into water are then dropped into the basket containing the meal. The rocks should be stirred constantly to prevent them from burning the basket, and when cooled they should be replaced with other hot rocks. This is repeated until the acorn meal takes on the consistency of porridge. Adding more water would result in a soupy concoction.

CHUCK-AHS
These are the granaries that acorns were stored in. A *chuck-ah* is basically a frame support made of four slender poles of incense-cedar about eight feet high around a center log or rock two feet high for the bottom. The basketlike interior is of interwoven branches of deer-brush tied at the ends with willow stems and fastened together with grapevine. This is then lined with dry pine needles and wormwood (to discourage insects and rodents). After the *chuck-ah* was filled with acorns, it was topped with pine needles, wormwood, and sections of incense-cedar bark bound down firmly with wild grapevine to withstand winds. Nearly every family had at least one of these caches that could hold almost a whole winter's supply of acorns or seeds.

Table 2 Chemical Composition of Hulled Acorns as Compared with Barley and Wheat

Species	Water	Protein	Fats	Fiber	Carbohydrates	Ash	Total Proteins, Fats, and Carbohydrates
					Chemical Composition (%)		
Tanbark oak	9	3.0	12	20.0	54.5	1.5	69.5
Canyon live oak	9	4.0	9	12.5	63.5	2	76.5
Black oak	9	4.5	18	11.0	54.5	2	77.0
Coast live oak	9	6.0	17	11.5	55.5	2	78.5
Barley	10	8.5	2	6.0	71.0	2.5	81.5
Wheat	12	12.5	2	2.0	69.5	1.5	84.0

From Baumhoff, M. A. 1963. "Ecological Determinants of Aboriginal California Populations." *University of California Publications in American Archaeology and Ethnology* 49(2):155–236.

This porridge or soup can be eaten as is, or additional flavoring such as mushrooms, roots, berries, greens, meat, or dried fish can be added. Acorn meal, before cooking, is approximately 18 to 20 percent fat, 5 to 6 percent protein, and 62 to 68 percent carbohydrates, with the remainder being water, ash, and fiber. As compared to corn and wheat, both of which contain 1 to 2 percent fat, 10 percent protein, and 75 percent carbohydrates, acorn meal is much higher in fat and lower in protein and carbohydrates (Merriam 1918; Harrington 1967).

Quercus in general—Some Natives discovered the forerunner to modern penicillin-type drugs. The ground acorn meal was allowed to accumulate mold that was then scraped off and used to treat boils, sores, and inflammations. Oak galls were mashed and boiled and used as an emetic.

Q. kelloggii and *Q. wizlesinii*—A decoction from the inner bark was used to relieve rheumatic pain. Bark was pulverized and used to treat burns.

Q. chrysolepis—The acorns of this oak were sucked on when one had a cough or sore throat. The tannin soothes the throat and acts as a cough drop.

The bark of many oaks provided dye and tannin used to color, soften, and cure buckskin. Oak galls have been used throughout history as a dye source. They were crushed, steeped in water overnight, gently simmered, then thoroughly strained to produce a rich tan color.

Warning: Unprocessed acorns may be toxic if eaten in quantity.

Tanbark Oak (*Lithocarpus densiflora*)

Description: This tree grows up to 130 feet tall and has red-brown branches and densely woolly young twigs. The leaves have prominent lateral, parallel veins. The male catkins have a foul odor, and the acorn forms at the base of the catkins are cylindric and tomentose when young. The acorn cup is shallow with bristly, spreading scales and tomentose insides. Tanbark oak occurs on wooded slopes below 4,500 feet.

Interesting Facts: Where available, the acorns of tanbark oak were considered to be the best tasting. It has been estimated that 100 pounds of tanbark oak acorns provided about 70 pounds of actual food (meal) that contains about 2,250 calories per pound. For more information on how to prepare the acorns for food, see the discussion under oaks (*Quercus*). An infusion of the bark was used as a wash to soothe skin sores or to treat colds and stomachaches.

BIRCH FAMILY (Betulaceae)

Of the 6 genera and 150 species in the birch family, 5 genera are native to the United States. Members of this family are trees and shrubs with deciduous, simple, alternately arranged leaves that have toothed margins. The unisexual minute flowers are arranged in catkins. Both male and female flowers are borne on the same plant. The male catkins are soft and pendant. After releasing abun-

GALLS

There are more than two thousand kinds of galls in the world. These are made by wasps, ants, moths, beetles, flies, and mites. In all cases, the insect injects a substance into the plant in one way or another, which causes an abnormal growth. Usually, this growth is used by the insect to house its eggs and larvae. The gall provides food as well as shelter, and some growths produce large amounts of sugar.

Additionally, some galls are eaten by humans. In Mexico, for example, a large growth caused by a gall wasp is gathered and sold in the fruit stands. In the United States, the catmint gall has been eaten in the past, and it takes on the pleasant taste and aroma of the catmint. Other galls contain lots of tannin (up to 65 percent) and have been harvested commercially all over the world for use in the manufacture of inks and dyes. From the time of the Greeks, Aleppo oak of Asia Minor provided the galls that were used in making ink for important documents. Even in modern times, this ink was specified for use in the printing done for the United States Treasury and the Bank of England.

Quick Key to the Birch Family

Leaves are heart shaped at the base, and there is a soft pubescent on both sides; the fruit is a nut	Hazelnut (*Corylus*)
Leaf bases not heart shaped, and there is little or no hair on the leaves; fruits are conelike	
The female cones are clustered, and each one falls intact; the bark is not aromatic	Alder (*Alnus*)
The female cone is solitary and disintegrates when mature; the bark is aromatic	Birch (*Betula*)

---

dant pale-yellow pollen in early spring, they drop off. The hard female catkin is either erect or pendant and appears conelike. Most members of this family grow in moist soil, particularly along streams. Economic products of this family include lumber, edible seeds, and oil of wintergreen.

Alder (*Alnus*)

Description: The two species of *Alnus* in the Sierra Nevada are small trees or shrubs with smooth, reddish or gray-brown bark. Leaves are egg shaped and have serrate edges. The male catkins are grouped near the end of branches and drop off after pollen is shed. The female catkin is conelike and persistent. These plants are usually associated with riparian and wetland sites at low to midelevations.

Quick Key to the Alders	
Plants are trees more than forty feet tall	White Alder (*A. rhombifolia*)—This tree is found along streams throughout the Sierra Nevada. It flowers from May to April.
Plants are shrubs that grow in thickets	Mountain Alder (*A. incana*)—This shrub occurs in moist places between 4,500 and 8,000 feet. It flowers from April to June.

Interesting Facts: The edible catkins are high in protein but generally do not taste very good. The catkins are more tolerable if they are nibbled raw, added to soups, or dried and powdered and used as a spice. The inner bark is palatable only for a short time in the spring when it is less bitter. A patch of bark is removed from the tree and the tissue scraped off and eaten fresh or dried in cakes.

The bitter leaves and inner bark act on the mucous membranes of the mouth and stomach to stimulate digestion. A tea made from the leaves was used as a wash or a soothing remedy for poison oak or poison ivy, insect bites, and other skin irritations. Used fresh, the inner bark is an emetic, taken to induce vomiting if poisonous substances are ingested. A decoction from alder bark was used to treat colds and stomach trouble. The decoction was also used to reduce pain from burns and scalds. Chewing alder bark is said to turn one's saliva red, which was then used to dye basketry material (Reagan 1934).

Alder is valued for its hardwood and is useful for open fires as it does not readily spark. It is used widely by aboriginal peoples for woodworking, including dishes, spoons, and platters. The wood is also used for making fire drill sets. The astringent bark and woody cones are used for tanning leather. A black-brown dye from the bark was used for coloring fishing nets to make them less visible. Because alders usually grow in the vicinity of free-flowing water, they are considered botanical indicators of water. The roots have small nitrogen nodules that improve the soil for other plants. Alders are good for controlling erosion and floods and for stabilizing stream banks.

Birch (*Betula occidentalis*)

Description: This species is a deciduous shrub with simple, alternate, and sharply toothed leaves. Birches can be found along streams and in wet meadows and bogs from the foothills to upper montane zone.

Interesting Facts: Young birch leaves can be added to salads. The inner bark can be dried and ground into flour, and the twigs can be steeped in hot water for a tea. The juice of birch leaves makes a good mouthwash. A tea was made from the leaves of a related species, *B. papyrifera* (paper birch), and a poultice of the boiled bark was used to treat bruises, wounds, and burns. Birch contains a significant amount of methyl salicylate and is often used in teas for headaches and rheumatic pain. Birch is highly regarded as a medicinal plant in Russia and Siberia for treating arthritis (Tilford 1997).

Because birch bark burns even when wet, it makes a good tinder. Some Native Americans took the new soft wood, chopped it very fine, and mixed it with tobacco. The sap, collected in much the same manner as maple, was sometimes made into vinegar. The best time for tapping is early spring, before the leaves unfurl. The hard wood of some species is used for veneers, cabinetwork, and interior finish. It is also used for paper pulp, woodenware, and novelties. Twigs of some species are distilled to produce oil of wintergreen.

Hazelnut, Filbert (*Corylus cornuta*)

Description: This is a shrub up to ten feet tall with thin, downy, double-toothed leaves. The nuts are wrapped in an "envelope" of leafy bracts that extend to form a fringed tube longer than the nut. This is a widespread species found at the lower elevations on well-drained soils. The genus name is Latin for hazelnut.

Interesting Facts: The sweet nuts of hazelnut ripen in late summer and can be ground into meal and made into bread. The best time to collect the nuts is in early autumn. They may be stored until ripe, then eaten raw or roasted. The long, flexible shoots can be twisted into crude cordage or used in basketry. The roots are said to produce a blue dye, but the method used to obtain the color is unknown.

OIL OF WINTERGREEN
Oil of wintergreen (methyl salicylate) is a folk remedy for body aches and pains and is known for its astringent, diuretic, and stimulant properties.
Caution: Oil of wintergreen, used externally for pain (such as muscular, joint, arthritic, or rheumatic pain), may cause irritation to the skin. As a precaution, test by using a few drops of wintergreen oil in a carrier oil or liniment.

BIRTHWORT FAMILY (Aristolochiaceae)

Only one genus and species occurs in the Sierra Nevada, but about six genera and four hundred species can be found in the Tropics. Plants in this family are herbs or woody vines with commonly heart-shaped entire leaves. The flowers often lack petals and may be carrion scented.

Wild-ginger (*Asarum*)

Description: Wild-ginger is a perennial plant with shiny heart-shaped leaves that forms dense mats on shaded forest floors and meadows. It is fairly common at the lower to middle elevations on the western side of the range. The flowers are brownish purple and bell shaped and are borne in the axil of the leaf. The petal-like sepals are broad at the base and taper into slender threads.

Quick Key to the Wild-gingers

Calyx lobes gradually narrowing to a long, slender tip	Hartweg's Wild-ginger (*A. hartwegii*)—This species occurs in shaded places between 2,500 and 7,000 feet throughout the Sierra Nevada. It blooms from May to June.
Calyx lobes sharp pointed but not long pointed	Lemmon's Wild-ginger (*A. lemmonii*)—This species is found in moist places from 3,000 to 6,000 feet throughout most of the Sierra Nevada. It flowers from May to June.

Interesting Facts: When crushed, the entire plant has a strong lemon-ginger smell. The roots can be eaten raw or dried and ground as a ginger substitute or used to make a tea. The rootstock may also be dried and kept for later use or candied by tenderizing the short pieces in boiling water and then boiling the pieces in heavy syrup.

A tea made from the root was drunk for stomach pains and to relieve gas. A poultice was made for headaches, intestinal pains, and knee pains. A fine powder of the dried roots was inhaled like snuff to relieve an aching head and eyes. Wild-ginger is said to have antibiotic properties. Some Native Americans boiled the leaves, crushed and put them in bathwater, or rubbed them directly on the painful limb.

Warning: The species contains asarone, a compound found in laboratory tests to cause tumors in rats, but it may not have the same effect on humans (Miller 1973). Additionally, wild-ginger is usually found in habitats that are slowly being destroyed by human activities. Overharvesting the plant is becoming a real concern.

BLADDERNUT FAMILY (Staphyleaceae)

The family consists of several genera that occur in the Northern Hemisphere. In general, they are shrubs or trees with pinnate leaves that are usually opposite. There are two species of *Staphylea* in the United States, one in California and one in the eastern portion of the continent.

Bladdernut (*Staphylea bolanderi*)

Description: There are eleven species of deciduous shrubs and trees with opposite pinnate leaves and panicles of white flowers. The fruit is a membranous, inflated capsule. The genus name is from the Greek *staphyle* (cluster), referring to the inflorescence.

Interesting Facts: A related species (*S. trifolia*) was used medicinally by Native Americans in the eastern United States (Foster and Duke 1990). The various species are used as ornamentals.

BLADDERWORT FAMILY (Lentibulariaceae)

There are five genera and three hundred species in this family worldwide. Two genera, *Utricularia* and *Pinguicula,* are native to the United States and occur in California. They are described as annual or perennial herbs of moist and aquatic habitats. The insectivorous species in this family, trapping by means of sticky leaves and bladders, are sometimes cultivated as oddities.

Bladderwort (*Utricularia*)

Description: About 250 species of bladderwort occur in the United States. These are aquatic or bog plants with submersed stems. The leaves are finely dissected, and the yellow flowers are strongly two-lipped with a spur at the base. They are found growing in ponds, lakes, and sluggish streams at the low to middle elevations.

Interesting Facts: Bladderworts are carnivorous plants that entrap small aquatic animals in their bladders. The bladders are closed at the narrow end by valvelike doors that have stiff trigger hairs on the outer surface. When set, the bladders have a partial vacuum, and when a passing animal touches the bristles, the doors open, the walls of the bladder immediately expand, and the sudden inrush of water captures the prey. The process has been timed at 1/460 of a second. Enzymes digest the trapped victim.

The edibility and uses of bladderworts are unknown, but several species in this genus are reputed to have diuretic values and have been used to treat dysentery (Coon 1974).

BLUEBELL FAMILY (Campanulaceae)
Worldwide, there are more than seventy genera and two thousand species in this
family. Twelve genera are native to the United States. These are annual or perennial
herbs usually with milky juice. The flowers are typically five-parted with the calyx
divided into separate sepals, and the corolla is five-lobed and bell shaped. The family
is of little economic importance, but some species are cultivated as ornamentals.

Bellflower, Harebell (*Campanula*)
Description: These are perennial herbs from a rhizome. The blue (occasionally
white) flowers are tubular, bell, or cup shaped. The genus name is from the Latin
bell, and the common name, harebell, may allude to an association with witches,
who were believed to transform themselves into hares, portents of bad luck when
they crossed a person's path. The two species in the Sierra Nevada can be found in
open, dry, or rocky areas from low elevations to above timberline.

Quick Key to the Bellflowers	
Flowers are pale blue to white in color; leaves have long petioles	Scouler's Harebell (*C. scouleri*)—Look for this harebell in wooded places below 5,000 feet from Sierra County northward. It flowers from June to August.
Flowers are bright blue; leaves appear to be sessile	California Harebell (*C. prenanthoides*)—This species occurs in dry, wooded areas below 6,000 feet from Tulare County northward. It flowers from June to September.

Interesting Facts: The leaves and shoots of at least *C. rotundifolia* (bluebell bellflower) can be used in salads or cooked as a potherb. The roots can also be boiled
and eaten, and they have a nutlike taste. *C. rapunculoides* (rampion bellflower) is
also edible (Harrington 1967).

Venus Looking-glass (*Triodanis*)
Description: These are annuals with fibrous roots. The flowers are solitary or
several in the axils of the leaflike bracts. The two species that may be encountered
in California include *T. biflora* and *T. perfoliata*. Both occur in disturbed areas
below 6,000 feet and are relatively uncommon.
    Interesting Facts: *T. perfoliata* was used by the Cherokee as a drug plant. For
example, an infusion of the roots was taken and used as a bath for dyspepsia
(Hamel and Chiltoskey 1975).

BORAGE FAMILY (Boraginaceae)
The borage family has approximately one hundred genera and two thousand
species, with twenty-two genera native to the United States. Members of the
Boraginaceae can be identified by their alternate leaves, round stems, coiled
racemes, and five-parted radially symmetrical flowers. The corolla has a narrow
tube that is abruptly flared at the top. The fruit, composed of four nutlets, is
helpful to correctly identify species in this family. The name borage comes from a
Middle Latin source, *burra*, meaning rough hair or short wool, as many of the
plants in the family are covered with stiff hairs.

Catseye, Cryptantha (*Cryptantha*)
Description: The many species of *Cryptantha* are annual or perennial herbs that
are rough to the touch. They have linear or spatulate leaves, and the inflorescence

is scorpion tail–like with small white or yellow flowers. *Cryptantha* are usually found in dry, open areas at various elevations. Many species of *Cryptantha* can be found in California.

Interesting Facts: The seeds of some *Cryptantha* may have been eaten by the Chumash Indians in California (Sweet 1976).

Fiddleneck
(*Amsinckia menziesii*)

Fiddleneck (*Amsinckia menziesii*)

Description: Fiddlenecks are coarse annual herbs with stiff hairs. The flowers are in a scorpion tail–like spike. This species is usually found below 6,000 feet in dry areas. It is named for W. Amsinck, an early-nineteenth-century patron of the botanic garden in Hamburg, Germany.

Interesting Facts: The seeds of this species were pounded into flour, then made into cakes and eaten without cooking. Coastal California tribes also ate the young fiddleneck leaves (Reagan 1934). A related species, *A. douglasiana,* was apparently used medicinally (unspecified) (Strike 1994).

Hound's Tongue (*Cynoglossum*)

Description: These are taprooted perennial herbs. The flowers are dull reddish purple, and the nutlets are densely covered with short, hooked prickles. They are introduced Eurasian weeds and are common in disturbed areas, especially along logging roads and heavily used pastures.

Quick Key to the Hound's Tongues	
Stems hairy	Western Hound's Tongue (*C. occidentale*)—This species is found in dry openings between 4,000 and 7,000 feet. It flowers from May to July.
Stems not hairy	Large Hound's Tongue (*C. grande*)—This species is found in dry openings between 4,000 and 7,000 feet. It flowers from May to July.

Interesting Facts: The plants contain cynoglossine, consolin, and allantoin, a highly effective healing substance that was sometimes used to relieve pain. The use and effect of hound's tongue is similar to comfrey in treating skin and intestinal ulcerations (Moore 1979). The root and leaves of the plant make an effective tea for a sore throat accompanied by dry, hot cough. A leaf poultice was used for insect bites and piles and other minor injuries such as bruises and burns.

Caution: Though the young leaves of this plant are edible in small amounts after boiling, internal use of hound's tongue is not recommended. The plant contains potentially carcinogenic alkaloids that may be harmful to the liver if ingested in large quantities.

Lungwort, Bluebell (*Mertensia ciliata*)

Description: This is a perennial herb with succulent, alternate, and entire leaves. The blue flowers are funnel-form or trumpet-shaped. The genus is named after the German botanist F. K. Mertens. The common name lungwort comes from a European species with spotted leaves, believed to be a remedy for lung disease.

Interesting Facts: Lungworts are often overlooked in many edible plant guides. The flowers can be nibbled upon raw or added to salads. Because the leaves are a bit hairy, we found them better when chopped up and added to soups. Lungworts may contain alkaloids and other constituents that can be toxic if consumed in large quantities (Schofield 1989).

Popcorn Flower (*Plagiobothrys*)

Description: The species are annuals with alternate or opposite leaves. The white, salver-form flowers are in scorpioid racemes. The various species occur in moist soil. There are many species in the Sierra Nevada.

Interesting Facts: A rouge was obtained from the stem base and roots of some species of *Plagiobothrys*. Many species have purple dye in the stems and roots. When pressed and dried in a folded sheet of clean paper, a mirror-image pattern results. The shoots and flowers of related species in California (*P. fulvus*) were eaten and leaves eaten as greens. *P. nothofulvus* was rubbed between the hands by the Kawaiisu to dye hands purple or to dye their faces (Zigmond 1981).

Stickseed (*Lappula redowski*)

Description: This is a taprooted annual herb with linear, alternate leaves. The plants are densely hairy throughout. The flowers are blue (rarely white), and the fruits have prickles about the edges. Stickseeds are usually found at the lower elevations in dry, disturbed habitats up to 8,000 feet. *Lappula* is Latin, diminutive of *lappa*, meaning bur.

Interesting Facts: Medicinally, a poultice from stickseed was applied to sores caused by biting insects, whereas cold infusion was used as a lotion for sores and swellings. Additionally, the roots of European stickseed placed on a hot stone and allowed to smoke were used as an inhalant, and a snuff was made from the root for headaches (Moerman 1977).

Stoneseed, Gromwell, Puccoon (*Lithospermum californicum*)

Description: This plant has few to several stems arising from the base that are often ribbed and hairy. The lower leaves are lance-linear to lanceolate, and the upper ones are lance-oblong to lance-ovate. The inflorescence is rather congested, and the nutlets are shiny white. Look for this species on dry slopes and ridges below 5,000 feet from Placer County northward. It flowers from May to June. The genus name means stone seed, referring to the hard nutlets.

Interesting Facts: *Lithospermum* was used by Native Americans throughout the West as a medicine and food (Craighead, Craighead, and Davis 1963). Although little is known about its chemistry, the effectiveness of several species as contraceptives and as depuratives for skin conditions warrants further investigation. Extracts of *L. ruderale* appear to contain a natural estrogen that interferes with hormonal balances in the female reproductive system. The Shoshone Indians used a tea made from the plant to treat diarrhea and as a female contraceptive, which caused permanent sterility after six months of continued use (Craighead, Craighead, and Davis 1963). A salve from powdered and moistened leaves and stems was used for rheumatic and other pains where the skin is not broken.

Wild Forget-me-not, Stickseed (*Hackelia*)

Description: These are mostly tall, taprooted perennials with numerous blue or white flowers with yellow centers. Each of the flowers gives rise to four small nutlets that possess rows of barbed prickles down their edges—hence the name stickseed. It is these prickles that readily enter clothing or the fur of animals but retard their being pulled out again. The seeds are then transported, and the plants become established a long distance from where they originated. The genus is named for Joseph Hackel, a Czech botanist who lived from 1783 to 1869. At least eight species are known to occur in the Sierra Nevada and grow in moist to medium-dry soils in the foothills to above 8,000 feet. One species, Sharsmith's stickseed (*H. sharsmithii*), was named for Dr. Carl Sharsmith.

Interesting Facts: There are no reports of any of the species being eaten by humans. However, the majority of the species are reported to be "noxious," in that their fruits tend to cling in great numbers to wool shirts, socks, and trousers. Additionally, related species, *H. floribunda* (many-flower stickseed) and *H. hispida* (showy stickseed), were reported to be used as medicinal plants by several Native American tribes. The prickles on the fruit of many-flower stickseed can also cause skin irritation and swelling.

BROOM-RAPE FAMILY (Orobanchaceae)

There are 13 genera and 180 species found around the world, with 4 of the genera being native to the United States. Members of the broom-rape family are herbaceous, lack chlorophyll, and are parasitic on the roots of other flowering plants. The family is of no direct economic importance. The family name comes from the Greek *orogos*, meaning a clinging plant, and *acho*, to strangle.

Boschniakia (*Boschniakia stobilacea*)

Description: The stem is unbranched, brown, and thick, and arises from a corm-like basal thickening. The spike is dark reddish brown. The plant is found below 10,000 feet throughout the Sierra Nevada. It flowers from May to June.

Interesting Facts: These young plants were eaten by the Karok (Baker 1981). Other species were also used as food and medicine. For example, the peeled roots of *B. hookeri* were eaten by the Luiseno.

Broom-rape (*Orobanche*)

Description: These species parasitize the roots of other plants. These fleshy annual plants are nearly white to brownish or purplish in color and lack chlorophyll. The leaves are reduced to scales. Broom-rape is usually found in dry soils, associated with such genera as *Artemisia* and *Eriogonum*.

Quick Key to the Broom-rapes	
Greater than twenty flowers present	California Broom-rape (*O. californica*)—This glandular, finely hairy herb grows in moist woods from 3,000 to 8,000 feet. It flowers from May to July.
Between one and twenty flowers present	
Flowers between five and twenty; corolla lobes not longer than the tube	Clustered Broom-rape (*O. fasciculata*)—This fleshy and somewhat sticky, pubescent herb is parasitic on plants such as sagebrush and buckwheat and grows in moist, woodsy areas between 4,000 and 10,500 feet. It flowers from July to August.
Flowers generally one to three; corolla lobes usually longer than the tube	Naked Broom-rape (*O. uniflora*)—This species mainly occurs between 3,000 and 8,500 feet throughout the Sierra Nevada.

Interesting Facts: The entire plant of broom-rape, roots and all, can be eaten raw. Being succulent plants, they answer for food and drink and are often called "sand food." We found them to be better tasting when roasted in the hot ashes of a campfire. Strike (1994) also indicates that the roots of *O. californica* (California broom-rape) and the entire plant of *O. fasciculata* (clustered broom-rape) were eaten.

"LIFE IS VERY PERSISTENT . . . IT COMES IN AT ALL CORNERS." Dr. Carl Sharsmith may be remembered as the oldest and longest-serving national park ranger, as an expert alpine botanist, as a professor of botany at San Jose State University, as the discoverer of previously unclassified wildflowers, and for establishing the herbarium at San Jose State University. However, to many of us who see these mountains as a second home, he will be remembered as the best-loved naturalist of Yosemite. Dr. Sharsmith was an inspiration to all and has influenced thousands of people. He had a magical and delightful way of encouraging those who walk with him to develop a greater appreciation for wilderness. He, like many of us, was greatly influenced by John Muir, having first discovered, as a boy, Muir's writings, which, he said, "set me afire."

Broom-rape (*Orobanche* spp.)

The decocted blanched or powdered seeds are said to ease joint and hip pain. They can also be used as a toothache remedy. Moore (1979) states that the whole plant is astringent and makes an excellent poultice. Broom-rape is also mildly laxative and has sedative properties. The stalks with the white inner portions removed have been used as pipes. *Orobanche uniflora* was used to treat numerous ailments, including bronchial problems, intestinal upset, toothaches, and rheumatic pain. A decoction of *O. fasciculata* was used as a skin wash to kill lice.

BUCKBEAN FAMILY (Menyanthaceae)
Members of this family are perennials with thick rhizomes, usually found in aquatic habitats. The leaves are simple or divided into three sessile leaflets. There are five genera and thirty to forty species distributed worldwide.

Buckbean (*Menyanthes trifoliata*)
Description: This is a perennial marsh herb with creeping rootstocks. The leaves have long petioles and are all basal and divided into three leaflets. The whitish flowers are small, star-shaped, and crowded into a short inflorescence. The species can be found in bogs and lakes. The species is sometimes placed in the gentian family (Gentianaceae).

Interesting Facts: The herbage and rhizome of the plant are bitter, but we found that the rhizome can be made palatable when collected in early season and boiled in several changes of water. A nutritious flour can also be made from the rhizome by drying, crushing, and leaching it thoroughly. The fresh plant eaten raw may cause vomiting.

The dried leaves are tonic and diuretic and are esteemed, due to high content of vitamin C, iron, and iodine. Buckbean tea was used to relieve fevers and migraine headaches, for indigestion, to promote a healthy appetite, and to eliminate intestinal worms. A poultice of leaves can be applied to skin sores, herpes, and glandular swelling and for sore muscles. Fresh leaves are an emetic; therefore dry them well before use unless you intend to induce vomiting. The plant contains a bitter glycoside, menyanthine, which stimulates gastric juices. The leaves have been used in facial steams for those troubled with acne. Add the tea to a bath or use as a rinse for oily hair. The fruit has no known use. Leaves are a common ingredient in herbal smoking blends.

Water Fringe (*Nymphoides peltata*)
Description: The genus name comes from *Nymphaea* and the Greek ending *oides*, indicating resemblance. There are about twenty species of aquatic, perennial, and rhizomatous herbs with long-stalked, round, floating leaves, and aerial yellow or white flowers.

Interesting Facts: Several species are grown as ornamentals, and some have edible tubers or medicinal seeds. For example, a related species root, *N. cordata*, was used medicinally as a complex infusion.

BUCKEYE FAMILY (Hippocastanaceae)
This family contains two genera and about fifteen species that occur in North and South America. In general, they are trees or shrubs with opposite leaves that are palmately compound. The fruit is a leathery one-seeded capsule. The family is of limited economic importance as a source of ornamentals.

Buckeye, Horse-chestnut (*Aesculus californica*)

Description: This is an erect large shrub or small tree. The crowns are flat-topped to rounded and very broad. Leaves are deciduous, opposite, and palmately compound, and there are about five to seven serrated, oblong to lance-shaped leaflets. The flowers are pinkish white, and the fruits are pear shaped and pendant, and each leathery capsule contains one or two large glossy, brown seeds. It is common on dry slopes and in canyons below 4,000 feet. It flowers from May to June.

Interesting Facts: This was a food staple among many Native Americans in California (Harvard 1895; Strike 1994). The starchy raw nuts contain the bitter principle aesculin, which has had some use as a fish stupefier. To remove the poison from the nuts, several variations on the theme of roasting and leaching were used. The nut was first placed in a pit lined with hot rocks, covered with willow leaves and hot ashes, and baked for several hours. They were then removed, shelled, sliced into pieces, and placed in a container in a stream for a few days. A shorter leaching process was effected by mashing the nuts with water and then soaking them for less than a day in a sandpit along the stream. The pulpy mass was eaten without further preparation. The nuts were not preserved for winter use because they would sprout rather quickly and lose their agreeable taste.

BUCKTHORN FAMILY (Rhamnaceae)

Of the approximately sixty genera and nine hundred species found worldwide, ten genera are native to the United States. Economically, they are of little importance, but several species have edible fruits, and others are used as ornamentals.

Quick Key to the Buckthorn Family

Leaves with three prominent veins running from common origin at base to leaf margins (palmately veined); fruit is a capsule	Snowbush, Buckbrush (*Ceanothus*)
Leaves with a prominent leaf midrib and the other prominent veins lateral and ascending from all along the midrib; fruit is a berry	Buckthorn, Cascara (*Rhamnus*)

Buckthorn, Cascara (*Rhamnus*)

Description: They are shrubs and small trees with flowers that are greenish yellow and four or five parted. They are found in wet or moist soils at low elevations.

1. Alderleaf Coffeeberry (*R. alnifolia*)—Look for this plant in swampy or boggy areas between 4,500 and 7,000 feet from Plumas to Placer County.
2. Hollyleaf Coffeeberry (*R. ilicifolia*)—This almost arborescent shrub occurs on dry slopes below 5,000 feet throughout the Sierra Nevada.
3. Cascara (*R. purshiana*)—This plant with gray bark is found in moist places below 5,000 feet from Placer County northward.
4. Sierra Coffeeberry (*R. rubra*)—This low shrub occurs on dry slopes between 2,000 and 8,000 feet from Fresno County north. It flowers from May to August.
5. Chaparral Coffeeberry (*R. tomentella*)—Chaparral coffeeberry is found on dry slopes and in canyons between 2,000 and 7,500 feet from Madera County south. It flowers from June to July.

Interesting Facts: A decoction of the leaves of these species was used by some Native Americans to soothe rashes caused by poison oak. The inner bark provided a purgative and a laxative. To relieve toothache, the Maidu heated a piece of coffeeberry root and held it against the aching tooth.

The Kawaiisu used crushed berries of coffeeberry as a salve on burns, wounds, and sores to prevent infection. The berries stopped hemorrhages, counteracted poisons, and had a laxative effect (Zigmond 1981). There were a number of plants used by Native Americans to produce a laxative or purgative effect. This may be related to the foods they ate and the resulting gastrointestinal problems they created. On the other hand, they may be related to a ceremonial or ritual requirement that required purging the body.

Snowbush, Buckbrush (*Ceanothus*)

Description: These are shrubs with leaves that are more or less leathery. One important distinguishing feature is that there are three prominent veins originating from near the base of the egg-shaped leaves on some species. The flowers are small, blue or white in color. Look for them on open and dry montane slopes.

Buckbrush, Deerbrush (*Ceanothus* spp.)

Quick Key to the Snowbushes and Buckbrushes

Leaves alternate
  Branches rigid and with spinelike endings
    Leaves entire; plant found mainly above 4,000 feet — Snowbush (*C. cordulatus*)—This spreading shrub grows on dry, open flats and slopes at higher elevations from 3,000 to 9,500 feet. It flowers from May to July. This is a rather easy plant to identify and is recognized by its gray appearance and spiny branchlets.

    Leaves with serrations; plant found mainly below 4,000 feet — Chaparral Whitethorn (*C. leucodermis*)—This shrub occurs on dry, rocky slopes below 6,000 feet. It flowers from April to June.

  Branches not rigid
    Leaves entire and deciduous
      Flowers usually white — Deer Brush (*C. integerrimus*)—This deciduous shrub grows on dry slopes from 1,000 to 5,000 feet.

      Flowers blue — Littleleaf Ceanothus (*C. parvifolius*)—This spreading shrub with slender greenish to reddish twigs is found on wooded slopes between 4,500 and 7,000 feet from Tulare to Plumas Counties.

    Leaves serrate
      Plants forming mats — Pine Mat (*C. diversifolius*)—This trailing shrub with long flexible branches is found on dry flats in pine forests between 3,000 and 6,000 feet throughout the Sierra Nevada.

      Plants erect
        Flowers white to blue; leaves one-half to one and a half inches long — Woollyleaf Ceanothus (*C. tomentosa*)—This evergreen shrub is found scattered on dry slopes below 5,000 feet from Mariposa to Placer Counties.

        Flowers white; leaves one to three and a half inches long — Tobacco Brush (*C. velutinus*)—This spreading, round-topped shrub is found on open, wooded slopes between 3,500 and 10,000 feet throughout the Sierra Nevada.

Leaves opposite
  Leaves mostly entire
    Flowers white — Buckbrush (*C. cuneatus*)—This shrub is common on dry slopes of the chaparral below 6,000 feet. It flowers from March to May.

Flowers mostly blue; plant often forming mats	Fresno Mat (*C. fresnensis*)—This mat-forming plant is found on dry ridges between 3,000 and 6,500 feet from Fresno to Tuolumne Counties. It flowers from May to June.
Leaf margins with definite spines Branchlets red-brown in color; found in Calaveras and Alpine Counties northward	Prostrate Mat (*C. prostratus*)—This prostrate plant is found in open flats between 3,000 and 7,800 feet.
Branchlets brown or gray in color; found in Tulare and Inyo Counties southward	Kern Ceanothus (*C. pinetorum*)—This species occurs on dry slopes between 5,400 and 9,000 feet.

Interesting Facts: The genus has been long recognized as a substitute for commercial black tea, and the leaves and flowers could be used to make tea. The seeds can also be used as food. An infusion of the bark may be used as a tonic. Many species contain saponin, which gives the flowers and fruits their soaplike qualities. When crushed and rubbed in water, the flowers produce a light lather for purposes of washing oneself. The leaves can also be used as a tobacco substitute. The long, flexible shoots were used in basketry. The red roots yield a red dye.

BUCKWHEAT FAMILY (Polygonaceae)

In the buckwheat family there are forty genera and eight hundred species, of which fifteen genera are native to the United States. The family is best represented in the western states. The economic products include food plants and a few ornamentals. Overall, this is a generally safe family when considering edibility.

Quick Key to the Buckwheat Family	
Plants lacking stipules; flowers one to several in bell- to tube-shaped involucres composed of three to ten bracts fused to various degrees	Wild Buckwheat (*Eriogonum*)
Well-developed, generally membranous stipules sheathing stems above each node; flowers not borne immediately above a whorl of bracts	
Leaf blades kidney shaped; tepals four; stamens six; pistil composed of two carpels; fruit lens shaped	Alpine Mountain Sorrel (*Oxyria digyna*)
Leaf blades not kidney shaped; tepals mostly four; three carpels compose pistil; fruit not as above	
Tepals six, outer ones not enlarging in fruit; leaves often jointed to stipule base	Dock (*Rumex*)
Tepals mostly five (rarely four or six), remaining similar in size; leaves jointed to stipulate base	Knotweed, Smartweed (*Polygonum*)

Alpine Mountain Sorrel (*Oxyria digyna*)

Description: Alpine mountain sorrel is a low perennial with simple, roundish leaves clustered at the base of the stem. The flowers are small, red or greenish. The plant is found in cold, wet places among rock crevices between 9,400 and 10,700 feet. It flowers in July to September. The plant resembles a miniature rhubarb, with small rounded leaves. It has always been highly esteemed in arctic regions as a "scurvy-grass" with an agreeable sour taste.

Interesting Facts: Perhaps one of the most refreshing plants one encounters in the high country is the alpine mountain sorrel. The new growth up to flowering time can be eaten raw, when it tastes like a mild rhubarb. The stems and leaves can be used in salads or prepared as a potherb. Some aboriginal peoples have been known to ferment mountain sorrel as a kind of sauerkraut. This is accomplished by simply letting the plant(s) sit in water for a while. This sauerkraut can then be stored for winter use (Schofield 1989). The plants were also dried in the sun for later travel. These plants are high in vitamin C and can be used to prevent and cure scurvy. Large amounts could, however, cause oxalate poisoning.

Alpine Mountain Sorrel (*Oxyria digyna*)

Dock (*Rumex*)

Description: These are annual or perennial herbs. They have small flowers that are greenish and aggregated in a large terminal inflorescence. They can be found in many habitats in the mountains.

1. Sheep Sorrel (*R. acetosella*)—This introduced weed from Eurasia is found between 3,000 and 7,000 feet. It flowers from March to August.
2. Curly Dock (*R. crispus*)—This weed is usually found at the lower elevations.
3. Alpine Sheep Sorrel (*R. paucifolius*)—This plant is found in damp places between 4,000 and 9,000 feet, typically only in Fresno and Lassen Counties.
4. Willow Dock (*R. salicifolius*)—This species occurs in moist places below 9,000 feet. It flowers from May to September.

Interesting Facts: The young leaves of dock can be used as greens, and we have found that the flavor varies from species to species. The young leaves are best when collected before the flower stalk emerges. Also, because the leaves become watery when cooked, use very little water and do not overcook them. The older leaves in most cases may be too bitter for use. Euell Gibbons (1966) found that the leaves of dock are high in vitamin C and contain more vitamin A than carrots. Native Americans ground dock seeds and used the meal to make breads. However, removing the papery seed cover involves a lot of work and, depending on the species, is probably more work than it is worth. The distinctive sour taste of these plants is due to oxalic acid. As with other species that contain oxalic acid, docks should be consumed in small portions as they can cause calcium deficiency.

Poisoning from *Rumex* has been recorded only in livestock after large quantities were eaten. Medicinally, the crushed leaves can be applied to boils and the juice of leaves used to treat ringworms and other skin parasites. The juice of the plant and a poultice of the leaves have also been applied to the rash and pain caused by stinging nettles. A poultice of leaves was used for nervous or allergic hives. The fresh roots were boiled in water to provide a decoction for use internally as a laxative. The powdered yellow roots have been used as a tooth cleanser, laxative, astringent, and antiseptic (Lewis and Elvin-Lewis 1977). Some *Rumex* roots contain as much as 35 percent tannin and were used for tanning animal hides.

Knotweed, Smartweed (*Polygonum*)

Description: The many species of knotweed are annual or perennial herbs with stems that are more or less swollen at the nodes. The flower colors include white, greenish, or pink. They can be found in various habitats up to the higher elevations.

1. Water Smartweed (*P. amphibium*)—This glabrous perennial with simple, elongate stems grows in ponds and lakes below 10,000 feet. It flowers from July to September.

2. Bistort (*P. bistortoides*)—This slender, erect perennial with glabrous stems grows in wet places such as wet meadows and along streams from 5,000 to 10,000 feet. It flowers from June to August.

3. California Knotweed (*P. californicum*)—Look for this plant in dry, sandy, and gravelly flats and bars. It flowers from May to October.

4. Davis' Polygonum (*P. davisiae*)—This species occurs on talus and rocky slopes between 5,000 and 9,000 feet from Alpine County north. It flowers from June to September.

5. Polygonum (*P. douglasii*)—This plant can be found in fairly dry areas as at the edges of meadows between 4,000 and 9,000 feet throughout the Sierra Nevada. It flowers from June to September.

6. Leafy Dwarf Knotweed (*P. minimum*)—This species is found in damp meadows and banks between 5,000 and 12,000 feet.

7. Parry's Knotweed (*P. parryi*)—Look for this plant in dry, sandy places.

8. Alpine Knotweed (*P. phytolacceafolium*)—This species occurs in moist, often rocky places between 5,000 and 9,000 feet from Yosemite northward. It flowers from June to September.

9. Knotweed (*P. polygaloides*)—This plant is found in damp, silty, or gravelly places from 4,500 to 11,500 feet.

10. Shasta Knotweed (*P. shastense*)—Look for this plant in rocky or gravelly places between 7,000 and 11,000 feet. It flowers from June to September.

Interesting Facts: Experimentation may be the rule for *Polygonum,* as none of the species is known to be poisonous. The species do, however, vary in degrees of palatability. Tannins are found in the plants, and large amounts might cause digestive upset and possible kidney damage. In moderate quantities, however, the genus is generally regarded as safe. Based on our experiments with various species, some have peppery-tasting leaves that can be used in flavoring foods. Others have starchy roots that may be eaten raw or boiled and roasted. Still others have young foliage made into good salads or potherbs. In our opinion, of all the species, *Polygonum bistortoides* (bistort) tastes the best. This species is very common in mountain meadows.

The seeds have been used whole or ground into flour. The seeds of *Polygonum* are described as a prehistoric food source and are frequently found in archaeological remains.

A decoction of the roots can be made for a sore mouth or gums. The root can also be used as an astringent, diuretic, antiseptic, and alterative. The roots were eaten by maritime explorers to prevent scurvy. There is a traditional European "Easter pudding" made up of bistort, nettle, and dock, all of which are high in vitamin C (Schofield 1989).

Wild Buckwheat, Eriogonum (*Eriogonum*)

Description: These are annual or perennial herbs; some species are woody at the base. The flowers are small and usually brightly colored. The many species of *Eriogonum* can be found in various habitats at all elevations. The genus name is from the Greek *erion,* meaning wool, and *gony,* meaning knee or joint, referring to the hairy stems of many species. At least twenty-two species of *Eriogonum* are known to occur in the Sierra Nevada.

Interesting Facts: None of the species is known to be poisonous. The flowering stems can be eaten raw or cooked before they have flowered. Seeds can be collected (though tedious) and ground into flour. A tea from the root of *Eriogonum* was used to treat headaches and stomach problems. The plants are mildly astringent and were used as a gargle for sore throats.

BUTTERCUP FAMILY (Ranunculaceae)

Worldwide, there are thirty-five to seventy genera and two thousand species in the cooler regions of the Northern Hemisphere. Twenty-one genera are native to the United States. Many plants in this family are poisonous, some are grown as ornamentals, and others provide drugs.

Note: Only a few plants in this family were eaten. Most contain an irritating compound, protoanemonin, in the fresh leaves, stems, roots, flowers, and seeds. All must be cooked before eating.

Aconite, Monkshood (*Aconitum columbianum*)

Description: This is a perennial herb with palmately divided or lobed leaves. Its flowers are usually deep blue or purple but may also be pale to white. It is usually found in moist, densely shaded places often with streamside vegetation up to timberline.

Interesting Facts: Some species of monkshood have been a source of drugs, as a pain killer or as a sedative for nervous disorders. However, all parts of the plant are *poisonous* and should be considered dangerous if ingested. The drug acotine from *A. columbianum* was used to treat pain from neuralgia, toothache, and sciatica. Acotine is one of the most toxic plant compounds known. Nevertheless, a number of different species are used medicinally in various parts of the world, with apparently beneficial therapeutic effects. In China, processing is the key. For example, used raw it was an arrow poison, whereas steamed, it was an internal medicine to help improve the digestive system.

Anemone, Windflower (*Anemone*)

Description: The three species in the Sierra Nevada are described as perennial herbs from a rootstock or rhizome. The basal leaves are palmately lobed or divided, whereas the stem leaves are in whorls, each with two to four compound or simple leaves. There are no petals in the flowers, but the five to many sepals resemble petals. The plants can be found in dry to moist meadow areas from the foothills to the alpine zone.

1. Drummond's Anemone (*A. drummundii*)—This species grows on talus and gravelly or rocky slopes between 5,000 and 10,600 feet from Inyo County north. This plant is named for Thomas Drummond, an early botanical collector in the Southwest.
2. Pasque Flower (*A. occidentalis*)—This plant is found on dry, rocky slopes between 5,500 and 10,000 feet throughout the Sierra Nevada. It flowers from July to August.
3. Windflower (*A. oregana*)—Look for this species in shaded places, on the forest floor, below 5,000 feet from Plumas County north.

Interesting Facts: The Native Americans made a poultice from the leaves of a related species, *A. cylindrica* (candle windflower), to treat rheumatism and burns (Coffey 1993). The roots of *A. globosa* (= *A. multifida*) were used for treating wounds. *Anemone patens* contains a volatile oil used in medicine as an irritant (Craighead, Craighead, and Davis 1963).

BUTTERCUP CHEMISTRY
This family of plants contains a chemical compound known as ranunculin. This is an innocuous chemical by itself, but it readily hydrolyzes to form an extremely irritating oil, protoanemonin. This chemical blisters the tongue and throat if eaten, an obvious deterrent to grazing animals. Plants in the buttercup family have long been used in folk medicine to raise hot blisters when applied as poultices. The cultivated mustards that are also used as poultices have similar irritant oils.

Buttercup (*Ranunculus*)

Description: The many species in California are either perennial or occasionally annual herbs with simple to compound leaves. The flowers are solitary or borne in a small inflorescence. The five petals are normally yellow or white and have a nectar gland at the base. They can be found in many different habitats from the lower elevations to the alpine zone. The genus name is from the Latin *rana* for frog and refers to the wet habitat of some species.

1. Water Plantain Buttercup (*R. alismaefolius*)—This glabrous perennial grows on wet banks and in meadows from 4,500 to 6,500 feet. It flowers from June to July.
2. Water Buttercup (*R. aquatilus*)—This aquatic perennial with leaves and stems submerged in water grows in ponds or in slow-moving streams below 10,000 feet. It flowers from April to August.
3. Desert Buttercup (*R. cymbalaria*)—This glabrous or slightly hairy perennial plant has erect flowering stems and is found in muddy places below 10,000 feet. It blooms from June to August.
4. Eschscholtz's Buttercup (*R. eschscholtzii*)—This glabrous perennial with mostly leafless stems grows from 9,000 to 13,000 feet. It flowers from July to August. Eschscholtz was a naturalist on an early Russian scientific expedition that explored the West Coast.
5. Yellow Water Buttercup (*R. flabellaris*)—This plant is found in mud or shallow waters below 6,000 feet.
6. Creeping Buttercup (*R. flammula*)—Look for this species in marshy places below 7,500 feet from Fresno County north.
7. Sagebrush Buttercup (*R. glaberrimus*)—This plant is found in sandy places or meadows at about 5,000 feet from Mono and Nevada Counties northward. It flowers from April to June.
8. Western Buttercup (*R. occidentalis*)—This species is found in meadows and moist places below 6,500 feet throughout the Sierra Nevada.
9. Straight-beaked Buttercup (*R. orthohynchus*)—Look for this species in meadows below 7,500 feet.
10. Cusick's Buttercup (*R. populago*)—This species occurs in meadows and boggy places between 5,000 and 6,000 feet in Butte County.
11. Bongard's Buttercup (*R. uncintus*)—This species is occasionally found in moist, shaded places below 8,000 feet throughout the Sierra Nevada. It flowers from May to July.

Interesting Facts: All species are more or less *poisonous* when raw. The leaves and stems should be boiled in several changes of water to remove the poisonous compounds. The volatile toxin is also rendered harmless by drying. The seeds can be parched and ground into meal for bread or pinole. The roots can also be boiled and eaten and were an important part of some Native American diets. A yellow dye can be obtained by crushing and washing the flowers.

Clematis, Virgin's-bower (*Clematis*)

Description: These are herbaceous perennials with erect stems or woody vines. The leaves are opposite or whorled and simple to pinnately compound. The flowers, lacking petals, are solitary or borne in an open, pyramid-shaped inflorescence. Sepals are petal-like. The various species can be found from brushy slopes above creek bottoms to open areas from the low to high elevations.

1. California Clematis (*C. lasiantha*)—This woody vine climbs over other vegetation and is common in canyons and in moist areas below 6,000 feet. It flowers from March to June.
2. Western Virgin's Bower (*C. ligusticifolia*)—This vine is woody at the base and climbs over other plants. It grows in moist areas below 7,000 feet. It flowers from March to August.

Interesting Facts: The genus is essentially composed of *poisonous* species. Many references list western virgin's bower as poisonous even though the stems and leaves have been chewed by Native Americans as a remedy for colds and sore throats. The plants have a peppery taste and may cause lightheadedness. Tilford (1993) also indicates that western virgin's bower is diaphorhetic, diuretic, and offers unique vasoconstrictory or dilating action that makes it useful in the treatment of migraine headaches. The Thompson Indians used the plant to make a head wash for scabs and eczema, and a mild decoction was drunk as a tonic (Teit 1930). Sweet (1976) states that the white portion of the bark was used for fever, the leaves and bark were used for shampoo, and a decoction of the leaves was used on horses for sores and cuts. The fibers in the bark were used for snares and carrying nets. The dried stalks were used in fire-by-friction sets and the feathery seed tails for tinder.

Caution: The consumption of *Clematis* may cause internal bleeding. The entire genus contains strong chemical constituents that can irritate skin and mucous membranes.

Columbine (*Aquilegia*)

Description: These are perennials with ternately compound leaves. There are five petaloid sepals and five petals, each ending into a spur. The genus name stems from the Latin name for eagle. It may also stem from the Latin *aqua* (water) and *legere* (to collect), perhaps referring to the nectar that collects at the tips of the spurs.

Columbine
(*Aquilegia formosa*)

Quick Key to the Columbines	
Flowers red and yellow and nodding; usually occurs below 9,000 feet	Western or Crimson Columbine (*A. formosa*)—Look for this columbine in moist woods between 3,000 and 9,700 feet throughout the Sierra Nevada.
Flowers cream to yellow or pink in color; usually seen above 9,000 feet	Coville's Columbine (*A. pubescens*)—This plant occurs in rocky places and on talus between 9,000 and 12,000 feet from Tulare to Tuolumne Counties. Coville's columbine is named for Frederick Coville, who worked for the Smithsonian Institution and explored the West for plants. He headed the Death Valley Expedition in 1891.

Interesting Facts: The flowers of western columbine are edible and have a sweet taste. They can be added to salads in small amounts. Weedon (1996) indicates that the leaves of this species are also edible but grow bitter with age.

A tea made from the roots of western columbine is said to stop diarrhea, and the fresh roots can be mashed and rubbed on aching joints. Aboriginal peoples used various parts of the plants in medicinal preparations for diarrhea, dizziness, aching joints, and possibly venereal disease. The root, boiled with *Ipomopsis aggregata* (scarlet gilia), resulted in a brew that induced vomiting (Strike 1994). Ripe seeds can be mashed and rubbed into the hair to discourage lice.

It is believed that the cream-colored flowers of Coville's columbine attract certain moth species seeking the nectar in the long flower spurs, which are the same length as the moth's proboscis. Smaller moths are effectively kept from using the same nectar source due to the size of their proboscis. In the process of obtaining nectar, the moth pollinates the flower or carries the pollen to the next flower or both.

Warning: The seeds can be fatal if eaten, and most parts of the columbine contain cyanogenic glycosides. Any therapeutic use of columbine is strongly discouraged.

False Bugbane (*Trautvetteria caroliniensis*)

Description: This erect, nearly glabrous plant is much branched above and has lower palmately lobed leaves that are long petioled. The flowers have no petals. It is found in swamps and along streams between 4,000 and 5,000 feet from Placer County north.

Interesting Facts: This species was used by the Bella Coola as a dermatological aid, and a poultice of the roots was applied to boils (Moerman 1986).

Larkspur (*Delphinium*)

Description: These are all perennial herbs with tuberous or fibrous roots and erect stems. The leaves are roundish in outline and deeply lobed or divided. The flowers are showy, blue to partly white, containing five petal-like sepals with the uppermost prolonged into a spur. There are also four petals, two partly enclosed by the upper sepals, the lower two often hairy and lobed at the tip. They can be found in various habitats, including meadows, thickets, and open woods from the lower to high elevations. Larkspurs are easy to recognize, although the correct determination of the various species is sometimes quite difficult.

1. Anderson's Larkspur (*D. andersonii*)—This plant occurs in sandy and volcanic soils, usually among shrubs and pines between 5,000 and 7,000 feet from Mono County northward. It flowers from April to June.
2. Dwarf Larkspur (*D. depauperatum*)—Look for this species in damp thickets and along the edge of woods between 4,000 and 9,000 feet. It flowers from May to July.
3. Mountain Larkspur (*D. glaucum*)—This glabrous perennial with coarse, hollow stems grows in wet meadows and near streams from 5,000 to 10,500 feet. It flowers from July to September.
4. Slender Larkspur (*D. gracilentum*)—This species is found in shady and damp places below 8,000 feet from Butte County south.
5. Hansen's Larkspur (*D. hansenii*)—This species is found in open, grassy places below 4,000 feet from Butte County south.
6. Red Larkspur (*D. nudicaule*)—Look for this species on dry slopes among shrubs and in woods below 6,500 feet from Mariposa, Butte, and Plumas Counties northward. It flowers from April to June.
7. Nuttall's Larkspur (*D. nuttallianum*)—This species is found in grassy and brushy places and open woods between 5,000 and 10,000 feet.
8. Dissected-leaf Larkspur (*D. polycladon*)—Look for this species in moist to wet habitats from 7,500 to 11,000 feet. It flowers from July to September.

Interesting Facts: Cattle and horses can contract the usually fatal disease of delphinosis from eating delphiniums (Muenscher 1962). Plants should therefore be regarded as *poisonous*. Strike (1994) indicates that some *Delphinium* roots were dried, pulverized, mixed with water, and used by the Kawaiisu Indians in California as a salve on swollen limbs.

Marsh-marigold (*Caltha leptosepala*)

Description: The leafless flowering stem of this plant has one or two flowers with five to fifteen white or blue-tinged, petal-like sepals. There are no petals. The leaves are dark green and basal. It can be found in marshes and wet meadows to above timberline. Marsh-marigolds bloom close to receding snowbanks. Marsh-marigold is not related to the cultivated marigold, a member of the sunflower family. Its nickname dates back to the Middle Ages when a plant (*Caltha*) was dedicated to the Virgin Mary and widely used in church celebrations. The flowers were fermented for making wine. The genus name is from the Greek *calathos*, meaning a cup.

Interesting Facts: The young leaves can be used as a potherb, and the spaghetti-like roots can be dug up during the winter and boiled as a pasta substitute. Though the plant is poisonous when raw (the plant contains the poisonous glucoside protoanemonin), cooking appears to destroy the poison. It also contains the deadly glucoside hellebrin, which breaks down with boiling (Mitchell and Dean 1982).

The roots have diaphoretic, emetic, and expectorant properties. The leaves are diuretic and laxative, and a tea from the leaves mixed with maple sugar was used as a cough syrup by the Ojibwas (Willard 1992). The tea was also used as an antispasmodic and expectorant for treating cramps, and convulsions. In *Stalking the Healthful Herbs*, Euell Gibbons reports a drop of juice squeezed from the fresh leaves is caustic and will remove warts.

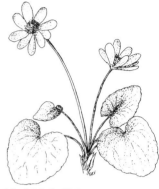

Marsh Marigold
(*Caltha lepsepala*)

Meadow-rue (*Thalictrum*)

Description: These are rhizomatous, erect perennial herbs. The alternate leaves are two to four times branched into ultimate leaflets that are petiolate and shallowly lobed or toothed and closely resemble the leaves of columbine (*Aquilegia*). There are no petals, and the four to five sepals fall soon after opening. The fruit is a ridged or nerved achene with a persistent style (beak). Mature achenes are important for identification.

1. Dwarf Meadow-rue (*T. alpinum*)—This glabrous plant is found in moist meadows and bogs in the Rock Creek Lake Basin and Convict Creek Lake Basin. It flowers from June to August.

2. Fendler's Meadow-rue (*T. fendleri*)—This species occurs in moist habitats between 4,000 and 10,000 feet throughout the Sierra Nevada.

3. Few-flowered Meadow-rue (*T. sparsiflorum*)—This plant occurs on moist stream banks and bogs, often in association with willow thickets between 5,000 and 11,000 feet throughout the Sierra Nevada.

Interesting Facts: The dried plant of Fendler's meadow-rue was rolled into a cigarette and smoked, or sprinkled on a fire, to treat headaches (Pojar and Mac-Kinnon 1994). The young leaves of a related species, *T. occidentale* (western meadow-rue), are said to be edible (Willard 1992). A tea was made from the roots as a cure for colds and venereal disease. When dried and powdered the roots can be used as a shampoo. Additionally, thalicarpine, a substance used in cancer treatment, has been isolated from *T. pubescens, T. revolutum*, and *T. dasycarpum*, species found in eastern North America (Mitchell and Dean 1982).

Red Baneberry (*Actaea rubra*)

Description: Red baneberry is a perennial herb with fibrous roots. The leaves have long petioles and are two to three times divided into sharply toothed, lance-shaped segments. The small flowers are white and borne in a branched, congested, hemispheric inflorescence. The fruits are shiny red or white. Red baneberry is common in moist, montane forests and riparian areas, usually with some partial shade.

Interesting Facts: The entire plant, especially the berries, is *poisonous*. The plant is sometimes confused with *Osmorhiza chilensis* (western sweetroot), which often shares the same habitat. However, unlike red baneberry, sweetroot has a strong licorice-like odor.

The roots are considered a laxative and can cause vomiting. The roots were also ground, mixed with grease or tobacco, and rubbed on the body to treat rheumatism (Bacon 1903). Ground seeds mixed with pine pitch were applied as a poultice for neuralgia. *A. arguta* is described by Moore (1979) as moderately *poisonous* when taken internally, with cardiac arrest possible from large doses. The powdered root was mixed with hot water and applied as a counterirritant.

CACAO FAMILY (Sterculiaceae)

This family comprises herbs, shrubs, and sometimes woody vines. The leaves and stems for the most part have stellate hairs. The family is distributed throughout the Tropics and subtropics. There are at least five genera in the United States. The most famous member of this family is *Theobroma cacao* (cocoa tree), which is native to tropical America. This plant has large reddish yellow fruits, and the seeds are the source of chocolate and cocoa. Another species, *Cola acuminata* (cola nut), found in Africa, is the source for cola drinks.

Flannel Bush (*Fremontodendron californicum*)

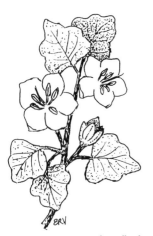

Flannelbush
(*Fremontodendron
californicum*)

Description: This is an evergreen shrub, five to twenty feet tall. The leaves are simple, alternate, round-ovate, palmately three lobed, dark green and slightly pubescent above, and yellowish and densely pubescent below. The showy flowers are bright yellow. There are no petals, and the calyx is petal-like with five lobes. The five stamens are attached to the style of the pistil. The fruit is a bristly hairy, ovoid capsule. Flannel bush grows on dry slopes between 3,000 and 6,000 feet. It blooms from May to June.

Interesting Facts: Medicinally, an infusion of the inner bark was taken as a physic. The inner bark was also used as a poultice for sores, and when soaked in water, the inner bark works as a purgative.

The fiber from the outer bark can be twisted into cordage, rope, or nets. The stems were used for cradle-board frames, and the smaller branches were used to make bows and arrows. One Native American tribe, the Yokuts (San Joaquin Valley area), considered *Fremontodendron* a "one-stop" weapons store. The bark was stripped and twisted into good bowstring, whereas the large branches were split into staves for bows and small straight branches were used for arrows (Romero 1954).

CACTUS FAMILY (Cactaceae)

There are more than 140 genera and 2,000 species within this family worldwide. Approximately 16 genera are native to the United States. Native to the Western Hemisphere, cacti have been spread all over the world, frequently carried by explorers and other travelers. Cacti are typically succulent spiny herbs of diverse form. One distinctive feature is the presence of areoles, which are round or elongated spots or openings that may be raised or pitted and usually arranged in rows or spirals over the surface of the plant. The spines grow from the areoles. The flowers have many sepals, petals, stamens, and an inferior ovary. Economic products from this family include ornamentals, edible fruits, nopalitas, and the hallucinogenic peyote. The spines of some cacti were once used as phonograph needles.

Prickly Pear (*Opuntia*)

Description: Prickly pear needs very little introduction. The various species in California are succulent herbs with fibrous roots, and the stems are flat or cylindric. The leaves when present are small, fleshy, and awl shaped. The genus name may be from the Papago name for this food plant, *opun*, or named for a spiny plant of Opus, Greece. Many *Opuntia* species have glochids—minute, nearly invisible barbed hairs that grow in clusters in areoles. They easily become embedded in skin or clothing and, because of their light tan or yellowish color and barbed surface, are almost impossible to remove. Prickly pear cacti occur in dry soils at the lower elevations.

Interesting Facts: Cacti have provided Native Americans with food, medicine, dyes, and a variety of other uses for thousands of years. They have also been attributed with saving many lives by supplying both food and water to people stranded in the desert. Because eating too many cactus fruits can cause constipation, they should be eaten in moderation. The cactus also contains oxalic acid, so watch your intake, as you may develop a deficiency of calcium.

The pads, especially younger pads, make an excellent cooked vegetable. Harvest the young pads by grasping them with tongs and slicing them at the stem joints. Hold over a flame to singe off the spines and glochids, and scrape off the remaining ones with a knife. Rinse well. Slice into thin strips and boil for at least ten minutes. Drain off water, rinse to remove the slippery gum, and they are ready to eat. To use the older pads, slice away the more fibrous section, then cook accordingly.

After removing the spines, cactus fruits (also known as prickly pears) can be peeled and the pulp eaten raw or boiled and then fried or stewed. One solution for removing the spines is to burn them off; another is to split the fruit into two halves and eat the insides. The pulp can also be sun or fire dried for future use. High in protein and oil, the nutritious seeds may be eaten or dried and ground into flour. Grind the seeds into flour or add them to soups.

Prickly pear cacti pads have been used as a soothing poultice that can be applied to wounds and bruises. A tea made from the flowers was said to increase urine flow, and a tea from stems was used as a wash to ease headaches, eye troubles, and insomnia. The fruits are high in calcium, potassium, and vitamin C. Additionally, the Chumash Indians in California used large baskets baited with crushed prickly pear pads to catch sardines. Another tribe roasted *Opuntia* stems, then soaked them. The resulting extract was used to improve the plasticity and cohesion of clay when making pottery.

Archaeologists in Texas have discovered purses made from prickly pear pads. According to Tull (1987), the dried pads were hollowed out to form a small container. A dye from the juice of the uncooked fruit can also be obtained. Bryan and Young (1940) suggest letting wool soak for about a week in the fermenting juice. The color ranges from pink to magenta and appears to fade when exposed to sunlight.

CALTROP FAMILY (Zygophyllaceae)

The caltrop family is composed of herbs, shrubs, and, rarely, trees that exhibit a xerophytic or halophytic life history. The leaves are usually opposite and pinnately compound. There are approximately 30 genera and 250 species distributed primarily in tropical and subtropical areas. Six genera are native to the United States. Guaiacum wood and lignum vitae (a valuable heavy hardwood and wood

Within the genus *Opuntia* are various species with common names like "prickly pear," "beavertail," and "cholla." Here is one way of distinguishing them: prickly pear—flat pads with spines; beavertail—flat pads with no spines; and cholla—cylindrical stems.

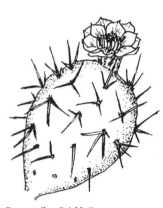

Beavertail or Prickly Pear Cactus (*Opuntia* spp.)

The most striking and characteristic part of this widely varied vegetation are the cactaceae—strange, leafless, old-fashioned plants with beautiful flowers and fruit, in every way able and admirable.
—John Muir, *Steep Trails*

resin used in some chemical tests) are the important economic products in this family.

Goathead, Puncture Vine (*Tribulus terrestris*)

Description: This is an introduced annual plant with trailing, hairy stems and an extensive root system. The leaves are opposite and pinnate with four to eight pairs of leaflets. The flowers are yellow and are borne singly in the leaf axils. The fruit is hard, consisting of five spiny nutlets or burrs that break apart into five "tacklike" sections upon maturity. These burs may injure livestock and are the bane of bicyclists. The plant is found in disturbed and waste places at the lower elevations.

Interesting Facts: Puncture vine has a history of medicinal uses, particularly in China and India, stretching back five thousand years. It was used for boosting the hormone production in men and women, urinary tract problems, itchy skin, and blood purification. The stems of the plant are considered to be astringent and act upon the mucous membrane of the urinary tract. A tea from the aboveground part of the plant is said to be good for arthritis.

Moore (1979) indicates that studies have been conducted on the seeds that show they are useful in the early treatment of elevated blood fats and cholesterol. It apparently helps prevent or lessen the severity of arteriosclerosis and atherosclerosis. Moreover, he says that the plant is useful in treating mild hypertension, contributing to a slower, stronger, more well-defined heart function, with greater relaxation between contractions and a lowering of the diastolic pressure.

CALYCANTHUS FAMILY (Calycanthaceae)

These are aromatic shrubs with simple, entire, and opposite leaves. One species occurs in California and three or four from Pennsylvania to Louisiana and Florida.

Sweet Shrub, Spice Bush (*Calycanthus occidentalis*)

Description: This shrub has large fragrant leaves and brown bark and grows to twelve feet tall. The fragrance is sweet and pleasant to some people and disagreeable to others. Leaves are deciduous, the blades are broad lance-shaped and rough to the touch, and the flowers occurring at the ends of branches are showy, reddish brown, and two inches in diameter. Fruits are about one inch long and consist of an urn-shaped receptacle that contains many velvety achenes. This species grows in moist, shady places, canyons, and stream banks in the Sierra Nevada foothills and the central and southern High Sierra.

Interesting Facts: The bark is peppery to the taste. Native people used the wood and bark from fresh shoots in basketry work (Chestnut 1902). The main use of this plant was of its pithy stems that were much prized for making arrows.

CAPER FAMILY (Capparaceae; formerly known as Capparidaceae)

Members of this family are shrubs, trees, or rarely herbs. They have simple or palmately compound leaves. Flowers have four sepals and four petals, with the fruit a capsule or berry. There are about forty-six genera and eight hundred species distributed in the tropical and subtropical areas of the world. Eight or nine genera are native to the United States. The family is of economic importance as a source of ornamentals and capers, a salad seasoning.

Bee Plant (*Cleome serrulata*)

Description: This is an erect, showy plant up to forty inches tall with alternate leaves divided into three lance-shaped, entire leaflets. The reddish purple to pink flowers are arranged in a dense, narrow, terminal inflorescence. The petals are separate, whereas the sepals are united. The fruits are long-stalked, pendulous capsules, linear to lance shaped in outline. Bee plant is found in disturbed areas (such as roadsides and railroad rights-of-way) at the lower elevations. A second species, *C. lutea* (yellow bee plant), also occurs on the eastern side of the range in dry, sandy flats and along roadsides.

Interesting Facts: An important food for many western Native Americans, bee plant was extensively used as a potherb. The young tender shoots and leaves and flowers are preferred. The plant has an unpleasant odor, especially when older, and a pungent taste much like the mustards. We found it necessary to cook the plants in at least two changes of water to remove the bitter taste. The seeds can also be collected and ground into flour.

The Blackfeet Indians used the whole plant to make a medicinal tea to alleviate fever (Hart 1996). Bee plant may have been used by Native Americans to treat stomachaches.

As a dye, plants are collected in quantity and boiled down for several hours until a thick, fluid residue is produced. The water is then drained off and the plants allowed to dry and harden into cakes. When black dye or paint is needed, a piece of the cake is soaked in hot water.

Bee Plant (*Cleome serrulata*)

CARPETWEED FAMILY (Molluginaceae)

These are often succulent and perennial herbs with whorled or opposite leaves. Petals are usually absent in the flowers, and the fruit is a many-seeded capsule. The family is chiefly distributed in South Africa with approximately 130 genera and 2,500 species. Other centers of distribution include Australia, Asia, South America, and the western United States. The family is of little economic importance, except that a few species are eaten as potherbs.

Green Carpetweed (*Mollugo verticillata*)

Description: This one species is an annual with whorled leaves. The flowers are whitish and without petals. Carpetweed is a weed of fields and waste places found at the lower elevations.

Interesting Facts: Kirk (1975) indicates that the plant may be used as a potherb.

CARROT FAMILY (Apiaceae: formerly Umbelliferae)

Approximately three hundred genera and three thousand species are found in this family. About one-quarter of the genera and 10 percent of the species are native to the United States. The carrot family is of considerable economic importance because of numerous food plants, condiments, ornamentals, and poisonous species. Some familiar members of this family include carrot (*Daucus*), parsnip (*Pastinaca*), celery (*Apium*), anise (*Pimpinella*), and parsley (*Petroselinum*).

The essential oil extracted from the fruits of several members of the Apiaceae can cause nervous disorders in high doses. The excessive use of these fruits as a condiment should be avoided. The fruits of this family are generally referred to as "seeds." However, although they are small and dry, they are botanically fruits, not seeds.

Caution: No wild members of this family should be eaten until they have been accurately identified. Correct identification usually requires the mature fruit, called a schizocarp (a dry fruit that splits into two halves).

Angelica (*Angelica*)

Description: These are herbaceous perennials from stout taproots. The leaves are compound, the leaflets broad, distinct, and dentate or lobed. Its flowers are white, pink, or purplish, and the fruit is strongly flattened dorsally.

1. Brewer's Angelica (*A. breweri*)—Brewer's angelica grows in open woods between 3,000 and 8,000 feet throughout the Sierra Nevada and flowers from June to August.
2. Call's Angelica (*A. calli*)—This is an uncommon species found growing along stream banks between 4,000 and 6,500 feet.
3. Sierra Angelica (*A. lineariloba*)—Sierra angelica may be found on gravelly open slopes between 6,000 and 10,600 feet in the central Sierra Nevada. It flowers from June to August.

Interesting Facts: Although there are a number of edible species of angelica, we have not found any information regarding the edibility of this species. Therefore, the internal use of any angelica species in the Sierra Nevada is *not recommended* until studies have been conducted concerning their toxicity, and because they superficially resemble the poisonous water hemlock (*Cicuta maculata*). All species of angelica contain furanocoumarins, which increase skin photosensitivity and may cause dermatitis (Muenscher 1962; Turner and Szczawinski 1991).

Very little is known about the medicinal aspects of this species. Angelica has been described, in general, as an antispasmodic, expectorant, diaphoretic, diuretic, effective astringent to the stomach lining, and menstrual stimulant that helps reduce cramps. Poultices from the mashed roots of angelica were applied for arthritis, chest discomfort, and pneumonia (Moerman 1986).

Warning: Angelica closely resembles the highly poisonous water hemlock (*Cicuta maculata*). Positive identification of the plants is paramount. The identification of young plants of angelica and water hemlock can be made by examining the leaf venation. The leaf edges of angelica are serrate and pinnately divided into opposing pairs, like water hemlock, but the leaf veins extend from the midribs to the outer tips of the serrations. Water hemlock has leaf veins terminating within the notches of the serrations. Also, keep in mind that you may still have poison hemlock (*Conium maculatum*) (Bomhard 1936).

California Orogenia (*Orogenia fusiformis*)

Description: This is a small perennial plant with fleshy roots. Its flowers are white and in compound umbels. Look for the species soon after snow melts in the mountains in spring and early summer. They are sometimes called "snow drops." It flowers from May to July.

Interesting Facts: The roots can be boiled, steamed, roasted, or baked in any way used for preparing potatoes. When small, they can be eaten raw. The roots can also be cooked and mashed into cakes for drying, and when protected from moisture they will keep a long time. The hard cakes can be soaked and cooked in stews (Schofield 1989).

Cow-bane (*Oxypolis occidentalis*)

Description: This erect glabrous perennial arises from a tuber. The leaf blades are oblong in outline, the leaflets serrate, and there are only a few stem leaves.

Its flowers are white. Cow-bane occurs in shallow waters and wet places between 4,000 and 8,500 feet throughout the Sierra Nevada. It blooms from July to August.

Interesting Facts: The Cherokee baked and ate the roots of a related species (*O. rigidior*) (Hamel and Chiltoskey 1975; Peterson 1978). It is unknown if our species was used by Native Americans in the Sierra Nevada. Therefore, the plant should be considered poisonous.

Cowparsnip (*Heracleum lanatum*)

Description: Cowparsnip is a stout perennial up to or more than seven feet tall. The lower leaves are three-lobed, resembling a maple leaf up to fourteen inches long. The white flowers are in compound umbels, and the fruits are egg shaped with only the marginal ribs winged. It is usually found in moist soils around streams, seeps, and avalanche chutes up to subalpine environments. The genus is named for Hercules (from the Greek *Heracles*), who is reputed to have used it as a medicine. It blooms from April to June.

Interesting Facts: The young stems of cowparsnip can be peeled and eaten raw but are best when cooked. The hollow base of the plant can be cut into short lengths and used as a substitute for salt by eating or cooking with other foods. The young leaves are also edible after cooking, but we find them not very tasty. The leaves can also be dried and burned and the ashes used as a salt substitute. Strong and bitter tasting, the cooked roots are said to be good for digestion, as well as for relieving gas and cramps. In our experience, some plants are much more palatable than others.

The seeds can be sparingly added to salads for seasoning. However, you should be aware that the mature green seeds have a mild anesthetic action on tissues in the mouth. When gently chewed and sucked, they will numb the tongue and gums in a manner similar to clove oil.

Medicinally, root pieces were placed in tooth cavities to stop pain. An infusion for a sore throat can be made by soaking the mashed root in water. Root tea was used for colic, cramps, headaches, sore throats, and flu.

The leaves of cowparsnip are large enough to be used as a toilet paper substitute and as a mild insect repellent. However, because furanocoumarin is present in the sap and the outer hairs, it may be a problem for people with sensitive skin. When the sap comes in contact with the skin in sensitive people, it causes a type of "sunburn" effect (that is, redness, blistering, and running sores) when exposed to light. As a poultice, the leaves were used externally for sores, bruises, and swellings (Gail 1916; Muenscher 1962; Pojar and MacKinnon 1994).

The older stems, before the flower cluster unfolds, can be peeled and the inner tissue eaten raw or cooked. Although it is edible, it does have an unpleasant taste. Cooking it in a couple changes of water usually helps the taste and digestibility. In any case, cowparsnip is considered to be an excellent survival plant in the mountains. Cowparsnip has been probably the most intensively used springtime green among Native Americans.

Warning: Do not confuse this plant with other species in the same family that are highly toxic (such as *Cicuta* spp. and *Conium maculatum*).

Coyote Thistle (*Eryngium alismaefolium*)

Description: This perennial with many stems has basal leaves that are lanceolate to ovate in shape. These basal leaves are also spinose-serrate to pinnatifid on short petioles. The stem leaves are reduced. The species is found in seasonally moist places (that is, vernal pools and flooded meadows) in the northern Sierra Nevada and flowers from July to August.

Interesting Facts: The Paiute made an infusion of the whole plant and used it for diarrhea. The herbage of related species is edible (Kelly 1932; Fowler 1989).

Cut-leaved Water-parsnip (*Berula erecta*)
Description: Cut-leaved water-parsnip superficially resembles a fern but is a perennial aquatic herb that grows up to about three feet tall. The leaves are pinnate with toothed segments. The inflorescences are erect and leafy, and they bear small umbels of white flowers.

Interesting Facts: An infusion of the whole plant was used as a wash for rashes and athlete's foot infection (Strike 1994). Although it's unknown how it was prepared, the leaves and blossoms were used as food by the Apache. The plant is known to produce coumarins. Therefore, it should be considered toxic.

COUMARINS
Coumarin is found in many plants and can be found in the seed coats, roots, leaves, flowers, fruits, and stems. It seems to play an important role in a plant's defense against herbivory and have some antimicrobial properties. Coumarin has a sweet smell, which will be familiar to anyone acquainted with new-mown sweetclover hay. Insects reject the bitter taste of coumarin. Dicoumarol is produced from coumarin and interferes with the body's ability to produce vitamin K (essential to blood clotting). Therefore, dicoumarol acts as an anticoagulant. When livestock eat sweetclover hay that has spoiled, the dicoumarol works to thin their blood, leading to internal (or external) bleeding. Derivatives of dicoumarol (such as warfarin) have been used in rat poisons.

Cymopteris (*Cymopteris*)
Description: The three species of *Cymopteris* in the Sierra Nevada are low perennial herbs with long, thick taproots. The leaves are two to four times pinnately divided into small ultimate segments, and the flowers are yellow or white in terminal, compound umbels. The round fruits have winged ribs on the outer faces. Most of the species occur in dry soils or gravelly slopes.

Quick Key to the Cymopterises	
Flowers yellow	Rocky Pteryxia (*C. terebinthinus*)—This species is found in rocky habitats below 11,000 feet throughout the Sierra Nevada. It flowers from May to June.
Flowers white or purple	
Fruits glabrous; flowers white	Gray's Cymopteris (*C. cinerarius*)—This species occurs on dry, open slopes. It flowers from June to July.
Fruit hairy; flowers purple	Ripley's Cymopteris (*C. ripleyi*)—Look for this rare species in sandy soils below 5,000 feet in the southern Sierra.

Interesting Facts: All species produce edible roots. We found the older roots more fibrous than the younger ones. The root can be used in stews, or it can be boiled or roasted in a pit, mashed, and dried into cakes. When dried, it will keep indefinitely. During the Lewis and Clark expedition, it was known as *kouse* (bread of cows) (Harrington 1967; Hart 1996). The old roots can also be used as an effective insect repellant when boiled. Just sprinkle the tea around camp and in sleeping areas.

The upper parts of the plants have been used raw or as potherbs. If cooked, they will require several changes of water. The seeds of some species are edible when ground and used as flour.

Gray's Lovage (*Ligusticum grayi*)
Description: The stems of this plant are erect and glabrous, and the leaflet blades are ovate to oblong in outline. It is found growing in meadows and on slopes between 4,000 and 10,500 feet throughout the Sierra Nevada. It flowers from June to September.

Interesting Facts: Some of the species contain alkaloids, but the green stems and roots of a related species, *L. filicinum* (fernleaf licoriceroot), may be eaten raw or cooked. Several different species have been used medicinally. They contain volatile and fixed oils and a very bitter alkaloid that has been shown to increase blood flow to coronary arteries and the brain.

L. canbyi (Canby's licoriceroot) has been described as an antiviral, expectorant, diaphoretic, and anesthetic to the throat and is useful in the treatment of upper respiratory infections (Tilford 1993). Native Americans commonly chewed the dried roots of *L. canbyi* for relief from sore throats, colds, toothaches, headaches, stomachaches, fevers, and heart problems (Hart 1996).

Lomatium (*Lomatium*)
Description: These are perennial plants with thick roots and leaves that are divided several times from the base. The white, yellow, pink, or purplish flowers are in compound umbels. The fruits are flattened and elliptical to oval in shape, and the margins may or may not be winged. Most species are found in dry ground or rocky situations. The genus name means "small border," alluding to the wings of the fruit. The following species are known to occur in the Sierra Nevada.

Biscuitroot (*Lomatium* spp.)

1. Fernleaf Lomatium (*L. dissectum*)—This perennial herb occurs in rocky soils from the chaparral and woodlands to the subalpine (500 to 9,000 feet).
2. Lomatium (*L. foeniculaceum*)—This plant occurs in subalpine scrub, below 11,000 feet.
3. Nevada Lomatium (*L. nevadensis*)—This lomatium grows on dry slopes between 5,000 and 9,000 feet. It flowers from April to July.
4. Piper's Lomatium (*L. piperi*)—This plant from a small rounded tuber is found in dry, stony places. It flowers from March to May.
5. Plummer's Lomatium (*L. plummerae*)—This plant with sometimes purple flowers grows on sandy slopes and flats between 3,000 and 5,000 feet. It flowers from May to June.
6. Owen's Peak Lomatium (*L. shevockii*)—This rare plant is found on rocky slopes and forests between 7,000 and 8,000 feet.
7. Stebbin's Lomatium (*L. stebbinsii*)—This rather rare species occurs on gravelly, volcanic soils in yellow pine forests below 5,000 feet in the northern Sierra Nevada.
8. Torrey's Lomatium (*L. torreyi*)—This species occurs on granite rocks and slopes. It blooms from May to August.

Interesting Facts: All *Lomatium* species have edible roots and were an important staple among many Native Americans (Kirk 1975). They can be eaten raw or cooked or dried and ground into flour. The flour can then be kneaded into dough, flattened into cakes, and dried in the sun or baked. Some of the species we have tried were too resinous to enjoy. Personal taste will guide one to choose the more palatable species.

The green stems can be eaten after boiling in the springtime, but as summer progresses they become tough and fibrous. A tea can be brewed from leaves, stems, and flowers. The tiny seeds are nutritious raw or roasted and can be dried and ground into meal.

The plants are rich in vitamin C. Seeds were chewed for colds and sore throats, and sap from the roots was used to treat cuts and sores. A poultice of pulverized roots was applied to a newborn baby's umbilical cord to facilitate healing. Roots were also chewed to relieve sore throats. Recent studies have shown that fernleaf biscuitroot has an ability to kill certain forms of influenza virus, especially those that infect the respiratory tract. It also has other antimicrobial and immuno-stimulating qualities (Moore 1989).

Caution: As with any member of the carrot family, positive identification is important before consumption. Strike (1994) indicates that some indigenous peoples considered the mature stalks, leaves, roots, and flowers of *L. dissectum* as poisonous. In fact, the roots were used as a fish poison and insecticide by some Native people in the West. The plant contains phototoxic compounds of the furanocoumarin group, and one or more of these compounds is responsible for the fish poison and insecticidal properties found in the chocolate-tipped roots.

Poison Hemlock (*Conium maculatum*)

Description: Poison hemlock is a biennial with a stout taproot and a disagreeable odor when crushed. The stems are purple-blotched and hollow, and the leaves are pinnately dissected with a lacy appearance to them. The flowers are white in compound umbels, and the fruits are egg shaped, flattened with prominent, wavy ribs. The plant is usually found in disturbed sites and waste places at low elevations (below 5,000 feet). It blooms from April to September.

Interesting Facts: This is an *extremely poisonous* plant. Death is said to result from the ingestion of the leaves, roots, or seeds. The most famous use of poison hemlock was by the ancient Greeks as a humane method of capital punishment. It is said to be quite painless, and the recipient's mind remains clear to the end. Introduced from Europe, poison hemlock has established itself as a common weed. Socrates was said to be killed by the plant in 399 BC when he was forced to drink it (Hardin and Arena 1974).

Ranger's Button (*Sphenosciadium capitellatum*)

Description: This is a stout perennial plant that grows twenty to sixty-four inches tall. The leaves are large, one to two times pinnate, on swollen petioles. The leaflets are linear, oblong, three-eighths to three-quarters of an inch long. The white flowers occur in compact umbels that are ball-like in appearance. The fruits are flattened dorsally, about one-quarter-inch long, woolly, and with lateral ribs winged. Ranger's button can be found in moist or wet meadows, bogs, or stream banks between 3,000 and 10,000 feet. It flowers in July and August.

Interesting Facts: The roots were chewed by the Maidu Indians to relieve sore throats, and a root decoction was used to treat bronchial problems (Strike 1994). An infusion of the roots was used by the Maidu and Paiute Indians to repel lice (Steward 1933).

Rattlesnake Weed (*Daucus pusillus*)

Description: This is a pubescent annual with pinnately compound leaves. The flowers are white, and the fruit is oblong to ovoid and flattened dorsally. These plants resemble poison hemlock (*Conium maculatum*) but are readily distinguished from each other in that *Daucus* has stems and leaves that are distinctly hairy. This is a common species on dry slopes below 5,000 feet, especially after fire or disturbance. It flowers from April to June.

Interesting Facts: The crushed seeds have been used as a contraceptive and herbal "morning-after pill" for at least 2,500 years (Chamberlain 1901). Rattlesnake weed has been shown to be successful for use as a contraceptive in laboratory trials and is used in some areas today.

The Costanoan Indians used *D. pusillus* to reduce fevers, heal snakebites, cure colds, and as a general blood medicine. As a poultice, the plant was used on bruises and swellings (Strike 1994).

Another species that may occur in the Sierra Nevada is Queen Anne's lace (*D. carota*). It is common at the lower elevations along roadsides and in fields, pastures, waste places, and moist clearings. Introduced from Europe, it is the ancestor of the cultivated carrot. The first year's roots can be prepared like garden carrots. We found the older roots tough and stringy. The roots can also be dried and roasted and then ground for use as a coffee substitute. The plant was used extensively by many Native Americans and should be kept in mind as an emergency food (Harvard 1895; Weiner 1972).

A tea made from the root has been traditionally used as a diuretic to prevent and eliminate urinary stones and worms. Laboratory studies confirm the bactericidal, diuretic, and worm-expelling properties.

Snakeroot (*Sanicula*)

Description: These are erect perennials with three to five lobed leaves. The flowers are in compound umbels, and the fruits are flattened laterally, densely covered by bristles. Two species occur in the Sierra Nevada.

Quick Key to the Snakeroots	
Basal leaves one to five inches long; stems from a globose- or irregular-shaped tuber	Tuberous Sanicle (*S. tuberosa*)—This species occurs on open or wooded slopes below 8,000 feet throughout the Sierra Nevada. It flowers from April to June.
Basal leaves one-half to one and a half inches long; stems from a taproot	Sierra Sanicle (*S. graveolens*)—This species grows in open forests from 4,000 to 8,000 feet throughout the Sierra Nevada. Sanicle flowers from April to June.

Interesting Facts: The herbage of both species contain various alkaloids and should therefore be regarded as inedible (Kingsbury 1964).

Sweet Cicely (*Osmorhiza*)

Description: These two species are herbaceous perennials from stout roots, with leaves twice divided into threes. The flowers are borne in open, compound umbels that arise from leaf axils. The fruit is spindle shaped and compressed along the sides. In general, they occur on moist slopes, open areas, and forests.

Quick Key to the Sweet Cicelys	
Fruit glabrous	Western Sweet Cicely (*O. occidentalis*)—This species occurs on wooded slopes between 2,500 and 8,700 feet in the central Sierra Nevada northward. It flowers from May to July.
Fruit densely bristly-hairy Involucels conspicuous	California Sweet Cicely (*O. brachypoda*)— This species grows in shaded woods below 8,500 feet and flowers from March to May. Its flowers are greenish yellow in color.
Involucels lacking or rudimentary	Sweet Cicely (*O. chilensis*)—Sweet cicely grows in woods below 8,000 feet throughout the Sierra Nevada. It flowers from April to July. Its flowers are greenish white in color.

Interesting Facts: The leaves of sweet cicely, also known as "dryland" parsnip, can be boiled and then eaten. The roots were dug in the spring and either pit cooked or boiled as a vegetable. To us the taste is reminiscent of baby carrots.

The roots of *O. occidentalis* taste and smell like licorice or anise and can overwhelm the taste buds if eaten in large amounts. In small quantities, this species can liven up the taste of teas (or meals) that are otherwise bland or unpleasant. Uses of other related species include a poultice from roots for boils, cuts, sores, and wounds and root tea for sore throats and upset stomachs.

Caution: Western sweetroot resembles the very poisonous water hemlock (*Cicuta maculata*), but the strong smell of anise gives it away as sweetroot. Also, the venation of water hemlock is unique among the Apiaceae (carrot family). Sweet cicely can be confused with baneberry (*Actaea rubra*). However, baneberry is easy to distinguish from sweet cicely by the cluster of red or white berries.

Tauschia (*Tauschia*)

Description: These perennial herbs from taproots or tubers have leaves that are pinnately compound. The yellow, white, or purplish flowers occur in loose compound umbels. The genus name honors I. F. Tausch, a nineteenth-century botanist. Two species of tauschia can be found in the Sierra Nevada.

Quick Key to the Tauschias

Plants twelve to forty inches tall; leaflets seven to twenty-four inches long	Hartweg's Tauschia (*T. hartwegii*)—This species grows on dry, sandy, or gravelly areas below 5,000 feet. It flowers from April to June.
Plants four to fifteen inches tall; leaflets two to five inches long	Parish's Tauschia (*T. parishii*)—This species grows on dry slopes (mainly eastern Sierra Nevada) from 4,000 to 9,000 feet. It flowers from May to July.

Interesting Facts: The boiled root of *T. parishii* was used to relieve internal pains by the Kawaiisu. For toothaches, the fresh root was mashed, or dried roots were pulverized and smeared on hot rocks. By placing the cheek of the toothache sufferer directly on the mashed root, the toothache was thought to be cured (Bean and Saubel 1972).

Poison Hemlock
(*Cicuta douglasii*)

Water Hemlock (*Cicuta maculata*)

Description: This species is found in marshes and along the edges of streams and ponds from low to midelevations (below 8,000 feet). Water hemlock is a stout perennial from fleshy, fascicled roots. Leaves are one to two times pinnately divided into narrowly lance-shaped, sharply toothed leaflets. The veins of the leaflets terminate at the notches between the teeth. Numerous white to greenish flowers are arranged in compound umbels. Fruits are slightly flattened with thickened ribs on the faces. The bruised foliage produces a musky odor. It flowers from June to September.

Interesting Facts: These are extremely *poisonous* plants if ingested. In fact, water hemlock has been described as *the most violently poisonous vascular plant in North America.* The whole plant contains cicutoxin, a resinlike substance that depresses the respiratory system, with the root being particularly dangerous. A single mouthful of the plant is capable of killing an adult. This species, as well as poison hemlock, has been used throughout the ages to execute criminals and kings. Many children have been fatally poisoned by blowing into whistles made from hollow stems of water hemlock.

In Oregon, Native Americans soaked arrows in *Cicuta* juice, rattlesnake venom, and decayed deer liver to poison arrow tips for hunting (Schofield 1989; Pojar and MacKinnon 1994). Water hemlock roots were mashed and smeared on a hot stone to relieve pain in sore arms or legs. The mashed root was then pressed against the sore arm or leg (Strike 1994).

Following is a graphic description of poisoning due to the ingestion of water hemlock in Europe. If anything, it should instill into the minds of wild-food gatherers the need to positively identify a plant species before eating it.

> When about the end of March 1670, the cattle were being led from the village to water at the spring, in treading the river banks they exposed the roots of this Cicuta (water hemlock) whose stems and leaf buds were now coming forth. At that time two boys and six girls, a little before noon, ran out to the spring and the meadow through which the river flows, and seeing a root and thinking that was a golden parsnip, not through the bidding of any evil appetite, but at the behest of wayward frolicsomeness, ate greedily of it, and certain of the girls among them commended the root to others for its sweetness and pleasantness, wherefore the boys, especially, ate quite abundantly of it and joyfully hastened home; and one of the girls tearfully complained to her mother she had been supplied meagerly by her comrades, with the root.
>
> Jacob Maeder, a boy of six years, possessed of white locks, and delicate though active, returned home happy and smiling, as if things had gone well. A little while afterwards he complained of pain in his abdomen, and, scarcely uttering a word, fell prostrate to the ground, and urinated with great violence to the height of a man. Presently he was a terrible sight to see, being seized with convulsions, with the loss of all his senses. His mouth was shut most tightly so that it could not be opened by any means. He grated his teeth; he twisted his eyes about strangely and blood flowed from his ears. In the region of his abdomen a certain swollen body of the size of a man's fist struck the hand of the afflicted father with the greatest force, particularly in the neighborhood of the ensiform cartilage. He frequently hiccupped; at times he seemed to be about to vomit, but he could force nothing from his mouth, which was most tightly closed. He tossed his limbs about marvelously and twisted them; frequently his head was drawn backward and his whole back was curved in the form of a bow, so that a small child could have crept beneath him in the space between his back and the bed without touching him. When the convulsions ceased momentarily, he implored the assistance of his mother. Presently, when they returned with equal violence, he could not be aroused by no pinching, by no talking, or by no other means, until his strength failed and he grew pale; and when a hand was placed on his breast he breathed his last. These symptoms continued scarcely beyond a half hour. After his death, his abdomen and face swelled without lividness except that a little was noticeable about the eyes. From the mouth of the corpse even to the hour of his burial green froth flowed very abundantly, and although it was wiped away frequently by his grieving father, nevertheless new froth soon took its place (Jacobson 1915).

Water Parsnip (*Sium suave*)

Description: This is a stout plant up to five feet tall. The leaves are pinnately divided, the flowers are white, and the fruit is oval in shape. It is usually found in water or swampy areas in the mountains.

Interesting Facts: The long, fleshy root of water parsnip, which is edible raw or cooked, has a sweet, carrotlike taste. The leaves and younger stems are also edible after cooking, but we found them better when boiled until tender. The older plants and flowers should be avoided because they are toxic and have been suspected of poisoning a wide range of livestock.

Warning: The plant is very similar in form and habitat to *Cicuta maculata* (water hemlock), which is the most poisonous vascular plant in North America. Both plants produce white flowers in umbrella-like clusters, and both grow in swampy ground at lake or pond edges. Water parsnip has leaves that are once compound, whereas the leaves of water hemlock are three times compound. Water hemlock also has a distinctive turniplike swelling at the base of the stem, which is usually chambered when cut open vertically and exudes a yellowish liquid along the cut surface. Therefore, when in doubt, *leave it alone!*

Wild Parsnip (*Pastinaca sativa*)

Description: Wild parsnip is a biennial with stout, leafy stems that arises from a large taproot. The basal leaves are once pinnately compound into usually nine to thirteen lance- to egg-shaped leaflets. Its flowers are yellow and in compound umbels. The fruits have fine ribs on the outer face. It can be found in damp disturbed areas at the lower elevations. It is apparently localized as a naturalized plant in some places in the Sierra Nevada. It blooms from June to July.

Interesting Facts: This introduced plant from Europe is the wild form of the cultivated parsnip and can be prepared and eaten in the same way (Saunders 1976). The taproot from first-year plants can be eaten raw or boiled until tender. The root of the plant was used by many people, from the ancient Romans to Native Americans. In Holland, the plant was used in soups. The Irish made a beer by boiling the roots with water and hops, then allowed the mixture to ferment. A tea from the roots was used by Native Americans to treat sharp pains. The roots were also poulticed and used on inflammations and sores (Saunders 1976).

Caution: Due to the presence of xanthotoxin, the plant may cause photodermatitis. The symptoms are much like those from exposure to poison ivy, but of longer duration. Xanthotoxin is used to treat psoriasis and vitiligo. Therefore, avoid contact and exposure to sunlight.

Yampah (*Perideridia*)

Description: These are biennial or perennial herbs with fascicled tuberous roots and pinnate leaves. The calyx teeth are well developed. The petals are white or pinkish, the stylopodium conic. The fruit is nearly terete or somewhat flattened laterally.

Yampah (*Perideridia* spp.)

Plants coarse and stout, arising from numerous fibrous or thickened roots	Howell's Yampah (*P. howellii*)—Look for this species in meadows below 5,000 feet in the northern Sierra Nevada.	77 *Major Plant Groups*
Plants slender, from solitary tubers of tuberous roots		
Basal leaves with ultimate segments less than one and a half inches long	Bolander's Yampah (*P. bolanderi*)—This species occurs in open forests below 7,000 feet throughout the Sierra Nevada.	
Basal leaves ternate, with ultimate segments one and a half to three inches long		
Rays not equal in size in fruit	Lemmon's Yampah (*P. lemmonii*)—This species can be found in moist habitats below 8,000 feet throughout the Sierra Nevada.	
Rays about equal in size in fruit	Perideridia (*P. parishii*)—Perideridia grows in moist meadows from 3,500 to 11,000 feet throughout the Sierra Nevada. It flowers from July to September.	

Interesting Facts: All of the species within this genus are edible. They were an important food of many indigenous peoples from British Columbia to California and the Great Basin region (Elias and Dykeman 1982; Coffey 1993). The raw fingerlike roots have a pleasant, nutty flavor when eaten raw and resemble carrots when cooked. They are best when dug up before the flowers appear. The roots should be washed and peeled before cooking. They can be easily dried and will keep well for future use. When dried, the roots can be pounded and ground into flour or mashed into cakes. The seeds may be parched and ground or eaten whole.

CROSSOSOMA FAMILY (Crossosomataceae)
Members of this family are shrubs with simple, entire, and coriaceous leaves. The flowers are white or sometimes rose tinged, and the fruit (follicle) is green at maturity. There are three genera and about eight species in the western United States and Mexico.

Nevada Greasewood (*Glossopetalion spinescens*)
Description: This shrub has stems that are more or less spiny and yellowish in age. The oblong- to lanceolate-shaped leaves are pubescent. This species occurs from 3,500 to 7,000 feet in the southern Sierra Nevada. It flowers from April to May.

    Interesting Facts: The Shoshone made a decoction of the shrub that was used for tuberculosis (Moerman 1986).

CUCUMBER OR GOURD FAMILY (Cucurbitaceae)
Members of this family are annual or perennial herbs that are climbing or prostrate, with spirally coiled tendrils. The leaves are alternate, often palmately lobed. The fruit is a berry (often referred to as a pepo) with a leathery or hard exocarp. There are approximately 100 genera and 850 species distributed in the warmer regions of the Old and New Worlds. Fourteen genera are native to the United States. The family is economically important as a source of many food plants and ornamentals.

California Man-root (*Marah fabaceus*)

Description: These are climbing perennials from swollen, woody roots. The leaves are palmately lobed and heart shaped at the base. The flowers are white and bell shaped. The fruits are football shaped, fleshy, and weakly spined. They tend to spread along the ground and climb onto other vegetation. The genus name is from the Latin, meaning bitter, referring to the taste of all plant parts.

Interesting Facts: Strike (1994) indicates that *Marah* seeds were roasted and eaten by several Native American tribes. The seeds are supposedly high in protein and oil and were crushed into a flour. The leaves may have been eaten by the Chumash.

Marah roots were brewed into a tea and used as a strong purgative by the Luiseno. California man-root was used by the Pomo as a dermatological aid. They pounded the nuts and grease and rubbed the concoction on the head for falling hair (Barrett 1952).

Marah roots, in general, are poisonous. The roots were mashed and put in streams to stupefy fish to make them easier to catch.

CURRANT AND GOOSEBERRY FAMILY (Grossulariaceae)

This family consists of a single genus (*Ribes*) with approximately 150 species. All are shrubs with palmately lobed leaves. Some species are armed with spines. The family is a source of ornamentals and edible fruits. In many old field guides, *Ribes* is sometimes included as a member of the saxifrage family (Saxifragaceae).

Currant, Gooseberry (*Ribes*)

Description: The many species of *Ribes* are shrubs. The species that have prickles on the stems and bristles on the fruit are commonly called gooseberries. Those without prickles on the stem or bristles on the fruit are currants. Leaves are palmately veined and shallowly or deeply lobed. The five petals are smaller than the sepals and usually narrowed to a clawlike base. The fruit is a berry.

Gooseberry (*Ribes* spp.)

Quick Key to the Currants and Gooseberries

Plant and berries without spines (currants)	
Berry red	Wax Currant (*R. cereum*)—This erect, fragrant shrub occurs in dry, rocky areas between 5,000 and 12,500 feet. It flowers from June to July.
Berry blue or black	
Herbage glandular	Sticky Currant (*R. viscosissimum*)—This species occurs in shaded woods and rocky places between 5,000 and 9,500 feet. It flowers from June to July.
Herbage may be only slightly hairy but not glandular	Mountain Pink Currant (*R. nevadense*)—This deciduous shrub grows in moist places between 3,000 and 10,000 feet. It flowers from May to July.
Plant and sometimes berries with spines (gooseberries)	
Berry without spines or prickles	
Berry with soft pubescence, may be glandular	Plateau Gooseberry (*R. velutinum*)—This stout, rigid branched shrub with yellow to white flowers grows on dry slopes between 2,500 and 8,500 feet.

Berry without pubescence		
Flowers lemon-yellow in color; leaves glandular-pubescent on both sides	Alpine Gooseberry (*R. lasianthum*)—Look for this species in rocky places between 7,000 and 10,000 feet from Tulare to Nevada Counties. It flowers from June to August.	
Flowers green except for white petals	White-stemmed Gooseberry (*R. inerme*)—This species occurs in moist, cool habitats between 3,500 and 11,000 feet throughout the Sierra Nevada. It flowers from May to July.	
Berry with spines and bristles		
Plants trailing; twigs are often prickly		
Petals purple in color; berry about one-quarter inch in diameter	Mountain Gooseberry (*R. montigenum*)—This straggly shrub may be found on dry, rocky areas between 7,000 and 12,500 feet. It flowers from June to July.	
Petals white; berry about one-half inch in diameter	Sequoia Gooseberry (*R. tularense*)—Look for this species in wooded forest situations. It flowers in May.	
Plants not trailing, more erect; twigs without any spines		
Leaves glandular below; berry with short gland-tipped bristles	Bitter Gooseberry (*R. amarum*)—This erect, deciduous shrub grows in wooded canyons below 5,000 feet from Eldorado County southward. It flowers from March to April.	
Leaves not glandular; berry pubescent and long spines	Sierra Gooseberry (*R. roezlii*)—This stout shrub grows on dry, open slopes between 3,500 and 8,500 feet. It flowers from May to June.	

79
Major Plant Groups

Interesting Facts: The berries of almost all species of *Ribes* are edible raw, and none are known to be poisonous. However, we have come across some unpalatable species, berries with an unpleasant odor and a taste to match. The berries are high in vitamin C and one of the richest plant sources for copper. One method of collecting them in bulk is by shaking the bushes over sheets of plastic or blankets. Those that are too sour or spiny become more palatable if they are cooked or dried. In regard to the fruits with bristles, one can also roll the berries on hot coals in a basket until the bristles have been singed off. When dried, the berries are a great trail snack. The dried berries can also be mixed with meat to make pemmican. The berries contain enough natural pectin to make jelly. The seeds also contain large quantities of gamma-linolenic acid, and many herbalists use this oil to treat skin conditions, asthma, arthritis, and premenstrual syndrome. The nectar-filled flowers are considered good trail snacks. The wood makes good arrow shafts.

Leaves of currants and gooseberries may be added to herbal tea blends. The leaves should be fresh or thoroughly dried, not wilted, as they may be toxic.

GAMMA-LINOLENIC ACID
Gamma-linolenic acid is an essential fatty acid (EFA) in the omega-6 family that is found primarily in plant-based oils. EFAs are essential to human health but cannot be made in the body. For this reason, they must be obtained from food. EFAS are needed for normal brain function, growth and development, bone health, stimulation of skin and hair growth, regulation of metabolism, and maintenance of reproductive processes.

DODDER FAMILY (Cuscutaceae)
Members of the dodder family are leafless, rootless, parasitic herbs that lack chlorophyll. The stems are threadlike and often yellowish in color. The small flowers have four or five distinct sepals and four or five united petals. The fruit is a dry or fleshy globose capsule. The family has one genus (*Cuscuta*) with approximately 170 species. The genus is native to the United States and may cause great losses to crop plants. The family was once included in the morning glory family (Convolvulaceae).

Dodder (*Cuscuta*)

Description: Dodders are leafless, twining perennials with slender stems that are colored pink, whitish, or yellowish, never green. Both the leaves and pink to white flowers are highly reduced. Dodder can normally be identified only with a microscope or hand lens. The many species of *Cuscuta* parasitize different flowering plant hosts at low elevations.

Dodders have a very unique life cycle. The small seeds usually germinate in the soil and produce slender stems without seed leaves (cotyledons). Unless the slowly rotating plant encounters a host plant within a short period of time, the dodder seedling will wither and die. However, if the seedling encounters the living stem of a susceptible host plant, the dodder will twine around it and at certain points develop suckers that penetrate the tissue of the host. Nutrition is received through these suckers. Dodder then loses all contact with the soil. After a period of growth, small flowers develop, and large amounts of seeds are produced to start the process all over again (Frankton and Mulligan 1987).

Interesting Facts: Dodder was often called "love vine" and "vegetable spaghetti" by some Native Americans, but it is generally not considered edible and may cause digestive upset. However, the seeds of *C. californica* (chaparral dodder) were parched and eaten by Maidu Indians.

Chaparral dodder, when brewed as a tea, was considered to be an antidote for black widow bites. However, only the dodder from *Eriogonum fasciculatum* (California buckwheat) was used for this purpose. Other Native Americans chewed a mass of dodder and stuffed it in their nose or pulverized the plant and sniffed the powder to stop nosebleeds.

Dodder stems were used by the Cherokees as a poultice for bruises. In China, the stems of some species of dodder are used in lotions for inflamed eyes. Moore (1979) indicates that a rounded teaspoon of chopped dodder is a good laxative-cathartic. In smaller quantities, and drunk every few hours, it is said that it will aid in spleen inflammations, lymph-node swellings, and "liver torpor." Additionally, handfuls of dodder can be gathered and used as scouring pads for cleaning.

Dogbane Family (Apocynaceae)

This is a large family of about two hundred genera and two thousand species that are mostly found in the Tropics. Nearly all of the members within this family are poisonous and usually have milky juice. Some of the well-known genera are ornamentals such as *Vinca minor* (periwinkle) and *Nerium oleander* (oleander). In recent years, *Rauwolfia serpentina* (Indian snakeroot), a tropical tree, was found to yield a wonder drug used in the treatment of high blood pressure, and periwinkle is a source of an antitumor drug.

Dogbane (*Apocynum*)

Description: This is perennial herb with milky juice that has leaves that are opposite, and the pink, bell-shaped flowers are borne in cymes. There is considerable hybridization between species. *Apocynum* is Greek, meaning noxious to dogs.

Dogbane (*Apocynum* spp.)

Flowers pinkish in color	Spreading Dogbane (*A. androsaemifolium*)—Spreading dogbane grows on dry flats and slopes from 5,000 to 9,500 feet. It flowers from June to August.
Flowers greenish in color	Indian Hemp (*A. cannabinum*)—This species is occasionally found in damp places throughout the Sierra Nevada. It flowers from June to August.

Interesting Facts: The primary use of dogbanes is for fiber. The stem fibers are strong and can be used for rope making, mats, baskets, bowstrings, fishing lines and nets, sewing, animal-trap triggers, snares, cordage for bow and drill fire making, and general weaving. One of the easiest ways to isolate the fibers is to soak the stems in water. Archaeologists in Utah have discovered nets made with the fiber dating back to about 5000 BC (Tull 1987). Many Native American tribes used dogbane to make rabbit-catching nets. Some of these nets were about two hundred feet long, three to four feet high, with a three-inch opening. The nets were propped on sticks across level ground. The men formed a line some distance away and advanced toward the nets, beating the brush with sticks, and driving the rabbits into the net (Ebeling 1986).

Dogbanes should be considered *poisonous* to humans if ingested. However, some authorities have indicated that the small seeds can be parched and ground into a meal to make fried cakes. For example, Strike (1994) believes the seeds eaten were that of *A. pumilum* (mountain dogbane). *A. pumilum* is considered to be a subspecies of *A. androsaemifolium*.

Dogbanes were extensively used as medicine by aboriginal peoples. They contain highly toxic glycosides and resins, with cymarin and apocannocide being major medicinal constituents found throughout the plants. Recent research indicates that the cardiac glycosides may be useful in treating malignant tumors. Millspaugh (1974) describes spreading dogbane as an emetic without causing nausea, a cathartic, and a quite powerful diuretic and sudorific; it is also an expectorant and antisyphilitic.

Fiber plants are second only to food plants in terms of their usefulness to humans and their influence on the advancement of civilization. Tropical people use plant fibers for housing, clothing, hammocks, nets, baskets, fishing lines, and bow strings. Even in our industrial society, we use a variety of natural plant fibers. . . . In fact, the so-called synthetic fibers now providing much of our clothing are only reconstituted cellulose of plant origin.
—Mark Plotkin, *Biodiversity*

Dogwood Family (Cornaceae)

There are twelve genera and one hundred species in the dogwood family, including many ornamentals. A single genus, *Cornus*, occurs in the United States. Dogwoods are trees or shrubs, often with tiny flowers surrounded by petal-like bracts that resemble a single large flower. The leaves are opposite and simple.

Dogwood (*Cornus*)

Description: Dogwoods are shrubs or semiwoody perennials with simple leaves that are opposite or whorled. The flowers mature into red or white drupes. The species can be found in moist mountain and foothill forests, preferring partial shade, up to the subalpine zone.

Dogwood (*Cornus* spp.)

Flowers occurring in compact heads that are subtended by petal-like bracts	Mountain Dogwood (*C. nuttallii*) -This species is found in mountain woods below 6,000 feet. It flowers from April to June.
Flowers occurring in open cymes or umbels Leaves are pinnately eight to fourteen veined	Red Osier Dogwood (*C. sericea*)—This shrub occurs in many moist habitats below 8,000 feet. The species is highly variable with many local forms.
Leaves are pinnately six to eight veined; plants are usually found below 5,000 feet Inflorescence is an umbel subtended by bracts; fruit is black when mature	Blackfruit Dogwood (*C. sessilis*)—This species grows along stream banks from Calaveras County and northward. It flowers from April to June.
Inflorescence is a cyme without bracts subtending; fruit is white when mature	Brown Dogwood (*C. glabrata*)—This species occurs in moist places throughout the Sierra Nevada. It flowers from May to June.

Interesting Facts: The fruits of *C. nuttallii* can be eaten raw or cooked. Strike (1994) suggests that the fruit contains enough protein, carbohydrate, and fat to sustain life when other food sources are not available. The inner bark of mountain dogwood twigs can be scraped off and used as an additive to tobacco mixes for smoking. Pounded twigs can be used as a toothbrush. The wood for both species was used for bows and arrows, fishing hooks, and other implements. The bark was boiled and used to make a brown dye.

The fruits of *C. sericea* were sometimes consumed by Native Americans. We have found them to be extremely bitter. In fact, in large quantities they may be toxic. The Blackfeet Indians used the bark of red osier dogwood as a laxative. Other Native Americans smoked the inner bark as part of a ceremonial herb blend.

The fruits of a related species, *C. canadensis* (bunchberry dogwood), may be eaten raw or cooked. Because the berries are rather bland, we like to mix them with other more tasty fruits. The unripe berries may cause stomachaches or act as a laxative. The chewed berries have been used as a poultice to treat local burns. A cold and fever remedy can be made by boiling dried root or bark (the root is more potent). The feathered bark can be used as a toothbrush. Fresh bark is a cathartic. Leaf tea was used for aches and pains, kidney and lung ailments, coughs, and fevers and as an eyewash. Dogwood has earned a reputation as an anti-inflammatory and general analgesic due to the presence of cornine and other flavonoid compounds. Researchers are studying these properties as an anticancer agent. The current interest by pharmaceutical companies may stem from the fact that Native Americans used dogwood as an antidote for a variety of poisons (Pojar and MacKinnon 1994).

Durango Root Family (Datiscaceae)
This family comprises three genera with a total of four species. A single species of *Datisca* occurs in California and ranges into Mexico. The other members of this family occur in eastern Asia and on the adjacent Pacific islands.

Durango Root (*Datisca glomerata*)

Description: This is a glabrous erect perennial plant with ovate lanceolate leaves that are pinnately divided into lanceolate-toothed segments. Flowers occur in axillary clusters, and there are no petals. Durango root grows in dry streambeds and washes below 6,500 feet. It flowers from May to July.

Interesting Facts: The pulverized root was used as a decoction to wash sores and heal rheumatic pain. The Costanoan Natives used a decoction of the plant to relieve sore throats and swollen tonsils (Bean and Saubel 1972).

Warning: All parts of durango root should be considered toxic.

ELM FAMILY (Ulmaceae)

Members of this family are trees or shrubs with simple, alternate leaves. The flowers lack petals, and the fruit is a samara, nut, or drupe. There are approximately fifteen genera and two hundred species in this family distributed through the temperate and subtropical regions of the Northern Hemisphere. *Ulmus, Planera,* and *Celtis* are native to the United States. The family is of economic importance as a source of wood.

Net-leaved Hackberry (*Celtis reticulata*)

Description: This is a small tree or shrub with leaves that are ovate to lanceolate in shape, with entire to serrate edges. The fruit is a drupe, and the plants can usually be found growing along streams or on dry canyon slopes at the lower elevations between 2,800 and 5,500 feet.

Interesting Facts: The small orange, red, or yellow fruits are edible raw and have a sweet taste to them. The entire fruits can also be dried and then ground into a flour.

Net-leaved Hackberry
(*Celtis reticulata*)

EVENING PRIMROSE FAMILY (Onagraceae)

There are about 20 genera and 650 species worldwide, with a dozen genera native to the United States. The family is of little economic importance, but a few are considered to be ornamentals. However, oil of evening primrose is obtained from this family and is said to be the world's richest source of natural unsaturated fatty acids. The oil is helpful in cases of obesity, mental illness, heart disease, and arthritis and is advertised widely in natural food publications.

Clarkia (*Clarkia*)

Description: These are annuals with brittle stems and purple or red showy flowers. They are usually found on dry slopes at the lower to middle elevations. The genus honors Captain William Clark of the Lewis and Clark expeditions to the Northwest in 1806.

The diversity of life forms, so numerous that we have yet to identify most of them, is the greatest wonder of this planet.
—E.O. Wilson, *Biodiversity*

1. Two-lobed Clarkia (*C. biloba*)—This species grows on open, dry slopes below 4,000 feet from El Dorado to Mariposa Counties. It flowers from May to July.
2. Dudley's Clarkia (*C. dudleyana*)—Look for this species in open woodlands below 4,500 feet.
3. California Clarkia (*C. heterandra*)—This species is found in dry, shaded places between 2,000 and 5,000 feet from Placer County south.
4. Lassen Clarkia (*C. lassenensis*)—This species is occasional on open slopes below 7,000 feet. It flowers from May to June.
5. Purple Clarkia (*C. purpurea*)—This erect annual has lavender to red-purple petals that are not clawed. It is common below 6,000 feet and grows in open areas. It flowers from April to July.

6. Rhomboid-leaved Clarkia (*C. rhomboidea*)—This erect annual with clawed, pinkish lavender petals grows on dry slopes below 8,000 feet. It flowers from May to July.

7. Elegant Clarkia (*C. unguiculata*)—This erect, glabrous annual has clawed, lavender to salmon or dark red-purple flowers. It is quite common below 5,000 feet. It flowers from May to June.

8. Williamson's Clarkia (*C. williamsonii*)—This species occurs in dry places below 5,000 feet from Fresno to Nevada Counties.

Interesting Facts: The seeds of *Clarkia* were among the most highly prized foods of some Native Americans in the Sierra Nevada, especially the Miwok (Barrett and Gifford 1933; Chatfield 1997; Vizgirdas 2003a). When ripe, the tops of the plants were tied in bundles and dried on rocks. After drying, plants were unbundled and the seeds were dislodged by beating with a stick. Seeds were then parched and ground into meal that was eaten dry or mixed with acorn meal. The roots of many *Clarkia* species were also eaten.

Enchanter's Nightshade (*Circaea alpina*)
Description: This is a low, slender perennial with opposite leaves. Flowers occur in a bractless raceme. There are two sepals that are turned back, and the petals are notched and white to pink in color. This species grows in deep woods below 8,000 feet in the Sierra Nevada. It flowers in June and August.

Interesting Facts: A related species, *C. lutetiana* (broadleaf enchanter's nightshade), was used by the Iroquois as a dermatological aid (on wounds). They also made an infusion as a wash on injured parts (Chamberlain 1901).

Evening Primrose (*Oenothera*)
Description: There are many species in this genus that are annual, biennial, and perennial herbs. The flowers are white or yellow, often opening at night. There are eight stamens, four petals, four sepals, and the stigma is globe shaped to deeply four lobed. The various species can be found in a variety of habitats up to the subalpine zone.

1. Evening Primrose (*O. flava*)—This plant is green and occurs mainly in sagebrush and juniper woodlands below 5,000 feet.

2. Woody-fruited Evening Primrose (*O. xylocarpa*)—Look for this gray-green plant on dry benches between 7,000 and 9,800 feet.

Interesting Facts: Most handbooks on edible plants indicate that at least *O. hookeri* (= *O. elata* ssp. *hookeri*) (Hooker's evening primrose) and *O. biennis* (common evening primrose) have edible roots. These are cooked and eaten as a vegetable when young, becoming tough and somewhat spicy or peppery with age. The leaves of *O. biennis* are also edible as cooked greens but are not exceptional unless mixed with bland greens to make a more acceptable salad. Harrington (1967) suggests that all species would stand a trial, as none are known to be poisonous. The various species are known to hybridize easily, making identification at times challenging.

We have cooked and eaten the young seedpods of several species and found them to have an acceptable taste. Olsen (1990) also suggests that many species have seeds that are edible after being parched or ground into meal. Strike (1994) states that the seeds and leaves of *Oenothera* were eaten by California Natives.

The leaves, stems, and crushed seeds have an astringent quality to them and can be used as a poultice to heal wounds and for bruises and piles. The seeds are also high in essential oils and have been shown in clinical studies to be effective for heart disease, asthma, arthritis, alcoholism, and other fatty acid problems. The medicinal uses of the oil in these plants are a recent discovery following scientific research in the 1980s that demonstrated their effectiveness for a wide range of intractable complaints. The oil contains gamma-linolenic acid, an unsaturated fatty acid, which assists in the production of hormonelike substances. Evening primrose oil, in the form of gel caps, is becoming popular in the natural supplements marketplace (Tilford 1997). Additionally, the stringy bark makes good cordage material.

Fireweed, Willow-herb (*Epilobium*)

Description: There are many species of willow-herb. The genus includes annual and perennial plants that have willowlike leaves. The flowers are white or lavender in color with petals that are often notched. Fruits are long, narrow pods that open by four slits to release the numerous small, densely hairy seeds. The roots and pods are often needed to make positive identification of the many species. The genus name is from the Greek, meaning on a pod, describing the elongated ovary bearing the other flower parts on its top. The common name refers to the tufts of hairs at the end of the seed, which is similar to that on willow seeds.

Fireweed
(*Epilobium angustifolium*)

1. Alpine Willow-herb (*E. anagallidifolium*)—This species occurs in moist rock slides and stony places between 8,000 and 11,500 feet.
2. Fireweed (*E. angustifolium*)—This robust perennial is common in dry areas following a burn. It also grows in moist areas in the mountains below 9,000 feet and north to Alaska. At higher elevations, the plant is shorter with shorter leaves. It flowers from July to September.
3. Willow-herb (*E. brachycarpum*)—This erect annual grows in open, dry disturbed places below 7,500 feet. It flowers from June to September.
4. California Fuschia (*E. canum*)—Look for this species on dry slopes and ridges below 10,000 feet.
5. Northern Willow-herb (*E. ciliatum*)—This erect perennial grows in moist places below 11,000 feet. Several varieties of the species also occur in the area. It flowers from July to September.
6. Willow-herb (*E. clavatum*)—This species grows in rocky places between 4,000 and 12,000 feet in the central Sierra Nevada.
7. Dense-flowered Epilobium (*E. densiflorum*)—This species grows in moist places below 8,500 feet throughout the Sierra Nevada.
8. Epilobium (*E. foliosum*)—This species usually grows in dry, disturbed habitats below 5,000 feet.
9. Glaucous Willow-herb (*E. glaberriumum*)—This slender, erect perennial grows along streams and in wet areas from 3,000 to 11,500 feet. It flowers from July to August.
10. Hall's Willow-herb (*E. halleanum*)—Look for this species in wet places between 5,000 and 9,000 feet.
11. Hornemann's Willow-herb (*E. hornemannii*)—This species occurs in rocky and mossy places between 6,000 and 11,500 feet.
12. Howell's Willow-herb (*E. howellii*)—Look for this species in wet habitats between 6,500 and 7,500 feet in the central and northern Sierra Nevada (for example, Yuba Pass).
13. White-flowered Willow-herb (*E. lactiflorum*)—This species is found in moist habitats between 6,000 and 11,000 feet.
14. Dwarf Willow-herb (*E. latifolium*)—This species occurs in wet, stony places between 7,600 to 11,400 feet in Inyo, Fresno, Mono, and Tuolumne Counties.

15. Rock-fringe (*E. obcordatum*)—This species grows on dry ridges and flats above 7,000 feet.

16. Oregon Willow-herb (*E. oregonense*)—This slender, erect perennial occurs in wet, boggy places from 5,000 to 11,500 feet. It flowers from July to August.

17. Pale Boisduvalia (*E. pallidum*)—This species is found in damp places between 4,000 and 6,000 feet from Fresno County north.

18. Narrow-leaved Boisduvalia (*E. torreyi*)—Look for this species in moist places below 8,500 feet.

Interesting Facts: The dozens of species of *Epilobium* are all reported to be edible for people caught in survival situations, but *E. angustifolium* and *E. latifolium* are the best-known and most commonly consumed species. Food, drink, tinder, twine, and medicine are all provided by these abundant herbs. There are many small and "weedy" species found. In general, they are survivors in landscapes that have been ravaged by man-made and natural forces (such as fires and clear-cuts). Soil conditions do appear to affect their flavor. Many Native Americans "owned" good patches of fireweed, and these were passed on to subsequent generations. The most distinctive identifying feature of fireweed is the unique leaf venation. Unlike other plants, the veins do not terminate at the edges of the leaves, but rather join together in loops inside the outer margins (Vizgirdas 1999c).

The young shoots and leaves of fireweed may be boiled like asparagus but are better when mixed with other raw greens for a salad. The leaves, green or dry, make a good tea and are useful in settling an upset stomach. Be careful, the leaves are slightly laxative. The unopened flower buds can be used in the same manner as leaves and stems. The young fruits can also be boiled like green beans and are tasty before the seed fibers form. Mature plants tend to become tough and bitter.

The pith of the stems can also be scraped out and eaten as a snack or as a thickener for soups. If consumed in large amounts, fireweed is a gentle but effective laxative. The plant contains a relatively high content of vitamin C and beta-carotene. Raw roots are a popular food of Siberian Eskimos. A poultice made from the roots of fireweed can be used on skin inflammations, boils, ulcers, and rashes.

BETA-CAROTENE
In addition to providing the body with a safe source of vitamin A, beta-carotene is an antioxidant that works with other natural protectors to defend your cells from harmful free-radical damage caused by highly reactive substances that either form in the body or are acquired from the environment (such as air pollution, cigarette smoke, smoke carcinogens, and so on). Dark-green leafy vegetables, yellow and orange vegetables, and fruits are excellent sources of beta-carotene.

The fibrous inner bark can be used as cordage and tinder material. For use in making cordage, I found the fibers brittle. The seeds have cottonlike hairs and are great for fire starting (tinder) and insulation. Many Northwest Indians used the fluffy seed cotton as a wool substitute, mixing it with mountain goat wool or duck feathers. Willow-herb fluff, however, lacks the qualities of a really fine fiber. The flowers can also be rubbed into rawhide to repel water.

The seeds of *E. densiflorum* can be gathered, parched, pulverized, and then eaten. The Maidu also rubbed the plant on their head to relieve headaches.

Groundsmoke (*Gayophytum*)

Description: These are slender-stemmed annuals with alternate leaves, the lower ones often being opposite. The small flowers have distinct, reflexed sepals, white to pink petals, and eight stamens. Fruits are linear to club-shaped capsules. The various species are found on dry slopes and on the edges of meadows.

1. Gravel Gayophytum (*G. decipiens*)—This species occurs in dry, sandy, and gravelly places below 7,000 feet.

2. Diffuse Gayophytum (*G. diffusum*)—This species is found in dry to moist habitats between 3,000 and 9,000 feet.

3. Coville's Gayophytum (*G. eriospermum*)—This plant is found in open montane forests between 300 and 9,000 feet in the Sierra Nevada south of Placer County.

4. Varied-seeded Gayophytum (*G. heterozygum*)—This is an infrequently found species below 9,000 feet
5. Low Gayophytum (*G. humile*)—This slender, much branched annual grows on dry slopes between 3,000 and 11,000 feet. It flowers from June to August.
6. Black-foot Gayophytum (*G. racemosum*)—This species is found on drying slopes and flats between 5,000 and 11,000 feet.
7. Much-branched Gayophytum (*G. ramosissimum*)—Look for this species on dry slopes and ridges between 4,500 and 8,000 feet on the eastern side of the range.

Interesting Facts: An infusion of *G. ramosissimum* was used to soothe irritated skin.

Sun Cup (*Camissonia*)

Description: These are annual plants with basal or alternate leaves. The inflorescences are bracted and nodding, and the four sepals are reflexed. The white or yellow flowers usually fade to red.

Quick Key to the Sun Cups	
Flowers white	Booth's Sun Cup (*C. boothii*)—This is mainly a desert species but is occasionally found up to 7,000 feet in the southern Sierra Nevada.
Flowers yellow	
Stems appearing fleshy; basal leaves many	Sun Cup (*C. ignota*)—This species occurs in heavy soils below 3,500 feet from Madera County southward.
Stems not fleshy, slender; no basal leaves	Sierra Sun Cup (*C. sierrae*)—Look for this species in rocky outcrops below 8,000 feet from Madera and Mariposa Counties southward. It flowers from May to June.

Interesting Facts: Although there are no recorded uses for these species, other species have been utilized as food. The leaves were used as greens, eaten raw, boiled, or steamed.

False Mermaid Family (Limnanthaceae)

The two genera and twelve species in this family are restricted to North America. They are annual herbs found in wet places.

False Mermaid (*Floerkea proserpinacoides*)

Description: This is a slender annual with succulent stems. The leaves are pinnately divided into three to five oblong leaflets. The white flowers are stalked and borne singly in the leaf axils. The plant is found in moist, shaded habitats, especially under shrubs. The genus is named after a German botanist, H. G. Floerke. This plant is so inconspicuous that only those who are familiar with it are apt to notice it.

Interesting Facts: We have sampled the stems and leaves of this plant and found them spicy. They were an acceptable addition to our wild salad.

Meadow Foam (*Limnanthes*)

Description: The seven species are fragile annual herbs with pinnate leaves and solitary, regular white flowers. These plants have been considered a potential oilseed crop plant.

1. White Meadow Foam (*L. alba*)—This species occurs in moist habitats from Tuolumne County north. It flowers from May to June.
2. Mountain Meadow Foam (*L. montana*)—Look for this species in seeps and moist habitats. It flowers from March to May.

Interesting Facts: The oil contained in *Limnanthes* seeds is like the seed oil from jojoba (*Simmondsia chinensis*), which is similar to sperm whale oil. As such, the species are being tested as a source of oil to replace whale oil. Whale oil has the unique ability to cling to metal surfaces of gears and bearings while withstanding wide temperature variations and high pressure. The seeds of a *L. alba* were also eaten by Native Americans (Dunmire and Tierney 1997).

FLAX FAMILY (Linaceae)
There are about twelve genera and three hundred species in this family worldwide. The family is of some economic importance because of flax fibers, linseed oil, and the ornamentals obtained.

Flax (*Linum lewisii*)
Description: This much-branched annual has blue or rarely white flowers and alternate, sessile, and linear leaves. This species grows on open slopes from 1,000 to 9,500 feet. It flowers from May to July.

Interesting Facts: Flax has had value through the ages for its many uses, such as for thread, fabric, oil, paper money, and cigarette paper. It possesses excellent fibers for making cordage.

The seeds contain a cyanide compound but are edible after roasting. They have a high oil content that contains essential fatty acids that are very much needed in our daily lives, plus they add an agreeable flavor to cooked foods. The crushed seeds have been used as a poultice for irritation, boils, and pain. An infusion of stems is said to relieve stomachaches or intestinal disorders. The roots were also steeped to make an eye medicine. The stems are a source of linen, a fabric used for clothing.

Northwest Yellow Flax (*Sclerolinon digynum*)
Description: This annual plant with yellow flowers and oblong to elliptic and mainly opposite leaves occurs in moist, grassy meadows between 3,500 and 4,700 feet from Fresno County north. It flowers from June to July.

Interesting Facts: The seeds are probably edible after cooking.

Small-flowered Flax (*Hesperolinon micranthum*)
Description: This annual plant has very slender branches and filiform leaves. It is often found growing on serpentine, open slopes and ridges, mainly below 5,500 feet throughout the Sierra Nevada. It flowers from May to July.

Interesting Facts: The seeds are probably edible after cooking.

FOUR-O'CLOCK FAMILY (Nyctaginaceae)
There are thirty genera and about three hundred species in this family found in the Tropics and the New World. Fifteen genera are native to the United States, chiefly centered in Arizona and Texas. In general, the plants are trees, shrubs, and herbs with opposite leaves, and the flowers have bracts that mimic the sepals, and the sepals look like petals. Except as being a source of ornamentals, the family is of little economic importance.

Sand Verbena (*Abronia alpina*)

Description: This rather rare species is a matted perennial that is sticky-hairy and has opposite leaves. The flowers are lavender-pink in color. It occurs in sandy meadows between 8,000 and 9,000 feet in Tulare County. It flowers from July to August.

Interesting Facts: Although there are no recorded uses of this species, the Diegueno used an *Abronia* species as a diuretic (Ellis 1941; Burgess 1966). Other species have also been used as food and medicine.

GENTIAN FAMILY (Gentianaceae)

There are seventy genera and approximately eleven hundred species of this family worldwide. Thirteen genera are native to the United States. They are mostly annual or perennial herbs with bitter juice. Several species are cultivated as ornamentals.

Canchalagua (*Centaurium venestum*)

Description: The leaves of this plant are broadly ovate to lanceolate. The flowers are rose colored with red spots in the white throat; sometimes the flowers are all white. This species occurs in dry habitats below 5,000 feet throughout the Sierra Nevada. It flowers from May to August.

Interesting Facts: The genus name comes from the Greek *kentaur* (centaur). The centaur in Greek mythology was said to know the medicinal value of plants. The Luiseno used an infusion of this plant for fevers, whereas the Miwok used a decoction of the flowers for fevers (Bean and Saubel 1972). A related species (*C. scilloides*) was grown as an ornamental and is reported to have medicinal properties.

Fringed Gentian (*Gentianopsis*)

Description: The genus name is from *Gentiana* and the Greek *opsis* (resemblance). The genus consists of sixteen to twenty-five species of annual and perennial herbs with simple, opposite leaves and blue or white, four-lobed, funnel-shaped flowers.

Quick Key to the Fringed Gentians	
Plants with three to six pairs of leaves; calyx lobes are not dark ribbed	Hiker's Gentian (*G. simplex*)—This erect annual or biennial grows in damp meadows between 4,000 and 9,500 feet. It flowers from July to September.
Plants with mostly one to three pairs of leaves above the base; calyx lobes are dark ribbed	Sierra Gentian (*G. holopetala*)—This is an erect annual that grows in wet meadows from 6,000 to 11,000 feet. It flowers from July to September.

Interesting Facts: Gentian flowers close at night and during cloudy weather, thereby conserving pollen for the pollinators, which are most active during the fair-weather days. An infusion of roots of a related species (*G. crinita*) was used as a blood purifier and stomach strengthener.

Gentian (*Gentiana*)

Description: This is a large genus with most species occurring in moist or wet soil. Members are annual, biennial, or perennial herbs from fleshy roots or rhizomes. The flowers are four or five lobed, tubular or funnel shaped. The three species can be found from the foothills to alpine meadows. The genus honors King Gentius of Illyria, ruler of an ancient country on the east side of the Adriatic Sea, who is reputed to have discovered medicinal virtues in gentians.

Quick Keys to the Gentians	
Stems six to sixteen inches tall; corolla without dark bands	Explorer's Gentian (*G. calycosa*)—This species occurs in moist places between 4,000 and 10,000 feet throughout the Sierra Nevada. It flowers from July to September.
Stems one and a half to five inches tall; corolla with dark bands; subalpine	Alpine Gentian (*G. newberryi*)—Look for this species in moist meadows and banks between 7,000 and 12,000 feet throughout the Sierra Nevada. It flowers from July to September.

Interesting Facts: Moore (1979) suggests that gentians are perhaps the best stomach tonics. As a bitter, gentians excite the flow of gastric juices, thereby promoting an appetite and aiding in digestion. The root or chopped herb is steeped and drunk before a meal. Herbage and roots of most species are bitter. Craighead, Craighead, and Davis (1963), in discussing *G. calycosa*, make mention of medicinal uses of European and Asian gentians and that early settlers used them in much the same way as a tonic. Gentians contain some of the most bitter compounds known, against which the bitterness of other substances is scientifically measured.

Green Gentian (*Swertia*)

Description: Members of this genus are perennial herbs with opposite or whorled leaves. Its flowers are bell shaped and are densely aggregated into pyramid-shaped panicles. The species can be found in dry, open areas or meadows up to the subalpine zone.

Gentian (*Gentiana* spp.)

Quick Key to the Green Gentians	
Flowers five-merous; leaves mostly basal	Perennial Swertia (*S. perennis*)—This species is found in meadows and damp places between 7,000 and 10,500 feet throughout the Sierra Nevada.
Flowers four-merous; many stem leaves Stem leaves are opposite Flowers are in an open, broad panicle	Inyo Swertia (*S. puberulenta*)—Look for this species on dry slopes.
Flowers are in a narrow, spikelike inflorescence	White-stemmed Swertia (*S. albicaulis*)—This species grows in dry to moist places.
Stem leaves are whorled Leaves with narrow white margins	Kern Frasera (*S. tubulosa*)—Look for this species in dry, granitic, and volcanic gravels and slopes between 6,000 and 9,000 feet.
Leaves not white margined	Giant Frasera (*S. radiata* = *Frasera speciosa*)—This species grows on dry slopes between 6,800 and 9,800 feet. Giant frasera is an arresting and unique plant, not easily confused with any other.

Interesting Facts: The fleshy root of giant frasera can be eaten raw, roasted, or boiled. But, because the root is very bitter, we suggest mixing it with salad greens. An infusion of *S. albicaulis* was used to treat infected sores.

Little Gentian (*Gentianella*)

Description: These are glabrous annuals with basal and cauline leaves. The calyx tube is shorter than the lobes, and the corolla has spreading lobes that are shorter than the tube.

Quick Key to the Little Gentians	
Flowers five-merous and clustered	Felwort (*G. amarella*)—This common slender, branched annual grows in moist places from 4,500 to 11,000 feet. It flowers from June to September.
Flowers four-merous and solitary	Dane's Gentian (*G. tenella*)—Look for this rare species in meadows between 8,000 and 12,200 feet in the Rock Creek Lake basin and Whitney Meadows. It flowers from July to August.

Interesting Facts: A related species (*G. propinqua*) was used as a cold remedy. A decoction of the leaves, stems, and flowers was also taken for colds and cough. Another species (*G. quinquefolia*) provided an infusion for diarrhea, and the liquid from the root was used for hemorrhages (Gottesfeld 1992).

GERANIUM FAMILY (Geraniaceae)

There are eleven genera and eight hundred species in this family distributed worldwide. *Geranium* and *Erodium* are native to the United States. Members of this family are five-merous plants (five petals, five or ten stamens, and pistil of five parts). The seedpod resembles the head and beak of a stork or crane, hence the common name, with the seeds in the short, thickened "head" and the style elongated into the pointed "beak." The family is of no real economic importance, except as a source of ornamentals, primarily from the cultivated geranium (*Pelargonium*), a tropical genus well developed in South Africa.

Quick Key to the Geranium Family	
Leaves pinnately compound	Storksbill (*Erodium*)
Leaves palmately veined or divided	Geranium (*Geranium*)

Red-stem Storksbill (*Erodium cicutarium*)

Description: Red-stem storksbill is a low-growing annual with mostly basal, finely dissected, fernlike, pinnately divided leaves. The flowers are small and pink and mature into the distinctive "stork's bill" fruit. This is an introduced plant that is widespread on disturbed sites at low to middle elevations.

Interesting Facts: *Erodium* seed has a long tail that dries into a coil resembling a clock spring. At night the tail straightens when moistened by dew but recoils in the morning when dried by the sun. The sharp, pointed seeds thereby slowly plant themselves by the repeated straightening and coiling of the tail. An activity to do with kids is to breathe heavily on the well-coiled seed and then place it in the sun to watch the seed coil.

The leaves of red-stem storksbill can be eaten raw in salads or cooked as a potherb. They are particularly palatable when picked young and have a parsley-like taste. We find it nicely complements an otherwise bland wild salad and provides a good source of vitamin K. It is uncertain whether other species of *Erodium* are edible, thus experimentation is not recommended.

The species has a reputation of being a diuretic, astringent, and anti-inflammatory herb. The entire plant was used in a warm-water bath for persons suffering from the pains of rheumatism. Leaves were also used in a hot tea to increase urine flow and to increase perspiration.

Wild Geranium (*Geranium*)

Description: Geraniums are annual or perennial herbs that are hairy. The leaves are mostly basal, and the flowers are showy, with five petals and sepals and ten stamens. The mature fruits are spirally coiled. They can be found in wet meadows or dry, open forests. *Geranium* derived from the Greek word *gernion*, meaning crane.

Flower and Fruit of Geranium
(*Geranium* spp.)

Quick Key to the Geraniums

Plant annual arising from a slender base; leaves are usually less than one and a half inches wide	Carolina Geranium (*G. carolinianum*)—This species is common in grassy and shaded places below 5,000 feet. It flowers from April to July.
Plant perennial; leaves are usually more than one and a half inches wide	
Free tips of styles less than one-quarter inch long	Geranium (*G. californicum*)—This slender perennial occurs in moist places from 7,000 to 9,000 feet. It flowers from June to August.
Free tips of styles one-quarter to one-half inch long	Wild Geranium (*G. richardsonii*)—This glabrous or pubescent perennial grows in moist places between 4,000 and 9,000 feet. It flowers from July to August.

Interesting Facts: The leaves and flowers of most species can be eaten, but because of their astringent properties and texture, they are not a choice edible. We find that they are best when tossed in with other greens in salads or steamed as potherbs. In any case, the leaves are better treated as a filler to stretch supplies of other more tasty and less abundant greens. The leaves can also be chopped and added to soups, thereby blending flavors, making the leaves more acceptable. Leaves do toughen with age but are still palatable in stews. Geranium leaves are similar looking to monkshood (*Aconitum* spp.), so positive identification of the flowerless plants is important. Harvest leaves and roots from plants identified with flowers.

A leaf or root tea of *G. richardsonii*, which is one of the most widespread western species and frequently hybridizes with other species, can be used as a gargle for a sore throat. The root sliced fresh can be used as a first aid for gum or tooth infections when applied directly on the area of pain (Willard 1992).

The herbaceous part of a related species not found in the Sierra Nevada, *G. viscosissimum* (sticky geranium), was used as an astringent and styptic, internally for diarrhea and hemorrhages. The plant is high in tannins, providing astringent remedies important in traditional medicine for the emergency treatment of injuries and diarrhea. A hot poultice of boiled leaves was used for bruises and skin problems. The green crushed leaves can be applied to relieve pain and inflammation.

GINSENG FAMILY (Araliaceae)

There are about fifty genera and five hundred species worldwide. Though they are of limited economic importance, several species are used as ornamentals. One

species, American ginseng (*Panax quinquefolius*), is the famous medicinal panacea and "mind enhancer." It also used to be one of the most important exports in
the United States. Another species, *Panax ginseng* from China, is much more
famous and has been economically important in China for thousands of years.

Elk Clover, Wild Sarsaparilla (*Aralia californica*)
Description: This is a large perennial with stems that grows up to nine feet tall.
The roots have a milky juice, and the leaves are two times pinnate. The leaflets are
toothed and ovate. The small flowers occur in a many-flowered umbel with petals
less than one-eighth inch long. The berrylike fruit is red when young, becoming
black with age. This species grows on moist and shaded slopes below 5,500 feet
throughout the western side of the Sierra Nevada. It flowers from June to August.

Interesting Facts: A poultice from the rhizomes of a related species, *A. nudi-
caulis,* can be used to treat burns and sores. As a tonic, it is a diuretic that lowers
fever. As a tea, wild sarsaparilla is rather pleasant tasting. Internally, it was used
for coughs and purifying blood. The long rhizome is often used as an ingredient
of root beer. Native Americans relied for long periods of time on these roots
during wars or when hunting (Fernald and Kinsey 1958). It has been used as a
substitute for true sarsaparilla (*Smilax officinalis*). The roots and rhizomes of
another related species, *A. racemosa,* have been used to treat rheumatism, coughs,
and backaches.

GOOSEFOOT FAMILY (Chenopodiaceae)
Approximately 102 genera and 1,500 species in the goosefoot family are found
worldwide. Fourteen genera are native to the United States, mostly in the West.
This family includes several food plants (for example, beets and spinach) and
weeds (such as Russian thistle).

Some plants in the Chenopodiaceae family can accumulate high amounts of
nitrates, which, when ingested, are reduced to toxic nitrites. Nitrites cause the
hemoglobin (red pigment) to transform into methemoglobin, which is dark
brown in color and makes the blood unable to transport oxygen to the body
tissues. Additionally, some plants in this family can become toxic through the
accumulation of selenium absorbed from the soil.

Goosefoot, Lambs's-quarters
(*Chenopodium* spp.)

Goosefoot, Lambs's-quarters (*Chenopodium*)
Description: The genus name comes from *Cheno,* meaning goose, and *podium,*
meaning foot, because the triangular leaves resemble the shape of a goose's foot.
Oil of chenopodium, distilled from the fruits, contains a broad-spectrum vermifuge that is widely used in veterinary medicine.

1. Lambs's-quarters (*C. album*)—This is a common weed that occurs below 6,000 feet
   in a wide variety of habitats.
2. Pigweed (*C. atrovirens*)—This annual plant is found in open, dry places and
   coniferous forests between 4,000 and 11,000 feet.
3. Pitseed Goosefoot (*C. berlandieri*)—This annual is quite common in open, often
   disturbed areas up to 7,000 feet.
4. California Pigweed (*C. californicum*)—This is common on dry slopes below 5,500
   feet. It blooms from March to June.
5. Goosefoot (*C. chenopodioides*)—This species occurs in alkaline places above 5,000
   feet.
6. Pigweed (*C. desiocatum*)—This species occurs in dry places between 4,000 and
   11,000 feet in the montane forests.

7. Fremont's Pigweed (*C. fremontii*)—This erect, pale-green annual is common in dry places from 5,000 to 8,500 feet. It flowers from June to October.
8. Inconspicuous Goosefoot (*C. incognitum*)—This plant is found in dry places between 6,000 and 8,000 feet in the montane forests. It flowers from July to August.
9. Goosefoot (*C. leptophyllum*)—This species occurs in dry places between 5,000 and 8,000 feet in sagebrush, pinyon-juniper, and yellow-pine forests. It flowers from July to September.

Interesting Facts: Leaves, tops, and seeds of all species can be used as an emergency or basic food and are quite tasty and nutritious. High in protein, the greens are a good source of vitamins A and C, iron, and potassium and are extremely rich in calcium. Because they do not become bitter with age, both young and old plants can be used. Leaves may be used raw in salads or boiled in water like spinach. The water can be saved and used as a yellow dye. The leaves were also eaten to treat stomachaches and prevent scurvy. A leaf poultice was used on burns. The flower buds and flowers can be used as potherbs. A single plant can produce up to seventy thousand seeds. Seeds can be ground as flour for use in bread or cooked as mush. Seeds can also be eaten without grinding or incorporated into pinole (flour made from a mixture of seeds of small plants). The seeds contain about 15 percent protein and 55 percent carbohydrates, more than is found in corn (Van Etten et al. 1963). The seeds can also be used as a coffee substitute.

Large quantities of the plant should not be eaten, as many species contain high levels of oxalic acid, which tends to bind calcium and prevent its proper absorption into the body. Cooking or freezing *Chenopodium* apparently breaks down the oxalic acid. Additionally, *Chenopodium* has been known to accumulate toxic levels of nitrates and may cause livestock poisoning. But because large quantities of the plant must be consumed to cause problems, this type of poisoning may be unlikely.

The hard root of some *Chenopodium* species was stored until needed, then grated on a rock to make soap. The leaves were also used to make soap but are not as effective as the roots.

Kochia (*Kochia*)

Description: Three species of kochia may be found in California, but it is *K. scoparia* (common kochia) that is most commonly encountered. Common kochia is a bushy annual with stems up to three feet tall. The leaves are alternate, narrowly lance shaped, and tapered at both ends. The herbage may or may not be covered with hairs. Its flowers are solitary or in clusters in spikes. The species is common in open, disturbed habitats at low elevations.

Interesting Facts: Common kochia is native to Europe and was introduced into the United States as an ornamental. It has since escaped and has become well established. In Asia, Japan, and China, common kochia was cultivated for its seeds. The tips of the young shoots can be prepared as potherbs (Clarke 1977). The seeds can be eaten raw or cooked or ground into meal and used in bread making. The genus is named for William Koch, a late-eighteenth- and early-nineteenth-century German botanist.

Povertyweed (*Monolepis nuttalliana*)

Description: Povertyweed is a low-growing winter annual with prostrate or ascending stems. The leaves are somewhat succulent and lance shaped, broadened and lobed at the base. Its flowers are borne in dense clusters at the leaf bases, and the solitary sepal is reddish in color. The seeds are dark brown. The plant is found in open, disturbed habitats at the lower elevations.

Interesting Facts: The aboveground parts of povertyweed may be eaten as a potherb. The seeds are also edible. Another species reported to occur in the Sierra Nevada is *M. spathulata* (beaver monolepis), but its uses are unknown.

Prickly Russian Thistle, Tumbleweed (*Salsola tragus*)

Description: This is not a true thistle (*Cirsium*), but a many-branched annual with purplish striped stems up to three feet tall in a rounded form. The lower leaves are threadlike; the upper leaves are awl-like and spine-tipped. The plant may or may not be hairy. When mature, the whole plant becomes rigid, breaks off at ground level, and becomes a "tumbleweed," blowing across the open plain.

The flowers are solitary in the leaf axils and are subtended by spiny bracts. Russian thistle is common in open, disturbed habitats, particularly around agricultural areas at low elevations and sand dunes. It was introduced to the United States from Europe. Fortunately, this species is not an aggressive competitor and does not appear to replace native plant species. However, it is still considered a noxious weed because of its distributional pattern and spines.

Interesting Facts: This unsavory-looking plant is edible. The young parts of the plant may be boiled and eaten as a potherb or chopped raw into a salad. On older plants, clip the tender branch tips that are green. We find the taste of the plant greatly improves when cooked in butter and lemon. In Europe, the ashes of the plant were once used in the production of carbonate of soda known as Barilla (Fernald and Kinsey 1958).

Warning: The older parts of the plants contain significant quantities of nitrates and oxalates and may be toxic if eaten in quantity.

Saltbush (*Atriplex*)

Description: These are annual or perennial herbs or shrubs with alternate leaves, and glabrous or scaly herbage. The flowers are unisexual, and individual plants have one or both sexes. The various species are found at the lower elevations in valleys, disturbed areas, or in dry, alkaline soils.

1. Silverscale (*A. argentea*)—This species occurs in alkaline places below 6,000 feet, mainly on the eastern side of the range.
2. Tumbling Oracle (*A. rosea*)—This is a common species in disturbed places.
3. Bractscale (*A. serenana*)—This species is found in alkaline places.
4. Wedgescale (*A. truncata*)—This species is mainly found below 8,000 feet.

Interesting Facts: There are many uses for these plants, from food to medicine and dyes, as well as soap and spice. The young leaves of many species can be cooked and eaten as greens and have a very distinct salty taste. We have often added them to otherwise bland foods to make our wild meals less boring. Adding the leaves to meats while cooking will help spice them up. The seeds were parched, ground into flour, and made into mush. They can also be soaked in water for a few minutes to make a rather pleasant-tasting drink. The Navajo used the flowers to make puddings (Bailey 1940). The ashes of *A. canescens* (fourwing saltbush) make a good substitute for baking soda.

Medicinally, the various species provided the Native Americans with many uses. As an analgesic, the Navajo used the leaves of *A. argentea* (silverscale saltbush) as a fumigant for pain, and the Zuni made a poultice from the chewed root for application to sores and wounds. A warm poultice made from the pulverized root fourwing saltbush was used to treat toothaches.

The leaves and roots of many species were used as a soap. They were rubbed in water for lather and used in washing clothing and baskets. Many Native Americans also carved arrowheads from the wood for use as weapons and for hunting. The seeds of some species were also used in making a black dye.

Winterfat (*Krascheninnikovia lanata*)
Description: In many older references, winterfat is also known as *Eurotia lanata*. Winterfat is a small shrub found at the lower plains and foothill elevations, often in saline or alkaline areas. The leaves are alternate, narrow and entire, whereas the flowers occur in heads or spikes in the axils of the leaves. The genus name is from the Greek *eurotios,* meaning moldy, and refers to the dense hairiness of the plant.

Interesting Facts: Although the edibility of this species is unknown, it is an important forage plant for horses and other livestock. Medicinally, the plant has been used by many Native American tribes. For example, the Hopi Indians used the powdered root for burns, and a decoction of the leaves was used for fevers. The Navajo made a poultice of the chewed leaves and applied it to a poison ivy rash. The Navajo also incorporated the stems and leaves of this plant in sweat-house ceremonies by placing them on hot rocks for the "mountain chant" (Elmore 1944).

GRAPE FAMILY (Vitaceae)
There are fourteen genera and eight hundred species of climbing plants in this family. They all have tendrils, and only a few species are shrubs and trees. The leaves are simple, palmately or pinnately lobed. The fruit is a berry. The most economically important member of the family is *Vitis vinfera.*

California Wild Grape (*Vitis californica*)
Description: The stems of wild grape can grow up to forty feet long and climb over other plants and rocks by means of tendrils that are opposite the leaves. The leaves are roundish, three lobed, and cordate at the base. The flowers are yellow, five-merous, and the fruits are purplish in color with a white bloom. Look for wild grape in shady canyons below 5,000 feet on the western side of the range.

Interesting Facts: Although grapes covered vast areas, they were not a major food source. All parts of the plant are edible, except the root. In comparison to domestic grapes, these wild ones are smaller, are thicker skinned, and have more seeds. The seeds can be eaten raw or dried, and the tender green shoots can also be eaten.

Leaves were used as a poultice for snakebites, and the sap from the leaves was used to treat diarrhea. The split grape shoots were used as twining material for baskets. The stems were used as rope. It is possible to obtain water by cutting the ends off and allowing the watery sap to drain.

Warning: Stems and roots can cause allergic reactions in some sensitive people.

HEATH FAMILY (Ericaceae)
There are about fifty genera and twenty-five hundred species in the heath family, the majority occurring in acidic soils. In the United States, approximately twenty-five genera are indigenous. Economic products provided by this family

include food plants, oil of wintergreen, and many ornamentals. Some herbaceous members are mycotrophic; that is, they depend on fungi for nutrient uptake and lack chlorophyll. The list of genera and species here includes the parasitic families Monotropaceae and Pyrolaceae and hemiparasitic herbs in addition to the autotrophic plants.

Alpine Wintergreen (*Gaultheria humifusa*)

Description: Alpine wintergreen is a dwarf evergreen shrub that forms small mats with leaves that are broadly egg shaped to elliptical. The bell-shaped flowers are white to pink, and the berries are red.

Interesting Facts: The small, red fruits are edible raw or cooked and can be made into jams, wines, or pies. The young tender leaves are suitable as greens and have a wintergreen flavor. The fresh leaves of a related species, *G. hispidula* (creeping snowberry), can be used to make a tea, and the berries are also edible.

In Native American medicine, the plants were used for treating aches and pains and to help with breathing while hunting or carrying heavy loads. The leaves of these species yield an oil upon steam distillation. This "oil of wintergreen" (methyl salicylate) is a folk remedy for body aches and pains and is known for its astringent, diuretic, and stimulant properties.

Warning: The wintergreen flavor in the plants is due to the presence of oil of wintergreen, which, if taken in excess, can be toxic, especially to children. In small amounts, such as in wintergreen tea, there is little danger. Children who are allergic to aspirin (a related drug) should not eat the plant or berries or even handle the plant.

Bog Kalmia (*Kalmia polifolia*)

Description: This species is a branched, evergreen shrub that spreads by short rhizomes and layering. The leaves are opposite, lance shaped to elliptical, and have rolled-in margins. The flowers are rose colored and bowl shaped. The plant has ten stamens that, when triggered by an insect landing on the flower, spring out of pockets in the petals and discharge pollen. The species occurs in the mid- to upper elevations, usually in moist to wet acidic soils, often along creeks.

Interesting Facts: The toxicity of *Kalmia* is legendary. Some Native Americans used it as a suicide plant. Game birds and livestock may be poisonous to eat if they have ingested the leaves. According to Peter Kalm (1715–1779), after whom the genus is named, "sheep are especially susceptible, while deer are unharmed. Though the flesh of affected animals is apparently not contaminated, the intestines will cause poisoning if fed to dogs so that they become quite stupid and as it were intoxicated and often fall so sick that they seem to be at the point of death" (Clawson 1933; Jaynes 1997).

All *Kalmia* species should be considered *poisonous*. They contain andromedotoxin, which causes a slow pulse, low blood pressure, lack of coordination, convulsions, progressive paralysis, and death. The honey made by bees from these plants is also poisonous (Kingsbury 1965).

Huckleberry (*Vaccinium*)

Description: The three species of *Vaccinium* in the Sierra Nevada are small to midsize shrubs with deciduous leaves. The twigs are often angled. The small flowers are urn shaped, and fruits have many-seeded berries. They can be found on well-drained sites, from wet meadows and around lakes up to the timberline.

SAPROPHYTIC HEATHS
A number of species in this family possess little or no chlorophyll to capture sunlight to derive nutrition and water. To compensate, they have an intimate association with mychorrhizal fungi. The degree to which these saprophytes are associated with or depend upon fungi is variable. The saprophytic heaths in the Sierra Nevada include the snow plant (*Sarcodes sanguinea*), pinedrops (*Perospora andromedea*), and coralroots (*Corallorrhiza* spp.). Some species of orchids (for example, *Eburophyton austinae*) are also saprophytes.

ANDROMEDOTOXIN
The major toxic substance in the Ericaceae appears to be andromedotoxin. This compound is known to occur in *Rhododendron, Leucothoe, Menziesia, Ledum,* and *Kalmia* and is probably more widespread than is now known. The leaves, twigs, flowers, and pollen grains of these genera all contain andromedotoxin. The course of poisoning includes watering of the mouth, eyes, and nose; loss of energy; vomiting; slow pulse; low blood pressure; lack of coordination; convulsions; and slow and progressive paralysis of arms and legs until death. Humans can be poisoned by chewing on the leaves and twigs, brewing "tea" from the leaves, or by sucking nectar from the flowers of these plants. The honey produced from bees after visiting large stands of rhododendron has been known for hundreds of years to be poisonous. The honey is normally so bitter to the taste that very little of it can be eaten.

Quick Key to the Huckleberries

Leaves are entire; twigs are not angled	Western Blueberry (*V. uliginosum*)—Look for this species in wet meadows between 5,000 and 11,000 feet throughout the Sierra Nevada. It flowers from June to July.
Leaves are toothed or minutely so; twigs may or may not be angled	
Twigs are sharply angled; fruit is red drying purple	Red Huckleberry (*V. parvifolium*)—This species is occasionally found in shady, moist habitats below 7,000 feet from Fresno County northward. It flowers from May to June.
Twigs are not or weakly angled; fruit is not red	Dwarf Bilberry (*V. caespitosum*)—This species grows in wet meadows and near snowbanks between 6,700 and 12,000 feet throughout the Sierra Nevada. It flowers from May to July.

Interesting Facts: In general, all *Vaccinium* berries can be eaten raw or dried in the form of cakes for future use. The various species we have sampled range in taste from sweet to tart. Hybridization between the species is known to occur, but the fruits are still edible. The berries have also been used as fish bait because they look very similar to salmon eggs. The leaves can be dried to make a tea. The leaves and berries are high in vitamin C.

Manzanita (*Arctostaphylos*)

Description: These are shrubs with evergreen, alternate leaves. The flowers occur in terminal racemes, and there are five sepals. The corolla is urn shaped with four to five lobes. Stamens number eight to ten, and the fruit is drupaceous. The genus name means bear grape and refers to the fondness shown by bears for the fruits of these shrubs, many of which are known as bearberry.

Manzanita
(*Arctostaphylos* spp.)

Quick Key to the Manzanitas

Plant grows prostrate to the ground, usually less than two feet high; the branchlets often root when in contact with the ground	
Flowers pink	Greenleaf Manzanita (*A. patula*)—Look for this species in open forest situations between 2,000 and 11,000 feet. It flowers from April to June.
Flowers usually white	
Through a handlens, stomates appear to occur on both surfaces of the leaves	Pinemat Manzanita (*A. nevadensis*)—This species is found in moist to dry slopes between 5,000 and 10,000 feet. It flowers from May to July.
Through a handlens, stomates appear to occur on only the lower surface of the leaves	Bearberry, Sandberry (*A. uva-ursi*)—This species is found mostly in the northern Sierra Nevada northward.
Plants usually taller than two feet	
Flower stalks with glandular-pubescent hairs; the ovary may be glandular pubescent	Whiteleaf Manzanita (*A. viscida*)—This species occurs on dry slopes below 5,000 feet. It flowers from March to April.

Flower stalks are not hairy

Leaves are pale gray-green in color	Indian Manzanita (*A. mewukka*)—This species occurs on dry slopes, mainly between 2,500 and 6,000 feet from Butte County southward. It flowers from March to April.
Leaves are bright green in color	Greenleaf Manzanita (*A. patula*)—Look for this species in open forest situations between 2,000 and 11,000 feet. It flowers from April to June.

Interesting Facts: The berries of all *Arctostaphylos* are edible. They may be eaten raw, and it is suggested that they not be eaten in large quantities because they may be hard to digest. Constipation or indigestion are common maladies of eating too much. The berries can also be stewed or dried and ground into meal and cooked as mush. A cider can also be made from the berries. The seeds alone can be collected and ground into meal too.

The leaves of bearberry can be boiled in water, allowed to cool, and the decoction applied to stop the itching and spreading of poison oak. The internal consumption of the leaf tea often results in urine becoming alkaline and bright green. This is caused by the urinary antiseptic hydroquinone, and it is relatively harmless. These hydroquinones (particularly arbutin) are strongly antibacterial and are effective against *Klebsiella* and *E. coli*, which are often associated with urinary infections.

Many species were mixed with tobacco and smoked. Leaves are an astringent due to the tannic acid and have been used to tan hides. The leaves can also be chewed to stimulate saliva, particularly when one is thirsty.

MANZANITA CIDER
To make cider, Native Americans crushed the berries into coarse pulp with a rock in a basket. Then they added water that seeped through the mass and dripped into a watertight basket below, extracting some of the berry flavor in the process. Manzanita cider, which was served as an appetizer, was enjoyed by dipping into the beverage a small stick with several short feathers fastened to one end and then sucking the drink off the feathers. A small watertight basket was also used for drinking the cider or water.

One-flowered Wintergreen, Woodnymph (*Moneses uniflora*)
Description: The leaves of this plant are basal, thin, ovate, and sharply serrulate. Its flowers are solitary, and the petals are white to pink in color. The anthers are two horned and the stigma five lobed. This is an uncommon plant found in woods in Fresno County below 3,500 feet.

Interesting Facts: A poultice of the leaves was applied to draw out pus from boils. An infusion of the dried plants was used for coughs and colds, and the plants were chewed for sore throats. A poultice of chewed or pounded plant was applied to pains. The fruit was used as food (unspecified) by Montana Indians (Hart 1996).

One-sided Wintergreen (*Orthilia secunda*)
Description: This species was once in the genus *Pyrola* (*P. secunda*). This low perennial has flowers dangling on one side of the stem and grows in dry, shady woods from 3,000 to 10,500 feet. It flowers from July to September.

Interesting Facts: A strong decoction of the root was used as an eyewash.

Pacific Madrone (*Arbutus menziesii*)
Description: This evergreen tree grows up to 120 feet tall and has peeling bark and stems that are red and appear polished. The large leaves are alternate, leathery, entire or toothed, ovate, and glabrous. They are dark green above and paler

below. The flowers are urn shaped, white to pink in color, and occur in terminal clusters. The fruit is a globose, red-orange berry. It has a rough, granular surface. Madrone grows in wooded, moist areas below 5,000 feet. It blooms from March to May.

Interesting Facts: The ripe berries can be eaten raw or cooked. Some Native Americans steamed the berries in a basket with madrone leaves on top, then dried and stored the berries. To eat, the berries were then soaked in warm water to soften. The berries do not store well, as they will decay when bruised. We have found them best when mixed and crushed with manzanita berries. The berries can also be used in making a tea. Simply crush the berries in water like you would manzanita.

Caution: If eaten in large quantities, the berries could cause a person to vomit.

Pipsissewa (*Chimaphila umbellata*)

Description: This is a short, evergreen semishrub (woody only at the base) that originates from a long, creeping rootstock. The leaves are whorled and leathery. Pipsissewa grows in dry, shady forests from 1,000 to 10,000 feet. It flowers from June to August.

Interesting Facts: The roots and leaves of pipsissewa may be boiled and the liquid cooled for a refreshing drink that is high in vitamin C. The leaves may also be nibbled raw, but because of their astringency and tough texture we found them unappealing.

Pipsissewa was an important herb to Native Americans for treating rheumatism. A tea from the leaves was used for the purposes of treating rheumatism and kidney problems. The plant contains quinone glycosides, such as that found in *Arctostaphylos*, but is less astringent and more a diuretic, making it better for long-term use. The plant was also mixed with tobacco for smoking. Pipsissewa produces a natural antibiotic that can be used by humans. Hot infusions of pipsissewa can be taken to induce perspiration in the treatment of typhus, and the berries can be eaten for stomach disorders.

Pipsissewa is a "secret ingredient" in certain popular soft drinks. In the Pacific Northwest, these plants, as well as certain species of *Pyrola*, are under commercial harvesting pressure and may be slowly disappearing (H. Norton 1981).

How hard to realize that every camp of men or beast has this glorious starry firmament for a roof! In such places standing alone on the mountaintop it is easy to realize that whatever special nests we make—leaves and moss like the marmots and birds, or tents and piled stone—we all dwell in a house of one room—the world with the firmament for its roof—and are sailing the celestial spaces without leaving any track.
—John Muir, *John of the Mountains*

Red Mountain Heather (*Phyllodoce breweri*)

Description: This is a low shrub, one foot or less tall, with leaves that are alternate, crowded, linear, and needlelike with rolled-in margins. The flowers occur in a terminal raceme and are bright rose-purple. The corolla is bell shaped and hangs downward. Red mountain heather occurs in the higher elevations in moist areas and alpine meadows from 6,000 to 12,000 feet. It blooms in the early summer.

Interesting Facts: The species is named after William H. Brewer, first person to botanize extensively in the Sierra Nevada in the late 1860s and collect this plant in Yosemite in 1862. Brewer was California's first state botanist and worked with the California State Geological Survey.

A related species, *P. empetriformis* (pink mountainheath), was used by the Thompson Indians (southwestern British Columbia, Canada) as a tuberculosis remedy. Apparently, a decoction of the plant was taken over a period of time for tuberculosis and spitting up blood (Perry 1952).

Sierra-laurel (*Leucothoe davisiae*)

Description: The stems of this plant are erect and leafy. The leaves are oblong, entire to somewhat serrulate. The calyx consists of five almost distinct segments and is white. This species occurs in moist, springy places between 3,200 and 8,500 feet from Fresno to Plumas Counties. The genus name is derived from Leucothoe, daughter of Orchamus, king of Babylon, and Eurynome in Greek mythology, who is said to have been changed into a shrub by her lover, Apollo.

Interesting Facts: The herbage contains andromedotoxin. An infusion made from a related species (*L. axillaris*) was rubbed on to treat rheumatism.

Snow Plant (*Sarcodes sanguinea*)

Description: This red, fleshy, saprophytic plant with scalelike leaves grows up to twenty inches tall. This plant grows in thick humus of the shady forests between 4,000 and 8,000 feet. It flowers from May to July. Here is an early description of the snow plant as it was written by John Muir in 1912:

Snow Plant
(*Sarcodes sanguinea*)

> The snow plant (*Sarcodes sanguinea*) is more admired by tourists than any other in California. It is red, fleshy and watery and looks like a gigantic asparagus shoot. Soon after the snow is off the ground it rises through the dead needles and humus in the pine and fir woods like a bright glowing pillar of fire. In a week or so it grows to a height of eight or twelve inches with a diameter of an inch and a half or two inches; then its long fringed bracts curl aside, allowing the twenty- or thirty-five-lobed, bell-shaped flowers to open and look straight out from the axis. It is said to grow up through the snow; on the contrary, it always waits until the ground is warm, though with other early flowers it is occasionally buried or half-buried for a day or two by spring storms. The entire plant—flowers, bracts, stem, scales, and roots—is fiery red. Its color could appeal to one's blood. Nevertheless, it is a singularly cold and unsympathetic plant. Everybody admires it as a wonderful curiosity, but nobody loves it as lilies, violets, roses, daisies are loved. Without fragrance, it stands beneath the pines and firs lonely and silent, as if unacquainted with any other plant in the world; never moving in the wildest storms; rigid as if lifeless, though covered with beautiful rosy flowers.

Interesting Facts: The genus name is from the Greek for fleshlike, referring to the inflorescence. The species name is Latin for bloodred, referring to the plant's overall color. Snow plants do not photosynthesize but derive their nutrients through a relationship with soil fungi.

The fleshy plant is edible when prepared like asparagus. However, because this is a protected and rare plant, only in an emergency should this be even considered. A decoction of the leaves and stems was used to treat ulcerated sores, irritated skin, and toothaches. The decoction was also used as a blood tonic. Some Native Americans also dried and ground this plant into a powder to relieve toothaches and other mouth pains (Strike 1994).

Western Azalea (*Rhododendron occidentale*)

Description: This deciduous shrub with shredding bark and large white flowers grows up to ten feet tall. The flowers are very showy and fragrant. Western azalea occurs in moist places in the Sierra Nevada below 7,500 feet. It flowers from May to July.

Interesting Facts: Many plants in this family contain a poisonous compound called andromedotoxin. If consumed in large concentrations this could be harmful, causing vomiting, illness, and even death.

Western Labrador Tea (*Ledum glandulosum*)

Description: This is an evergreen shrub, with short, fine hairs and glands on young branches and lower leaf surfaces. The leaves are elliptical to egg shaped and are clustered near the stem tips, giving them a whorled effect. The white flowers are in rounded clusters at the tip of the stem. All parts of the plant smell like turpentine when crushed. Look for the plants in the mid- to upper subalpine elevations, particularly in permanently wet or moist, acidic soils.

Interesting Facts: Labrador tea contains ledol, which is a narcotic toxin that causes drowsiness, delirium, cramps, paralysis, heart palpations, and even death if taken in excess. Prolonged cooking extracts large doses of ledol. Otherwise, the tea is a slight laxative. Infusions are recommended for camper's distress (that is, constipation). The leaves are astringent and useful in facial creams.

Andromedotoxin is also found in the leaves of these plants (see *Kalmia*), and therefore they should be considered *poisonous*. The leaves of *L. groenlandicum* (hog labrador tea) make a mild but agreeable tea when steeped in hot water. Willard (1992), Densmore (1974), and Foster and Duke (1990) indicate that the tea was used for colds, rheumatism, scurvy, and stomach ailments but is *not recommended*. A strong decoction of the leaves was used as a wash to get rid of lice. As an insect repellent, it was said to be quite effective against mosquitoes. According to Elias and Dykeman (1982), "[S]teep 1 tablespoon dried leaves in cup of boiling hot water for 10 minutes. Do not boil water with leaves in it, as it may release the harmful alkaloids. Serve hot or cold."

Another common name for this species is "trapper's tea." The name apparently originated not because trappers drank it, but because they boiled their traps in it to de-scent them.

White Heather (*Cassiope mertensiana*)

Description: The low branches of this plant can be easily mistaken for clubmoss (*Selaginella*). The heather is very distinctive when the delicate snow-white, bell-shaped flowers are present. This species occurs on rocky ledges and crevices between 7,000 and 12,000 feet from Fresno County northward. It flowers from July to August.

Interesting Facts: The plant appears as a thick mat when not in bloom. The waxy white flowers bloom thickly in the manner of lily of the valley. Their starlike appearance no doubt inspired the name Cassiope, who, in Greek mythology, was set among the stars as a constellation.

Wintergreen (*Pyrola*)

Description: In general, wintergreen are low, smooth perennial herbs with shiny, leathery leaves that are clustered at the base. The flowers are waxy and nodding. *Pyrola* stems from *pyrus* for pear, probably because the leaves of many species resemble pear leaves.

Wintergreen (*Pyrola* spp.)

|---|---|
Leaves are elliptical and somewhat toothed or with white veins; plants grow on dry forest floor	White-veined Wintergreen (*P. picta*)—This low perennial with cream or slightly greenish flowers grows in the shaded forest on rich humus, between 3,000 and 9,500 feet. It flowers from June to August.
Leaves are rounded and entire; plants occur in moist habitats	
Leaves are one and a half to two and a half inches long; flowers are red to purple colored	Pink Wintergreen (*P. asarifolia*)—This low perennial plant with pink flowers that grow on both sides of the stem grows in moist, shaded woods from 4,000 to 9,000 feet. It flowers from July to September.
Leaves are one-half to one inch long; flowers are white to pink	Common Wintergreen (*P. minor*)—This species occurs occasionally in boggy, shaded places between 7,000 and 10,000 feet. It flowers from July to August.

Interesting Facts: A tea made from the whole plant was used to treat epileptic seizures in babies. A leaf tea was gargled for sore throats and canker sores, whereas a tea from the root was a tonic. A poultice from the mashed leaves was used for tumors, sores, and cuts and to relieve the itch of insect bites. The plant is also an excellent astringent and disinfectant for urinary tract infections. The plants contain ursolic acid and the glycosides arbutin and ericolin, which were used in the treatment of kidney problems and skin eruptions.

Pyrola is also used as an ingredient in popular soft drinks. It is said to be an excellent substitute for *Chimaphila umbellata* (pipsissewa). In some areas, *Pyrola* may be exploited (overharvested) for commercial purposes.

Woodland Pinedrops (*Pterospora andromedea*)
Description: This is a brownish red plant with sticky stems up to three feet tall with pale-yellow flowers. Found in deep humus of coniferous forests between 2,500 and 8,500 feet, it is usually associated with ponderosa pine (*Pinus ponderosa*). It flowers from June to August.

Interesting Facts: Foster and Duke (1990) indicate that Native Americans used a cold tea made from the pounded stems and fruits to treat bleeding from the lungs. As a dry powder, the plant was used as a snuff for nosebleeds.

HEMP FAMILY (Cannabaceae)
The hemp family consists of annual herbs or climbing perennial herbs. The leaves are opposite, simple, or compound. There are two genera (*Cannabis* and *Humulus*), with three to five species distributed in the north temperate zone, and they are widely cultivated. *Humulus* is native to the United States. The family is a source of hempen fiber, oils, edible seeds, hops, and tetrahydrocannabinols (THC), the psychoactive compound in *Cannabis*. Some references place the family in the mulberry family (Moraceae).

Common Hop (*Humulus lupulus*)
Description: The genus was formerly included in the mulberry family (Moraceae). Common hop is a strongly twining, herbaceous vine with stems up to fifteen feet long. The stems and leaves are rough to the touch. The leaves are opposite, serrate, three to seven lobed, with heart-shaped bases. The underside of

the leaves is glandular. The flowers are small, green, and unisexual. This is a widely cultivated plant from Europe and Asia.

Interesting Facts: Hops are primarily grown for their fruits, used in brewing to give ale and beer a distinctive bitter taste. Additionally, the young shoots can be prepared as potherbs. They can be boiled in water for about three to five minutes, then boiled again in freshwater until tender.

A tea from the fruits was traditionally used as a sedative, antispasmodic, and diuretic and for insomnia, cramps, coughs, and fevers. Externally, the tea was used for bruises, boils, inflammations, and rheumatism. Recently, clinical studies have disproved the sedative qualities of hops and found that they have no physiological activity on the nervous system, yet anyone who drinks much of the tea tends to fall asleep or become groggy (Moore 1979).

Caution: The plant is known to cause dermatitis when handled.

Marijuana (*Cannabis sativa*)
Description: This is an unbranched, coarse, aromatic annual. The lower stem leaves are opposite, and the upper ones are alternate. The leaves are palmately compound, with three to nine leaflets that are lance shaped to elliptic and coarsely toothed. The small green male and female flowers (staminate and pistillate flowers, respectively) are found on separate plants. The fruit is an achene that is enclosed in the calyx and covered by a persistent bract.

Interesting Facts: The plant is a native of Asia and was cultivated in Europe for fiber (hemp) and seeds (hemp butter and oil). The seeds can be used as food, parched and mixed into a batter, and fried into cakes. Kephart (1909) indicates that the young shoots were used as a substitute for asparagus in Belgium.

Although the plant is now illegally cultivated in the United States and elsewhere for its euphoria-inducing properties, marijuana has been legitimately used to treat glaucoma and relieve nausea following chemotherapy. Although much maligned, marijuana is potentially a very useful medicinal plant.

Warning: Cultivation of this species and use as a drug are forbidden by law. It is usually grown by permit only under very rigid controls.

HONEYSUCKLE FAMILY (Caprifoliaceae)
There are fifteen genera and four hundred species in the honeysuckle family. Of the fifteen genera, seven are native to the United States. They are woody plants with opposite leaves. The flowers are five-merous, with the petals fused and an inferior ovary. Many genera in this family are cultivated as ornamentals.

Elderberry (*Sambucus*)
Description: Elderberries are shrubs with pithy stems. The species here have large, compound leaves with serrated leaflets. The white flowers are arranged in dense clusters. The fruits may be red or blue-black. Elderberries can be found in open areas, hillsides, and riparian habitats in the montane zone. The genus name comes from the Greek *sambuke,* an instrument made from the hollow stem.

Berries are blue, sometimes white; the inflorescence appears flat topped		
There are usually five to nine leaflets	Blue Elderberry (*S. caerulea*)—Look for this species growing in open places up to 10,000 feet throughout the Sierra Nevada. It flowers from June to September.	105 *Major Plant Groups*
There are usually three to five leaflets	Desert Elderberry (*S. mexicana*)—This shrub grows in open places up to 11,000 feet. It blooms from June to August.	
Berries are black or red; the inflorescence is dome shaped		
Berry is black, and there are usually five leaflets	Black Elderberry (*S. melanocarpa*)—This species occurs in moist places between 6,000 and 12,000 feet. It flowers from July to August.	
Berry is red; plant is ill-smelling	Red-berried Elderberry (*S. racemosa*)—This shrub grows in moist places between 6,000 and 11,000 feet. It blooms from June to August.	

Interesting Facts: The blue or black elderberry berries of blue, black, and desert elderberries are edible raw, or they can be made into excellent jams, jellies, and wine. They can also be dried and stored for winter use. The seeds contain hydrocyanic acid and if eaten in quantity can cause diarrhea and nausea. It is best to cook the berries or strain the seeds before use. The red-berried species contains much higher concentrations of these compounds and should be considered poisonous.

The blossoms can be added to pancakes to lighten batter and add flavor. The dried flowers are also ground and added to flours and baking mixes. Flower buds can be pickled or steamed as a potherb. Both the flowers and fruits contain a rich source of vitamin C.

The fresh flowers can be used externally as a decoction for an antiseptic wash. Flower tea contains a natural estrogen and is often effective for relieving menstrual cramps. The leaves were used as poultices for sprains and skin irritations. The leaves and flowers were common ingredients in skin salves for piles, burns, and boils. Recent studies of elderberry have confirmed that the berries possess antiviral properties that may be useful against influenza (Tilford 1997).

Elderberry stems can be cut and dried for use as musical instruments. After drying, holes can be bored into the branches to make flutes. During the drying process, the poisons are said to dissipate. The stems can also be used in making bows and arrow shafts for hunting small game. The odorous leaves can be used in water and sprayed on plants to repel aphids. The pith of the stem is used by watchmakers to absorb grease and oil. The leaves, with chrome as a mordant, yield a green hue. The berries, with alum and cream of tarter, yield a crimson dye.

Caution: The seeds, leaves, bark, and roots contain hydrocyanic acid and an alkaloid sambucine. They are toxic and cause acute emetic and laxative effects. Berries should be consumed when ripe and used for food after cooking and removal of seeds.

HYDROCYANIC ACID
Hydrocyanic acid is a colorless, volatile liquid with a peach-blossom odor. It is an extremely deadly poison and is also known as prussic acid. Prussic acid does not occur freely in normal, healthy plants. Instead, certain sugar compounds called cyanogenic glycosides contain the cyanide ion and only form prussic acid when degraded by certain enzymes. Living plant tissues can contain both cyanogenic glycoside (called dhurrin in sorghum species) and enzymes (beta-glycosidase or emulsin) in separate cells. When plant tissues are damaged, such as by freezing, chopping, or chewing, enzymes can come in contact with the cyanogenic glycoside and produce prussic acid. Bacterial action in the rumen of cattle and sheep can also release prussic acid from glycosides.

There are approximately 1,000 plant species in 250 genera that are known to be cyanogenic. Perhaps the most commonly known ones are the sorghums, sudangrass, and Johnsongrass. Other commonly known plants include plants of the *Prunus* genus, such as wild cherry and choke cherry.

Honeysuckle (*Lonicera*)

Description: This genus is composed of shrubs and woody vines with entire, opposite leaves (upper leaf pair fused in vines). Inflorescences are borne either two-flowered axillary stalks or terminal clusters in which the uppermost flower blooms earliest. Fruits are fleshy, several-seeded berries. They can be found in a variety of habitats from the foothills up to the alpine zone. The genus is named for Adam Lonitzer, a German naturalist who lived from 1528 to 1586.

Quick Key to the Honeysuckles

Flowers occurring in pairs
    Ovaries of the flower pair are obviously
    free

    Ovaries of the flower pair appear fused
        Flowers are yellow

        Flowers are dark red

Flowers occurring in a spike
    Leaves two to four inches long; flowers
    one-half to one and a half inches long

    Leaves one-half to three and a half inches
    long; flowers less than three-quarters inch
    long
        Upper leaf pair not fused around the
        stem; flowers often hairy

        Upper leaf pairs fused around the
        stem; flowers may be hairy
            Flowers glandular hairy; leaves
            with obvious stipules

            Flowers not glandular hairy;
            leaves without stipules

Twinberry (*L. involucrata*)—This upright shrub grows in moist places between 6,000 and 10,800 feet throughout the Sierra Nevada. It flowers from June to August.

Mountain Fly Honeysuckle (*L. cauriana*)— This species occurs along moist banks between 5,000 and 10,500 feet from Tulare to Nevada Counties. It flowers from May to July.

Double Honeysuckle (*L. conjugialis*)—This species can be found on wooded slopes between 4,000 and 10,200 feet from Tulare County northward. It flowers from June to July.

Orange Honeysuckle (*L. ciliosa*)—Look for this species on dry slopes between 2,000 and 5,000 feet in Butte County northward. It flowers from May to June.

Southern Honeysuckle (*L. subspicata*)—This twinning and trailing vine or shrub is common on dry slopes below 5,000 feet. It flowers from April to July.

Honeysuckle (*L. hispidula*)—Look for this species in canyons and woodlands below 7,000 feet.

Chaparral Honeysuckle (*L. interrupta*)—This twinning and trailing vine or shrub grows on dry slopes from 1,000 to 6,000 feet. It flowers from May to July.

Interesting Facts: The berries of honeysuckle are seedy but can be eaten raw or dried for future use. The bark and twigs of *L. involucra* were used for a variety of medicinal preparations, ranging from digestive-tract problems to contraceptives. Additionally, the juice from the stems was used as an antidote for bee stings (Willard 1992).

The berries provide a black pigment. The long stems of honeysuckle were used as basket foundation material by a number of Native American tribes. They also peeled and split the hairy stems as wrapping material for coiled baskets.

Snowberry (*Symphoricarpus*)
Description: Snowberries are erect shrubs with elliptical to egg-shaped leaves. Its flowers are white to pink and bell shaped, accompanied by two small bracts. The fruits are berrylike and white. Two species of snowberry in the Sierra Nevada are found in dry soils at various elevations.

Flower lobes are hairy inside	Creeping Snowberry (*S. mollis*)—Look for this species in wooded areas between 3,500 and 8,000 feet throughout the Sierra Nevada. It flowers from June to August.
Flower lobes are not hairy inside	Mountain Snowberry (*S. rotundifolius*)—This species occurs on rocky slopes between 4,000 and 11,000 feet throughout the Sierra Nevada. It flowers from June to August.

Interesting Facts: The white, tasteless berries are edible raw or cooked and are said to be emetic and cathartic in large amounts. Saponins are found in the leaves and can be used as a natural cleaning agent. A decoction of the pounded roots has been used for colds and stomachaches. An infusion made from *S. rivularis* (= *S. albus* var. *laevigatus*) was used to cure sores and skin lesions, and a root decoction was used to alleviate colds (Moerman 1986).

Twinflower (*Linnaea borealis*)

Description: Twinflower is a slender, trailing, mat-forming evergreen with short, leafless branches that divide into two at the top. At the top of these small branches arise small bell-shaped pink or white flowers, hence the name twinflower. The evergreen leaves are oval or round and about one-half inch long. Twinflower is found in the middle to subalpine elevations and is associated with conifers and moss-covered sites that also support *Pyrola, Clintonia uniflora,* and *Chimaphila umbellata*.

Interesting Facts: The genus name honors Carolus Linnaeus of Sweden, the person largely responsible for developing the binomial (two-name) system of naming plants and animals. It is said that twinflower was Linnaeus's favorite flower. The species name, *borealis*, means northern.

Twinflower (*Linnaea borealis*)

Twinflower is known to some Native Americans for its medicinal uses. As an orthopedic aid, the plant was mashed for inflammation of the limbs, whereas a poultice of the whole plant was applied to the head for headaches. The Algonquin Indians of Quebec made an infusion of the entire plant for menstrual difficulties and as a means to ensure "good health of the child" for pregnant women (Turner and Kuhnlein 1991). The Iroquois made a decoction of the plant and used it as a sedative for crying children and for children with cramps or fever (Herrick 1977).

Western Viburnum (*Viburnum ellipticum*)

Description: This species is a slender-stemmed shrub with opposite leaves that are lobed or toothed. The flowers are white, showy, and borne in a large umbrella-like inflorescence. It is found occasionally below 4,500 feet in the central Sierra Nevada. It flowers from May to June.

Interesting Facts: The berries of related species, *V. edule* (mooseberry viburnum) and *V. trilobum* (= *V. opulus* var. *americanum*) (American cranberry viburnum), are edible and taste somewhat like cranberries. The Hupa made necklaces from the black fruits of this species (Moerman 1986).

HORNWORT FAMILY (Ceratophyllaceae)

The hornwort family has only one genus, *Ceratophyllum,* with approximately three species distributed worldwide. These aquatic herbs occur in lakes, ponds, and slow streams and have no roots.

Coon's-tail (*Ceratophyllum demersum*)

Description and Interesting Facts: This is a rootless, submersed, or free-floating aquatic forb with slender, lax, and much-branched stems. The sessile leaves are in whorls of five to twelve, and the blades are dissected into linear, filamentous segments whose shape varies with the position on the plant. The minute flowers have no petals and are borne in the axils of the leaves. Coon's-tail is a common plant in standing or slowly flowing water of rivers, sloughs, and ponds to about 7,000 feet. It flowers from June to August. The Maidu Indians used coon's-tail to make a soothing lotion that was used on sore or inflamed skin.

LAUREL FAMILY (Lauraceae)

These are usually aromatic trees or shrubs. The family is largely tropical and most abundant in southeastern Asia and tropical America. The only representative of the family on the Pacific Coast is California laurel (*Umbellularia californica*). In the Midwest and eastern United States, there are sassafras (*Sassafras*) and wild allspice (*Lindera*). The avocado (*Persea*) is also a member of this family. It is native to tropical America but can be found growing from Florida to California.

California Laurel (*Umbellularia californica*)

Description: This is a slender tree growing up to one hundred feet tall. The leaves are glabrous, alternate, entire, and oblong-lanceolate. The leaves have a strong pungent odor when crushed. The flowers are small, yellow-green, and occur in six- to ten-flowered clusters. There are no petals, and the six sepals are green. Tiny orange glands may be observed inside the flower at the base. California laurel is a fairly common tree at the lower elevations in canyons and more shaded areas below 5,000 feet. It flowers from December to May.

Interesting Facts: The flesh of the fruit and the ripe kernel may be eaten. Many Native groups gathered the ripe fruits and dried them in the sun until the thick outer covering loosened and split. The flesh was eaten, as it does not store well. The kernels are rich in oil and can be stored. When needed, they can be roasted and eaten or pounded and then formed into cakes. These cakes can then be dried in the sun and stored for future use. California laurel leaves may be used for flavoring cooking but are more potent than the commercially available European bay (*Laurus nobilis*).

California laurel leaves contain acrid oils that were used by Natives for various medicinal treatments (Callegari and Durand 1977). The oil was sometimes pressed from the leaves and used to relieve toothaches, earaches, or headaches. A tea from the leaves was used to treat colds, whereas a leaf poultice was used on sores and boils.

To repel insects, the fresh leaves or a leafy branch were used. The boiled leaves were used as a shampoo.

Warning: This species contains volatile oils in the leaves that may cause severe headaches or unconsciousness when inhaled. The oil may also cause skin irritation.

LIZARD'S-TAIL FAMILY (Saururaceae)

The family consists of only five genera and about seven species that occur in southeastern Asia and in the United States and Mexico. In California, the only genus is *Anemopsis* (yerba mansa). In general, members are perennial herbs of moist places.

Yerba Mansa (*Anemopsis californica*)

Description: This plant has a flowering stem with one clasping leaf in the middle of the stem and one to three petioled leaves in the axil. Flowers occur in dense terminal spikes and are subtended by five to eight white- to rose-streaked bracts. This is a common plant in marshy places below 6,500 feet, occurring into the yellow-pine forest. It flowers from March to September.

Interesting Facts: Aside from the fact that the seeds of yerba mansa were eaten by Native Americans, the species has many medicinal uses. For example, the aromatic peppery roots were peeled, cut, squeezed, boiled, and drunk for pleurisy. An infusion of the root was used for stomach ulcers, chest congestion, and colds. The tea from the root also provided a general pain reliever. Other uses include an infusion of the bark to wash open sores and a drink for stomach problems. The bark tea was also used as a laxative. Finally, the leaves, after allowed to wilt in the heat, were applied to swellings. Other medicinal uses include the roots being pounded and used as a poultice for wounds and bruises.

Yerba Mansa
(*Anemopsis californica*)

LOOSESTRIFE FAMILY (Lythraceae)

Members of this family are herbs, shrubs, or trees. The leaves are opposite or whorled. There are approximately 25 genera and 550 species widely distributed around the world. Seven genera are native to the United States. The family is a source of dyes and ornamentals.

Purple Loosestrife (*Lythrum salicaria*)

Description: This erect perennial grows 20 to 72 inches tall and has pale-green, glabrous stems. The leaves are alternate, linear to linear-oblong in shape, entire, and ⅜ to 1½ inches long. Its flowers are purple, with a cylindric base, and solitary in the leaf axils. The six petals are each ¼ inch long, and there are four to twelve stamens. Loosestrife grows in moist places below 6,000 feet. It flowers from April to October.

Purple loosestrife is a Eurasian species that has become a nasty wetland weed. It is also cultivated in gardens by some who are unaware of its potential. It is often called the "beautiful killer" because it can take over wetlands and displace native species. It appears to have some efficacy against gnats and flies and was reported to calm quarrelsome beasts of burden at the plow if placed upon the yoke (Pojar and MacKinnon 1994).

Interesting Facts: A tea made from whole flowering plant of purple loosestrife, fresh or dried, is a European folk remedy for diarrhea, dysentery, and a gargle for sore throats. It was also used as a cleansing wash for wounds. Experiments have shown that the plant extracts stop bleeding and kill some bacteria (Foster and Duke 1990). Other species appear to have been used by Native Americans. For example, *L. californicum* was used by the Kawaiisu Indians in California as a medicine and as a dermatologic aid. The method, however, is not reported (Zigmond 1981). Additionally, *L. hyssopifolia* was used by Maidu Indians in California to expedite healing and to reduce inflammation of mucous membranes. It was also used as a shampoo for the hair, but the method is not reported.

MADDER FAMILY (Rubiaceae)

The madder family consists of approximately five hundred genera and six to seven thousand species distributed worldwide. About twenty genera are native to the United States. The family is of economic importance because of coffee, quinine, and many ornamentals.

110 Bedstraw (*Galium*)

Description: Despite their small flowers, the various species of *Galium* are unmistakable. They are annual or perennial herbs with four-angled stems and whorled leaves. The small, four-parted flowers are white or greenish, and the fruits are smooth or bristly hairy. They can be found in various habitats from the low to higher elevations. *Galium* is from the Greek *gala*, meaning milk, referring to the herb's traditional use as a milk coagulant for making cheese. The rennet (a substance that curdles milk in making cheese and junket) for this use was obtained by blending the herb with an equal amount of salt, covering it with water, and then simmering away half of the fluid.

1. Goose Grass (*G. aparine*)—This weak-stemmed, scrambling, and straggly annual is common on shaded banks below 7,500 feet. It flowers from March to July.
2. Tall Rough Bedstraw (*G. asperrinum*)—This species is found in shaded places between 1,500 and 7,300 feet from Mariposa County northward.
3. Low Mountain Bedstraw (*G. bifolium*)—This slender, glabrous annual grows in moist, shady places from 5,000 to 10,500 feet. It flowers from June to September.
4. Bolander's Bedstraw (*G. bolanderi*)—This bedstraw grows on dry, rocky slopes.
5. Gray's Bedstraw (*G. grayanum*)—This species with broadly ovate leaves occurs on dry, rocky slopes between 6,000 and 10,000 feet.
6. Alpine Bedstraw (*G. hypotrichium*)—Alpine bedstraw occurs in dry places about rocks between 7,000 and 12,000 feet.
7. Sequoia Bedstraw (*G. sparsiflorum*)—This species occurs in Plumas County and south but is rather rare north of Fresno County.
8. Trifid Bedstraw (*G. trifidum*)—This slender perennial with angled stems grows in wet places below 8,000 feet. It flowers from June to September.
9. Fragrant Bedstraw (*G. triflorum*)—This species occurs in moist, shaded canyons and wooded places below 8,000 feet throughout the Sierra Nevada.

Interesting Facts: None of the species of *Galium* are known to be poisonous. Although *G. aparine* is the most commonly used species, Tilford (1993) believes, as we do, that all other species can be used similarly. The very young leaves and stems can be used as a potherb. The small hairs on the stems make the plant difficult to swallow raw; boiling or steaming, however, does soften them up. If the stems are too fibrous, use only the leaves. Slow roasted until dark brown and ground, the ripe fruit can be used as a coffee substitute.

Medicinally, the plants were used to increase urine flow, stimulate appetite, reduce fevers, and remedy vitamin C deficiencies. It has diuretic, anti-inflammatory, and astringent qualities and has been used as a lymphatic tonic. A wash made from the plant is said to remove freckles, whereas a cool tea is reported to cool sunburns. Many species of *Galium* contain asperuloside, which produces coumarin, giving it the sweet smell of new-mown hay as the foliage dries. Asperuloside can be converted to prostaglandins (hormonelike compounds that stimulate the uterus and affect the blood vessels), making the *Galium* species of great interest to the pharmaceutical industry.

Dried, the foliage of bedstraw has been used as a stuffing for mattresses or as a tinder for starting fires. The roots may yield a red dye, but because the roots are threadlike and produce little dye, collecting enough for a strong dye bath would be fairly laborious.

Mallow Family (Malvaceae)

There are some eighty-five genera and fifteen hundred species in this family, most of which occur in the Tropics. Twenty-seven genera are native to the United States. The distinctive feature of this family is the uniting of the numerous stamen stalks to form a tube around the pistil that resembles a tree trunk, with the anthers and nonfused filaments as the branches and leaves. This "stamen tree" in the center of the flowers is almost a "never fail" characteristic of this family. The family is of moderate economic importance because of cotton fibers derived from the seeds of *Gossypium,* several ornamentals, and a few food plants. Mallow is from the Greek word *malva,* meaning soft and may refer to the soft fuzzy leaves characteristic of so many plants in this family or to the sticky, soothing juice obtained from the roots of some species.

Alkali Mallow (*Malvella leprosa*)

Description: This spreading perennial has densely grayish pubescent herbage. The stems are four to sixteen inches long, and the leaves are alternate, round, triangular, or kidney shaped. Its flowers are axillary and have five cream or pale-yellow petals. Alkali mallow grows in alkaline or heavy soil, mostly below 6,000 feet. It flowers from May to October.

Interesting Facts: Although it is not reported how, this species was used by the Maidu Indians as a laxative (Strike 1994).

Checker Mallow (*Sidalcea*)

Description: Checker mallows are annual herbs with lobed or divided leaves. Its flowers are white to deep pinkish lavender, in terminal clusters. The genus name is a combination of two Latin words for mallow—*sida* and *alcea.*

1. Glaucous Sidalcea (*S. glaucescens*)—This species occurs in dry, grassy places or open woods between 3,000 and 11,000 feet. It flowers from May to July.
2. Checker Mallow (*S. malvaeflora*)—Look for this checker mallow in open woodlands below 8,000 feet from Mariposa County northward.
3. Checker Mallow (*S. multifida*)—This species grows in dry habitats between 6,500 and 8,000 feet from Mariposa and Mono Counties south.
4. Spicate Checker Mallow (*S. oregana*)—This species occurs in moist habitats below 8,500 feet. It flowers from June to August.
5. Creeping Checker Mallow (*S. reptans*)—Look for this species in moist meadows between 4,000 and 7,600 feet from Tulare to Amador Counties. It flowers from July to August.

Interesting Facts: Early settlers applied the whole plant of *S. glaucescens* as a poultice to ease the pain of insect stings and to draw out thorns and splinters. *S. malvaeflora* was used as greens. The dried, mashed leaves were used to flavor manzanita berries. Kirk (1975) says *S. neomexicana* (New Mexico checker mallow) is edible as greens after cooking. Additionally, the thick sap of a species in the eastern United States was mixed with sugar and used to make marshmallows (Peterson 1978).

Checker Mallow
(*Sidalcea* spp.)

Cheeseweed (*Malva parviflora*)

Description: These plants are distinguished by their distinctive fruit and seeds, rather than their leaves and flowers. They are introduced annual or biennial herbs that are usually found in waste places at the lower elevations.

Interesting Facts: Medicinally, the bruised leaves of cheeseweed can be rubbed on the skin to treat skin irritations. As a headache remedy, leaves or the whole plant can be mashed and placed on the forehead. Leaf or root tea can be used for angina, coughs, bronchitis, and stomachaches. The fresh or dried leaves were used as a soothing poultice.

The entire plant of another species, *M. neglecta* (dwarf mallow), is edible. The young leaves are particularly good in salads or cooked up as a potherb. The plant is, however, very mucilaginous and it is often used to thicken soup and may take a little getting used to. Eaten in large amounts, however, may result in digestive disorder. The immature fruits (which look like cheese) can also be eaten raw or added to soups.

Globemallow (*Sphaeralcea munroana*)

Description: This is a perennial herb with star-shaped hairs on the leaves and stems. Flower colors range from red to pink. It can be found in open areas at the lower elevations. The genus name comes from the Greek *sphaira* and *alkea*, meaning spherical mallow.

Interesting Facts: This species was pounded in water and made into a gummy paste and applied over the rough surfaces of earthen dishes.

A related species, *Sphaeralcea coccinea* (scarlet globemallow), was chewed and applied to inflamed sores and wounds as a cooling, healing salve. It was also used as a pharmaceutical aid. The entire plant was ground and steeped in water for a sweet-tasting tea that was mixed with other bad-tasting medicines to make them more palatable. The Navajo Indians used the plant as a lotion to treat skin diseases, as a tonic to improve appetite, and as a medicine for rabies (Wyman and Harris 1941).

MAPLE FAMILY (Aceraceae)

This family consists of two genera and about two hundred species distributed worldwide. In general, they are shrubs or trees with opposite leaves that may be simple or compound. The flowers are small, usually appearing before the leaves. The family is of economic importance as a source of timber, ornamentals, and sugar. Maple wood is considered to be heavy, tough, compact, and very hard. Its light brown color with a dense, even grain and fine texture makes it one of the best woods for furniture, veneers, and flooring. It is also used in making violins, tool handles, and pianos.

Maple (*Acer* sp.)
leaves and fruits

Maple (*Acer*)

Description: Maples are deciduous trees or shrubs with male and female flowers on the same or separate plants. Its flowers are arranged in racemes, corymbs, or umbels. Fruits are winged schizocarps that resemble tiny "helicopters" when blown by the wind. Maples are usually found in moist places in canyons, hills, and along streams from low to high elevations. Of the approximately fifteen species of *Acer* native to the United States, four are found in the Sierra Nevada.

Leaves pinnately compound, usually three to five leaflets	Box Elder (*A. negundo*)—Box elder grows along streams and in moist areas below 6,000 feet. It flowers from March to April.
Leaves are simple, usually palmately lobed	
Leaves less than one and a half inches wide	Mountain Maple (*A. glabrum*)—Mountain maple is found from 6,500 to 9,000 feet. It blooms from April to May.
Leaves greater than one and a half inches wide	
Leaves are deeply lobed, four to eight inches wide	Bigleaf Maple (*A. macrophyllum*)—Bigleaf maple is fairly common in canyons and along streams below 5,000 feet. It blooms from April to May.
Leaves lobed only about halfway to the center	Vine Maple (*A. circinatum*)—This species grows on shaded stream banks below 5,000 feet in the northern Sierra Nevada. It flowers in May.

Interesting Facts: Sap can be harvested in much the same way as the eastern sugar maple. To obtain sap, simply bore a small hole into the tree a couple feet above the ground. The sunny side of the tree is usually the ideal spot to bore. Insert a small grooved wooden peg into the hole. This peg will be the spigot. If the tree is ready to flow, sap will immediately begin to flow after drilling. Hang or place a container under the spigot to collect sap. After collecting sap, seal the hole to protect it from infection and further sap loss while it heals.

Next, the sap must be boiled down because the majority of it is water. Only a small fraction of the original volume collected will be left. You may boil the sap so far as your personal taste dictates. As an alternative to boiling, the collected syrup can be allowed to freeze overnight, which allows the water to separate from the sap. The frozen water can be easily discarded (Hart 1996).

The inner bark of all maples can be eaten in emergencies. A tea made from the inner bark of box elder is used to induce vomiting. The young shoots of mountain maple can be used as asparagus. The winged seeds of box elder can be roasted and eaten (Harrington 1967).

Native Americans used the young saplings for basketry work. The saplings were split into quarters and used as a white wrapping or sewing strand in coiled basketry work. In some places, maple thickets were intentionally manipulated by burning the old growth to promote new growth. These straight, uniform shoots were highly valued as good basketry material. Maple wood has been used to make snowshoe framing, mush paddles, and other household utensils. Knots and burls on tree trunks can be used for making bowls, dishes, and other items. Gum from the buds was mixed with animal fat and used as a hair tonic. Inner bark can be shredded and twisted into a coarse rope.

MARE'S-TAIL FAMILY (Hippuridaceae)

The family consists of a single genus, *Hippuris*. The genus was at one time assigned to the water-milfoil family (Haloragaceae), but it is not closely related.

Mare's-tail (*Hippuris vulgaris*)

Description: This is a common plant in the main mountain chains of the western United States. The plant at first glance resembles an immature horsetail (*Equi-*

setum spp.), but they are unrelated. Horsetails reproduce by spores and have stems that can be quickly pulled apart. The flowers of mare's-tail are small and inconspicuous. The plant is found in the margins of shallow waters from ponds and lakes to streams. It can also be found in marshy and swampy areas, along roadsides, and in irrigation ditches.

Interesting Facts: The whole plant is edible when prepared as a potherb. The plant parts are tender and can be gathered in any stage, even in winter. Ancient herbalists are said to have employed mare's-tail for internal and external bleeding (Harrington 1967).

MENTZELIA FAMILY (Loasaceae)

There are about 15 genera and 250 species of this family occurring chiefly in South America and the warmer parts of North America. Various species of *Mentzelia* are endemic in the western United States.

Blazing Star (*Mentzelia*)

Description: Members of this family have alternate, entire, or pinnately lobed leaves. The fruit is a capsule that opens at the top. There are many species of *Mentzelia* in the western United States. It is also called "stick-leaf" because of the barbed hairs on the leaves that readily cling to fabric.

Quick Key to the Blazing Stars	
Lower leaves are pinnately lobed or divided	
Flower petals are usually orange and blunt	Blazing Star (*M. veatchiana*)—Look for this species on open slopes and flats below 6,000 feet.
Flower petals are bright yellow and sharp pointed	Smoothstem Blazing Star (*M. laevicaulis*)—This coarse, stout biennial has shining white stems that are rough hairy above. This species grows on dry, stony, gravelly slopes below 8,500 feet. It flowers from June to October.
Lower leaves are toothed or entire but not lobed	
Bracts of the inflorescence are green	Nevada Stick-leaf (*M. dispersa*)—This species occurs in dry, disturbed, sandy places below 8,500 feet throughout the Sierra Nevada.
Bracts of the inflorescence are white at the base	Mountain Blazing Star (*M. montana*)—This species grows between 4,000 and 8,000 feet.

Interesting Facts: *Mentzelia* was considered an important food source in many places of the West. The seeds are edible after being parched and ground into flour. The seeds were often stored for future use. Murphey describes a type of "gravy" made from the seeds of *M. laevicaulis* (smoothstem blazing star): "[T]he red seed is put into a hot frying pan and when the seeds turn a darker red, warm water is added and it is stirred till it thickens" (1990, 27).

The Hopi Indians in the Southwest parched and ground the small, oily seeds of a related species, *M. albicaulis,* into a fine, sweet meal and ate it in pinches (Hough 1897).

MILKWEED FAMILY (Asclepiadaceae)

About 250 genera and 2,000 species are found in this family worldwide. Milky sap in the stems, leaves, and flowers inspired the common name for the milkweed family. The family is of moderate economic importance as a source of ornamentals, latex, fibers, poisonous plants, and a few food plants. The sap contains latex, and in a few species it may yield industrially important hydrocarbons. The flowers are five-parted. Pollination involves an insect literally pulling up the pollen mass from the anthers and directly depositing it on the stigma.

Milkweed (*Asclepias*)

Description: These are erect or decumbent herbs from deep perennial roots. The leaves are opposite or whorled, and the corolla is deeply five-parted with the segments reflexed. The corona hoods each have an incurved horn within. The name *Asclepias* refers to Asklepios, the Greek god of medicine. The larvae of the monarch butterfly (*Danaus plexippus*) feeds on the leaves of milkweeds.

CARDIAC GLYCOSIDES
Cardiac glycosides are an important class of naturally occurring drugs whose actions include both beneficial and toxic effects on the heart. Plants containing cardiac steroids have been used as poisons and heart drugs at least since 1500 BC. Throughout history these plants or their extracts have been variously used as arrow poisons, emetics, diuretics, and heart tonics. Cardiac steroids are widely used in the modern treatment of congestive heart failure and for treatment of atrial fibrillation and flutter, yet their toxicity remains a serious problem.

Quick Key to the Milkweeds

Upper stem, leaves, and flower stalk definitely white, hairy

    Leaves are opposite — Showy Milkweed (*A. speciosa*)—This species occurs in many habitats, including roadsides and disturbed areas below 6,000 feet.

    Leaves are mainly whorled — Milkweed (*A. eriocarpa*)—Milkweed is common on dry washes and slopes below 7,000 feet. It flowers from June to August.

Upper stem, leaves, and flower stalks not hairy or only very slightly

    Flowers dark purple; leaves broad and heart shaped — Purple Milkweed (*A. cordifolia*)—Purple milkweed is common on dry slopes below 7,500 feet in the chaparral and the yellow pine forest habitats. It flowers from April to July.

    Flowers greenish white; leaves narrow and not heart shaped — Narrow-leaf Milkweed (*A. fascicularis*)—This species occurs in dry places mostly below 7,000 feet in many plant communities, usually on the desert side of the mountains.

Milkweed (*Asclepias* sp.) Flower

Interesting Facts: Almost every book on edible plants in the United States lists milkweeds as being edible. It should be noted that in most cases, they are referring to the eastern species of *A. syriaca,* which does not occur in California, and the western species of *A. speciosa* (showy milkweed) and *A. asperula* (spider milkweed). These latter two species also contain the cardiac glycosides that can cause severe poisoning if not properly prepared or cooked.

With respect to the two western species mentioned above, Harrington (1967) suggests gathering plants when they are six inches tall and then boiling for fifteen to twenty minutes in at least two to three changes of water. We have tried five to seven changes of water, and the plants were still bitter! The unopened flower buds can be served like broccoli by boiling in at least three changes of water.

A strong fiber can be obtained from the inner bark to make rope, fishing line, clothing, and nets (see the sidebar on the next page). Archaeologists have discovered clothing that was made from the fibers more than ten thousand years ago. The silky floss found in mature milkweed seedpods was used in making candlewicks, and the fiber can be spun like cotton. The floss is buoyant and water resistant and makes a good insulator. During World War II, schoolchildren in Canada harvested milkweed floss from the wild for the United States Navy's use as a substitute for kapok in life vests (Turner and Kuhnlein 1991). The dried pods were used as utensils. The sap was used as an adhesive.

Milkweeds contain asclepain in their plant parts and sap. Asclepain is a proteolytic enzyme that gives credence to the old pioneer remedy of applying the white juice daily to get rid of warts. However, some Native American tribes used to collect the milk of *A. speciosa* and roll it in hand until it became firm enough to chew as gum, but it was not swallowed.

Milkweeds have been used in folk medicine for hundreds of years. The powdered root of several species is reported to have been used to treat wounds, pulmonary diseases, rheumatism, and gastrointestinal problems, among other ailments (Coffey 1993). Many modern medicines were originally derived from poisonous plants. Perhaps research will validate some of the medicinal uses of milkweeds and provide us with new medicines from the old (Lewis and Elvin-Lewis 1977).

Additionally, milkweeds have the potential to furnish an exciting array of products for industry and home. In the future, as petroleum products dwindle, perhaps we will find ourselves taking a closer look at the possibilities of cultivating milkweeds for fiber, hydrocarbons, and medicines.

Warning: Milkweeds can be confused with other plants producing milky juice such as dogbane (*Apocynum*), which is also considered poisonous. Additionally, some species of milkweed at certain stages are poisonous to animals and could affect humans when eaten raw.

MILKWORT FAMILY (Polygalaceae)

The family contains about thirteen genera and eight hundred species of herbs, shrubs, climbers, and small trees that are widely distributed in the tropical to temperate regions of the world. The leaves are alternate and simple, and the flowers are usually subtended by a bract and a pair of bracteoles. The flowers look something like a papilionoid legume, which is often confusing to the budding botanist. A few species are cultivated as ornamentals.

Milkwort (*Polygala cornuta*)

Description: This species occurs on rocky or gravelly slopes between 1,000 and 5,000 feet from Fresno County north. It flowers from June to August. The genus name is from the Greek *polys* (much) and *gala* (milk); it was thought to enhance milk production.

Interesting Facts: Although the plant may be considered poisonous, a decoction of the plant was used by the Maidu and Miwok as an emetic or as a remedy for coughs, colds, and diverse pains.

MINT FAMILY (Lamiaceae; formerly Labiatae)

There are approximately 180 genera and 3,500 species worldwide, with the Mediterranean region being the chief area of diversity. All have epidermal glands that exude an odor when rubbed, not necessarily pleasant and mintlike. Labiatae refers to corolla shape, two lips (*labia*). This family is of considerable economic importance as a source of numerous ornamentals, aromatic oils, and a few weedy genera as well as medicines.

Bluecurls (*Trichostema*)

Description: From the Greek *trichos* (hair) and *stemon* (stamen), referring to the long slender stamens, a characteristic of the genus.

WILDERNESS CORDAGE
The survival and continued existence of primitive humans was as much dependent on fiber as on food. The cordage made from the fibers of wild plants can be used to make blankets, sandals, baskets, clothing, nets for fishing, and snares for capturing small game animals. In a wilderness situation, you will be surprised how important a piece of string or cordage can be. There are many species of plants in California that have fiber in the stem, leaves, or bark that can in one way or another be used as cordage. Some of the species discussed in this handbook include milkweed (*Asclepias* spp.), dogbane (*Apocynum* spp.), sagebrush (*Artemisia* spp.), cottonwood (*Populus* spp.), willow (*Salix* spp.), juniper (*Juniperus* spp.), thistle (*Cirsium* spp.), sunflower (*Helianthus* spp.), yucca (*Yucca* spp.), flax (*Linum*), and nettle (*Urtica* spp.). There are probably other species that can be used, and finding those out will be a matter of experimentation.

1. Vinegarweed (*T. lanceolatum*)—This species occurs in disturbed areas at the lower elevations.
2. Mountain Bluecurls (*T. oblongum*)—Look for this species on the dry margins of meadows throughout the Sierra Nevada.
3. Siskiyou Bluecurls (*T. simulatum*)—This species is found in dry, open places between 2,000 and 6,000 feet.

Interesting Facts: Vinegarweed has a strong, pungent vinegar odor and was best known as a fish poison. The fresh plants would be mashed and thrown into pools or sluggish streams. The intoxicated fish would then float to the surface where they were easily caught by hand. Strike (1994) indicates that the fine hairs on the flowers would catch on the gills of the fish and interfere with respiration, making the fish easier to catch. Medicinally, the leaves were chewed and put in the cavity of an aching tooth. Moore (1979) says that a tea made from the flower tops is good for stomachaches and promotes sweating in dry fevers.

Common Self-heal (*Prunella vulgaris*)

Description: This perennial grows four to twenty inches tall and has opposite, lanceolate-ovate shaped leaves that are one to two inches long. The herbage is glabrous to short pubescent. Its flowers are in a dense, terminal spike in the axils of round, membranous, purple-tinged bracts. The calyx is purplish, and the corolla is two-lipped, violet, and three-eighths to three-quarters of an inch long. Self-heal is common in moist woods from May to September at elevations up to 7,500 feet. It is also called Hercules' all-heal, because it is supposed that Hercules learned the herb and its virtues from Chiron.

Interesting Facts: The entire plant is edible, raw or cooked. However, we found that it is the young and tender plants collected in the early spring that are best. The crushed leaves can be used fresh or dried to make a tea.

Historically, as the common name implies, the plant has been used as a medicine for almost everything. Herbal uses include an astringent, antispasmodic, tonic, and styptic. The tea of the dried plant was also used as a gargle for a sore throat. Fresh plants can be made into an antiseptic poultice for bruises and scrapes, because of the high tannin content (Scully 1970).

Giant Hyssop (*Agastache urticifolia*)

Description: This is a tall perennial growing 3 to 6 feet tall. The leaves are opposite, ovate in shape, 1 to 3 inches long and 1½ inches wide. They are also coarsely toothed on the margins. Flowers occur in dense whorls and form a terminal spike, 1½ to 6 inches long. The calyx is green or rose, and the corolla is two-lipped, rose or violet, and ⅜ to ⅝ inch long. Giant hyssop grows in moist places below 9,000 feet. It flowers from June to August. *Agastache* is from the Greek *agan*, meaning much, and *stachys*, meaning ear of grain, referring to the flower cluster.

Interesting Facts: The seeds of giant hyssop may be eaten raw or cooked, and the leaves can be used as a tea or for flavoring stews. The Miwok of California drank an infusion made from the leaves to relieve rheumatic pain and for indigestion and stomach pains. The plant is said to have mild sedative qualities. Mashed leaves were made into a poultice for swellings.

Hedge-nettle (*Stachys*)

Description: Hedge-nettle has no stinging hairs, as do the true nettles (*Urtica* spp.), but resembles them before flowering. The genus name is from the Greek

Giant Hyssop
(*Agastache urticifolia*)

stachys (spikelike) and refers to the inflorescence. There are about three hundred species of perennial herbs and shrubs with wrinkled leaves.

1. Hedge-nettle (*S. ajugoides*)—Look for this species in moist habitats.
2. White Hedge-nettle (*S. albens*)—This species is found in wet places below 9,000 feet throughout the Sierra Nevada.

Interesting Facts: The leaves and flowers are edible, but because of their fuzzy texture and bitter taste, we find them unpleasant. The tubers can also be eaten raw, cooked, or pickled and are best if collected in the autumn. Other species may be edible, but this has not been confirmed. Strike (1994) states generally that *Stachys* tubers were eaten.

The leaves may be soaked in water for a few minutes and used as a poultice (Meuninck 1988). Additionally, an infusion of fresh leaves can be used as a wash for sores and wounds.

Horehound (*Marrubium vulgare*)

Description: This is a woolly perennial herb with bitter sap. The leaves are wrinkled and toothed. The flowers are small, white, and occur in dense whorls. Horehound is a common weed of waste places and fields at the lower elevations.

Interesting Facts: Horehound is listed as a stimulant, tonic, expectorant, and diuretic. The plant was highly valued by ancient Egyptian priests and the Romans, with the former calling it the "Seed of Horus" and "eye of the star."

The most famous use of this plant is horehound candy, which is used to soothe sore throats and coughs. A tea from the dried leaves and flowers is also used, but because of the extreme bitterness of the herb, it is obvious why it tastes better in the form of a candy. Following is a recipe from Clarke that we have used on a number of occasions to make the horehound candy and cough syrup:

> The plant can be boiled or dried without losing its flavor, which is unusual for a member of the Mint Family. One cup of fresh leaves or 1/4 cup dried leaves boiled in two cups of water for 10 minutes will make a strong concentrate. This concentrate can then be diluted with 2 parts water to 1 part concentrate for a tea. One part concentrate may also be added to 2 parts sugar or honey and a pinch of cream of tarter, brought to hard crack (290 degrees Fahrenheit), and poured into a buttered plate for the old fashioned cough drop candy. A bit of lemon added at the last minute improves the flavor. A cough syrup can be made of 1 part concentrate and 2 parts honey. (1977, 196)

HERBAL TEAS
Plants are the source of practically all the beverages consumed by mankind. Water and milk are the exceptions. They provide flavor, color, and aroma for endless variety and pleasure. They also provide nutrients for health.
Most herbal teas are infusions made by pouring boiling water over herb leaves or flowers and allowing them to steep for five to ten minutes to release the herb's aromatic oils. A general rule is one teaspoon of dried herb, or three teaspoons of fresh crushed herb, per cup of water. To make a stronger tea, add more of the herb rather than steeping the tea longer (long steeping makes the tea bitter). Experiment by combining various herb teas for interesting flavor results.

Other medicinal uses of the plant include making a warm infusion that will promote perspiration and the flow of urine. When taken cold, this infusion will expel worms.

Mint (*Mentha*)

Description: All species are distinctly aromatic perennial herbs with rhizomes. The flowers are arranged in whorls. *Mentha* comes from Mintho, the mistress of Pluto, ruler of Hades. Pluto's jealous queen, Proserpine, upon learning of Mintho, trampled her, transforming her into a lowly plant forever to be walked upon. Pluto made this horrible fate more tolerable by willing that the more the plant was trampled, the sweeter it would smell.

1. Field Mint (*M. arvensis*)—This aromatic perennial with rose to violet flowers is found in moist places below 7,500 feet. It flowers from July to October.
2. Spearmint (*M. spicata*)—This strongly aromatic perennial with pale lavender flowers is found in moist fields and in marshes below 5,000 feet. It flowers from July to October.

Interesting Facts: The fresh or dried leaves of *M. arvensis* (wild mint) and *M. spicata* (spearmint) can be steeped in hot water for a tea. They have also been used as flavoring agents for soups, meat, and pemmican. The young leaves can also be added to salads and soups. The plants are high in vitamins A, C, and K and the minerals iron, calcium, and manganese. It is an appetite stimulant and digestive aid. The leaf tea is considered medicinal and was used for colds, stomachaches, fevers, headaches, insomnia, and nervous tension. Crushed leaves were used by some Native Americans to poultice swellings and bruises.

Pure mint oil is a multimillion-dollar industry. It is added to shampoos, massage oil, salves, and soaps, as well as medicines, foods, and liqueurs. It takes approximately three hundred pounds of mint to yield one pound of oil.

Monardella (*Monardella*)

Description: These are perennial herbs with leafy stems. Its flowers are commonly rose-purple in dense terminal, globose clusters. The bracts are leaflike but usually colored. The genus name honors the Spanish botanist and physician Nicholas Monardes (1493–1588).

Pennyroyal (*Monardella* spp.)

Quick Key to the Monardellas

Leaf margins wavy and more or less entire	Sweet-smelling Monardella (*M. beneolens*)—This species occurs in subalpine forests between 8,000 and 11,000 feet from Fresno County southward.
Leaf margins barely or not wavy and entire to serrate	
Plant annual	Mustang Mint (*M. lanceolata*)—This erect annual grows eight to twenty inches tall and has a pleasant mint odor. Mustang mint grows in dry places in the mountains below 8,000 feet. It flowers from May to August.
Plant perennial	
Stems are lying on the ground; outer bracts are reflexed and leaflike	Monardella (*M. sheltonii*)—This species is found in dry habitats below 6,000 feet throughout the Sierra Nevada. It flowers from June to August.
Stems are erect; outer bracts are erect and not leaflike	Mountain Pennyroyal (*M. odoratissima*)—This aromatic perennial is found between 4,500 and 9,000 feet. It flowers from June to August.

PENNYROYAL INSECT REPELLENT

You will need the following ingredients:

2 cups mountain pennyroyal leaves and flowers

1/2 cup sagebrush (*Artemisia*) leaves

1/2 cup fresh yarrow (*Achillea*) leaves and flowers

2 cups rubbing alcohol

1/2 cup water (preferably distilled)

1 tsp. jojoba oil

Place all ingredients except for the water and oil in a quart-size jar and allow it to stand for two to three weeks. Strain the liquid concoction through a coffee filter and discard the solid material. Then add the water and jojoba oil to the strained liquid. Pour the liquid into a spray bottle. To use, shake well before using and apply as needed.

Interesting Facts: The leaves and stalks of *M. odoratissima* were eaten, and a thirst-quenching tea was made from the leaves and flower heads. Medicinally, the tea made from the inflorescence of *M. odoratissima* was used for colds and fevers, relieved digestive upsets, and purified the blood. The tea made from the whole plant used in the first stages of a cold when fever is present is said to help relieve elevated temperatures and toxins through sweating. Leaves and flowers rubbed on exposed skin repel mosquitoes and other biting insects (Callegari and Durand 1977).

Mountain-mint (*Pycnanthemum californicum*)

Description: The stems of this plant are white-woolly and often few branched. The leaves are ovate to lance-ovate in shape and finely pubescent. The flowers are white. Mountain-mint occurs on moist slopes and canyons below 5,000 feet from Mariposa County northward. It flowers from June to September.

Interesting Facts: Herbage can be used to flavor stews or soups. The Miwok made a decoction of the plant and took it for colds (Strike 1994).

Northern Bugleweed (*Lycopus uniflorus*)

Description: This is a perennial herb with rhizomes. The small flowers are pinkish purple, whorled in the axils of the upper leaves. They occur in wet to moist areas of lakes and riverbanks. Unlike other mints, bugleweeds do not have the mintlike odor.

Interesting Facts: The leaves are edible raw but are usually tough and bitter. The tubers can be added to salads, pickled, or boiled and eaten. Folk uses of *Lycopus* include its use as a cough remedy (Coffey 1993). It is also believed to have been used as a diuretic, mild sedative, and astringent in a tea or tincture form.

Sage (*Salvia pachyphylla*)

Description: This sprawling shrub is strongly aromatic. The flowers are blue-violet in color. Sage grows on dry, rocky slopes from 5,000 to 10,000 feet. It blooms from July to September.

Interesting Facts: The seeds of perhaps all species of *Salvia* may be eaten raw or parched and ground into meal. The seeds can also be soaked in water for a flavorful drink. Leaves of any fragrant sage can be used as a tea or spice for soups and meats. Sage does contain moderate amounts of vitamins A and C and can be added fresh to salads and sandwiches; however, we advise you to do this sparingly.

Tilford (1993) indicates that the aboveground parts are antiseptic, astringent, hemostatic, alterative, and tonic and make a good strong topical disinfectant and cleansing wash for abrasions, contusions, and chafed skin. They are also an effective gargle for sore throats and congested sinuses.

Note: *Salvia* and some species of *Artemisia* are often mistaken for "sage." Both are aromatic but are plants in different families. Sage (*Salvia*) is a member of the mint family, whereas wormwood (*A. vulgaris*) and sagebrush (*A. tridentata*) are in the sunflower family. Mints have opposite leaves, whereas these *Artemisia* species have alternate leaves.

Skullcap (*Scutellaria*)

Description: The genus name is Latin (*scutella*) and refers to "small dish," referring to the appearance of the sepals in the fruit. There are about three hundred species of rhizomatous herbs and subshrubs. The flowers are curved, tubular, and two-lipped.

Interesting Facts: *Scutellaria galericulata* has nervine-related therapeutic properties and has been used as a remedy for general restlessness. It has been used in acute or chronic cases of nervous tension or anxiety. The calming effects are said to be mild but reliable. A strong tea was also made of *S. lateriflora* (blue skullcap) and used as a sedative, nerve tonic, and antispasmodic for all types of nervous conditions. All the species contain scutellarin, the primary active compound that has been confirmed to be a sedative and has antispasmodic qualities. Other species may have similar qualities.

MISTLETOE FAMILY (Viscaceae; formally Loranthaceae)
These are parasitic shrubs found on the branches of a variety of trees such as oaks, alders, conifers, and cottonwoods. Leaves are opposite and somewhat leathery. The flowers are small, composed of two to four tepals (petals and sepals not differentiated) inserted on a cuplike receptacle. The fruit is a drupe or berry. There are approximately 11 genera and 450 species in this family distributed worldwide. In the United States, *Arceuthobium* and *Phoradendron* are native.

Dwarf Mistletoe (*Arceuthobium*)
Description: These species grow as parasites on coniferous trees and have opposite leaves. The flowers are found in the axils of the leaves. The pulp of the berry is mucilaginous, which glues the seed to whatever it touches (to eventually germinate on the trees). These mistletoes are found on a variety of conifers. The genus name is from the Greek *arkeuthos*, meaning juniper, and *bios*, meaning life, as the plant is often a parasite on members of the pine family (Pinaceae).

1. Fir Dwarf Mistletoe (*A. abietinum*)—on white fir (*Abies concolor*).
2. Lodgepole Pine Dwarf Mistletoe (*A. americanum*)—on lodgepole pine (*Pinus contorta*)
3. Sugar Pine Dwarf Mistletoe (*A. californicum*)—on sugar pine (*P. lambertiana*).
4. Western Dwarf Mistletoe (*A. campylopodum*)—on Jeffrey pine (*P. jeffreyi*) and ponderosa pine (*P. ponderosa*).
5. Limber Pine Dwarf Mistletoe (*A. cyanocarpum*)—on limber pine (*P. flexilis*).
6. Pinyon Dwarf Mistletoe (*A. divaricatum*)—on pinyon pines (*P. edulis* and *P. monophylla*)

Interesting Facts: Strike (1994) indicates that *A. campylopodium* mixed with the pitch of *P. edulis* was used to cure coughs and colds and to relieve the associated aches. Additionally, the smoke from burning mistletoe relieved colds and coughs and a decoction was used as a contraceptive.

A decoction of lodgepole pine dwarf mistletoe was taken for lung and mouth hemorrhages, as well as for tuberculosis. A decoction of pine dwarf mistletoe was taken for stomachaches.

Juniper Mistletoe (*Phoradendron juniperinum*)
Description and Interesting Facts: This species is usually found growing on junipers (*Juniperus*). The berries were eaten fresh by the Cahuilla. Medicinally, a poultice from the mashed berries was used on wounds to aid in healing, and a decoction of the mashed berries and water was used to bathe the eyes (Bean and Saubel 1972). Two other species of *Phoradendron* occur on other conifers (*P. bolleanum*) and on broad-leaved trees and shrubs (*P. flavescens*).

Warning: All parts of *Phoradendron* are toxic and should be avoided.

MOCK ORANGE FAMILY (Philadelphaceae)
There are approximately 17 genera and 130 species in this family mostly found in the Northern Hemisphere, from the Himalayas to North America. Nine of the genera are native to the United States. The family is a source of a few ornamentals.

Cliff Bush (*Jamesia americana*)
Description: The stem of this plant has shreddy bark and pubescent twigs. The leaves are oblong to roundish, green and pubescent above, and densely gray-

hairy below. It occurs about rocks between 7,800 and 12,000 feet from Mono and Fresno Counties south. It flowers from July to August. The genus name honors Dr. Edwin James (1787–1861), who first collected the plant.

Interesting Facts: The Apache, Chiricahua, and Mescalero apparently ate the seeds of this plant (Castetter 1935). Otherwise, it is grown as an ornamental.

Mock Orange (*Philadelphus lewisii*)

Description: This is a loosely branched shrub or tree with white flowers. It grows on rocky slopes and canyons below 6,000 feet. The genus is named for the Egyptian king Ptolemy Philadelphus.

Interesting Facts: The wood of mock orange is strong and hard and does not crack or warp. It is an excellent wood for making bows and arrows. The leaves and flowers foam into lather when bruised and rubbed with hands and can be used for cleaning the skin. The plant is otherwise considered *poisonous.*

Mock Orange
(*Philadelphus lewisii*)

MORNING GLORY FAMILY (Convolvulaceae)

Members of this family are herbs, shrubs, or trees. In some species, a milky latex is present. The flowers are usually five-merous with five united petals. There are approximately fifty genera and fourteen to seventeen hundred species in this family, distributed in tropical and temperate regions. Nine genera are native to the United States. The family is of some economic importance because of the sweet potato (*Ipomoea batatas*), several weeds, and ornamentals.

Sierra Morning Glory (*Convolvulus malacophylla*)

Description: This twinning and trailing perennial with grayish, woolly herbage grows to twelve inches long. The leaves are triangular shaped and woolly on both sides. The flower is creamy white. This species grows on dry slopes and ridges from 1,500 to 6,000 feet. It flowers from June to August. The genus name is Latin *convolvere,* meaning to entwine.

Interesting Facts: This species was used by Native Americans as a cold leaf tea as a wash on spider bites. A tea from the flowers was used for fevers and wounds. A tea was also made from the leaves and stems by the Kashaya and Pomo women to stop excessive menstruation (Barrett 1952). In European folk use, the flower, leaf, and root teas were considered a laxative. The root is considered to be a strong purgative, cathartic, and diuretic. The powdered rootstock was used as a laxative in ancient and modern China.

MUSTARD FAMILY (Brassicaceae; formerly Cruciferae)

There are approximately 375 genera and 2,000 species in this family worldwide. Mustard flowers are easy to recognize, with four petals in the form of a cross, hence the name cruciform. There are usually six stamens, four of which are longer than the remaining two. The fruit is a pod, either long and thin (silique) or short and wide (silicle), with a partition down the middle dividing the seeds into two individual chambers.

Mustards are a friendly family, having a characteristic peppery taste. In general, the genera are safe to experiment with. Mustards are associated with the Roman practice of soaking seeds in newly fermented grape juice ("must"), drunk as a stimulant to prepare armies for battle. Cauliflower, turnip, radish, cabbage, rutabaga, and watercress are among the economically important plants in this family. Despite the high nutritional value of mustards and the tenacity of the plants, they are largely neglected (Vizgirdas 2003a). In the United States, mustards are often regarded as pests, but in Europe and Asia various species are widely cultivated.

Bittercress (*Cardamine*)

Description: The six species of *Cardamine* in the Sierra Nevada are annuals or perennials with entire or pinnate leaves. The flowers are white or purple. The pods are elongate and flattened. The genus name comes from the Greek *kardamon,* because some bittercress were thought to have heart-strengthening qualities.

Quick Key to the Bittercresses

Stems and leaves compound
  Leaves with five to nine leaflets

Pennsylvania Bittercress (*C. pennsylvanica*)—This species is infrequently found in moist places between 3,000 and 6,800 feet in Amador and Nevada Counties. It flowers from May to June.

  Leaves with three to five leaflets
    Plant of wet habitats; petals one-quarter inch

Brewer's Bittercress (*C. breweri*)—Look for this species along streams between 4,000 and 10,200 feet throughout the Sierra Nevada. It flowers from May to July.

    Plant of moist, shady habitats; petals greater than a half inch

California Milkmaids (*C. californica*)—This species occurs on shady slopes below 5,000 feet. It flowers from May to June.

Leaves simple
  Flowering stalk not leafy

Alpine Bittercress (*C. bellidifolia*)—This species is found at about 7,000 to 8,000 feet from Lassen County northward. It flowers from June to July.

  Flowering stalk leafy
    Leaves ovate in shape

Stout-beaked Bittercress (*C. pachystigma*)—Look for this species on wooded slopes between 5,000 and 9,500 feet throughout the Sierra Nevada. It flowers from May to June.

    Leaves heart or kidney shaped

Lyall's Bittercress (*C. cordifolia*)—This species grows along streams between 5,500 and 7,300 feet from Placer County northward. It flowers from June to July.

Interesting Facts: The plants can be eaten raw in salads; however, we suggest cooking them in at least a change of water to improve the taste. Some plants in this genus were reputed to have medicinal qualities that were used in the treatment of heart ailments.

Caulanthus (*Caulanthus coulteri*)

Description: These are mostly annual or sometimes perennial herbs. Flowers occur in racemes and are white, purple, or yellow in color. There are approximately eighteen species in western North America. The genus name is from the Greek *kaulos,* meaning stem, and *anthos,* for flower—referring to cauliflower because some species can be used like it. This species grows on dry slopes below 5,000 feet. It flowers from March to May.

Interesting Facts: The leaves of *C. coulteri* can be eaten. The upper stem section of *C. inflatus* (desert candle) can be roasted and then eaten. This later species occurs in the southern Sierra in open, sandy places.

Hedge Mustard (*Sisymbrium*)

Description: These are annual, biennial, or perennial herbs. The small flowers are yellow or white, and the fruits are linear. The six species in California are introduced from Europe and are widespread throughout the United States. They are usually found in waste places and disturbed habitats at low elevations.

Interesting Facts: The seeds of hedge mustard can be parched and then ground into a flour. The plants also make fine potherbs. As with other mustards, it is best to cook the plants in a couple of changes of water.

Mustard (*Brassica*)

Description: *Brassicas* are large annuals with showy yellow flowers. The pods are round or four-sided in cross section, with a conspicuous beak. In California, the various species can be found in waste places and fields at lower elevations. *Brassica* is the Latin name for cabbage.

Interesting Facts: All species of *Brassica* have leaves that are edible as greens—an excellent source of vitamins A, B, and C and calcium and potassium. The older leaves should be boiled in at least one change of water. The seeds contain thiocyanate and may cause goiter if consumed in large amounts. The seed can be ground or crushed to a flour and applied as a mustard plaster. The plaster is a long-used remedy for aches and pains. Mustard oil is also a caustic irritant and can discolor and blister the skin if left on too long. The flower buds are rich in protein. The table mustard comes from *B. nigra* (black mustard). In China, mustard oil from *B. rapa* (rape mustard) was used for illumination until the introduction of kerosene.

Peppergrass (*Lepidium*)

Description: The three species of *Lepidium* are annual or perennial plants that are widely distributed. One species is grown for salad. The genus name is Greek (*Lepidion*) and refers to a little scale with reference to the shape of the pods.

Quick Key to the Peppergrasses	
Upper leaves perfoliate	Shield Cress (*L. perfoliatum*)—This peppergrass is found occasionally below 7,000 feet. It flowers from April to June.
Upper leaves not perfoliate No petals	Common Peppergrass (*L. densiflorum*)—This species is widespread and has no petals.
Petals are present and about as long as the sepals	Virginia Pepperweed (*L. virginicum*)—This is a widespread species found in waste places below 7,000 feet. It flowers from March to August.

Interesting Facts: The young stems and leaves may be eaten raw or dried for future use. The plants contain vitamins A and C, iron, and protein. The seedpods and seeds can be used as a flavoring. Fresh *L. virginicum* (Virginia pepperweed) plants were bruised or a tea made from leaves was used for poison ivy rash and scurvy.

Prince's Plume (*Stanleya pinnata*)

Description: This species has flowers that occur in elongated racemes. The fruits (siliques) are borne on a long stipe. The plants are usually found in sagebrush habitats at low elevations. The genus is named for Lord Edward Stanley, a British ornithologist who lived from 1775 to 1851.

Interesting Facts: The tender leaves and stems of all four species can be prepared in much the same way as cabbage. They are bitter, but boiling in several changes of water removes some of the astringent properties. The seeds can be collected, parched, and then ground into a flour. They can be eaten as a mush or used in making breads.

Rock Cress (*Arabis*)

Description: These are biennial or perennial herbs with stellate hairs. Their flowers are in racemes, usually white to purple in color. The fruits are linear siliques, usually flattened parallel to the partition. The many species of *Arabis* in the Sierra Nevada are found in a variety of habitats at various elevations.

1. Brewer's Rock Cress (*A. breweri*)—Brewer's rock cress is found on dry, rocky slopes and summits below 7,400 feet from Yuba County north. It flowers from May to July.

2. Davidson's Rock Cress (*A. davidsonii*)—Look for this rock cress in rocky places between 5,000 and 11,500 feet from Tulare to Plumas Counties. It blooms from July to August.

3. Bent-pod Rock Cress (*A. divaricarpa*)—This species grows on dry slopes in open woods from 7,000 to 11,000 feet throughout the Sierra Nevada. It flowers from July to August.

4. Drummond's Rock Cress (*A. drummondii*)—Drummond's rock cress occurs in dry or dampish benches and slopes between 5,000 and 10,900 feet throughout the Sierra Nevada. It blooms from June to July.

5. Tower Mustard (*A. glabra*)—Look for this species in shaded canyons and mountains up to 9,400 feet. It flowers from May to July.

6. Hairy Rock Cress (*A. hirsuta*)—Look for this rock cress in moist places between 4,000 and 8,200 feet throughout the Sierra Nevada. It flowers from May to July.

7. Holboell's Rock Cress (*A. holboellii*)—This rock cress grows in dry, rocky places from 6,000 to 11,000 feet throughout the Sierra Nevada. It flowers from May to July.

8. Inyo Rock Cress (*A. inyoensis*)—Inyo rock cress is found dry, rocky places between 5,000 and 12,000 feet in Tulare, Inyo, and Mono Counties. It blooms from May to July.

9. Lemmon's Rock Cress (*A. lemmonii*)—Look for this species in dry, rocky places between 8,000 and 14,000 feet from Tulare County to Lassen Peak. It flowers from June to August.

10. Lyall's Rock Cress (*A. lyalli*)—This species grows in rocky places between 8,000 and 12,000 feet. It flowers from July to August.

11. Small-leaved Rock Cress (*A. microphylla*)—Small-leaved rock cress occurs on rocky outcrops from 4,000 to 8,000 feet in the northern Sierra Nevada. It flowers from May to July.

12. Flat-seeded Rock Cress (*A. platysperma*)—This species can be found on dry benches and slopes from 5,000 to 12,750 feet throughout the Sierra Nevada. It flowers from June to August.

13. Blue Mountain Rock Cress (*A. puberula*)—This rock cress is found in dry, stony places between 4,000 and 10,300 feet on the eastern side of the Sierra Nevada. It blooms from June to July.

14. Dwarf Rock Cress (*A. pygmaea*)—Look for this species on dry flats of volcanic sand or gravel from 8,500 to 11,000 feet in Tulare County from Rock Creek to Templeton Meadows. It flowers from June to July.

15. Bristly-leaved Rock Cress (*A. rectissima*)—This species is found on dry slopes and benches from 4,000 to 9,000 feet. It blooms from June to July.

16. Repand's Rock Cress (*A. repanda*)—This species occurs on dry slopes between 4,600 and 11,600 feet from Tulare to Nevada Counties. It flowers from June to August.

17. Elegant Rock Cress (*A. sparsiflora*)—This species is found on dry slopes from 2,500 to 6,000 feet on the western side of the Sierra Nevada and below 9,000 feet on the eastern side in Mono County. It flowers from April to July.

18. Woody Rock Cress (*A. suffrutescens*)—Woody rock cress occurs in dry places between 5,500 and 9,000 feet from Fresno County north. It flowers from June to July.

19. Tiehm's Rock Cress (*A. tiehmii*)—This species is rare on rocky formations and decomposed granite between 10,000 and 11,000 feet in the northern Sierra Nevada (east slope). It flowers from July to August.

Interesting Facts: The crushed plant of *A. puberula* serves as a liniment or mustard plaster. Some species of rock cress were eaten by California tribes, and an infusion was used to cure colds (Strike 1994). A decoction of *A. drummondii* (whole plant) was taken for pains in the lower back and kidney pains by some northwestern Indians. The leaves of *A. holboellii* were chewed as a medicine for toothaches. The chewed roots and subsequent juice of *A. sparsiflora* were swallowed by the Okanagan-Colville for the treatment of diarrhea (Moerman 1986).

Sand Lacepod (*Thysanocarpus curvipes*)
Description: This is a slender, branched annual with stem leaves and basal leaves arranged in a rosette. The flowers are purplish, and the circular, flattened pods are surrounded by a flat, nearly circular wing. The species occurs in open areas at low to midelevations.

Interesting Facts: Sand lacepod seeds may be parched and eaten or ground into flour. A tea made from the plant is said to cure a stomachache; a drink made from the leaf can be used to relieve colic.

Shepherd's Purse (*Capsella bursa-pastoris*)

BIOLOGICAL CONTROL OF MOSQUITOES
The seeds of shepherd's purse have been used against mosquitoes. By sprinkling the seeds on water where mosquitoes breed, the mucilage on the seeds will kill the larvae. One pound of seeds is said to destroy ten million larvae. In water, the seeds release a mosquito attractant and gummy substance that bind the mouth of larva to the seed. In addition, the seed releases a substance that destroys larvae. Despite their pesky nature to humans, mosquitoes do have an important ecological role in nature, providing food for many fish, birds, and other insects.

Description: This is a pubescent annual with leaves mostly in a basal rosette. The petals are white and the pods obcordate (heart-shaped) and strongly flattened, contrary to the narrow septum. *Capsella* means little box, referring to the fruit, as does *bursa-pastoris*. Collectively, the name means "purse of the shepherd." This is a common weed on dry or disturbed soil. It blooms most of the year.

Interesting Facts: Shepherd's purse has been used as food for thousands of years. The seeds have been found in the stomach of Tollund man (approximately 500 BC–AD 400) and during excavations of the Catal Huyuk site, approximately 5950 BC (Tull 1987). The seeds of shepherd's purse may be parched and eaten or ground into flour. The whole pod, with the seeds beaten out, can be added to salads or soups or dried for winter use. The young leaves can be prepared as a potherb and are a good source of vitamin C. With age, they develop a peppery taste. The entire herb (leaves, stems, and green seedpods) can be chopped and added to soups. The roots may be ground or chopped and used as a ginger substitute. The seeds are known to cause blistering of the skin.

The plant is extremely high in vitamin K, the blood-clotting vitamin. Mash or chew the leaves and hold them on a cut. The juice of the plant on a ball of cotton was used to stop a nosebleed. Shepherd's purse also contains significant amounts of calcium, potassium, sulfur, and ascorbic acid. Used as a decoction, shepherd's purse has been used to treat hemorrhoids, diarrhea, and bloody urine. The decoction has a gentle detergent action and is very cleansing to the skin.

Tansy Mustard (*Descurainia*)
Description: Tansy mustards are annual or biennial herbs with leaves that are one to three times pinnately divided. The foliage is covered with simple, branched, or short gland-tipped hairs. The flowers are cream-colored or light yellow, and the pods are long, narrow, and three-sided to nearly round in cross section. These are weedy species occurring in disturbed soils at the lower elevations.

Style inconspicuous	Western Tansy Mustard (*D. pinnata*)—This species grows on dry, sandy areas below 8,000 feet. It flowers from March to June.
Style conspicuous Fruit one-half to one and a quarter inches long	Mountain Tansy Mustard (*D. incana*)—This tansy mustard grows on dry slopes from 5,000 to 11,000 feet. It flowers from May to August.
Fruits about one-quarter inch long	Sierra Tansy Mustard (*D. californica*)—This tansy mustard occurs on dry slopes between 7,000 and 11,000 feet, especially on the eastern Sierra Nevada, from Nevada and Mono Counties southward. It flowers from May to August.

Interesting Facts: All species of tansy mustard are edible as greens but are bitter. The seeds can be parched, ground, and prepared as mush. The seeds were parched by tossing in a basket with hot stones or live coals, then ground into a fine flour and made into mush. Because of its peppery taste, the mush was often mixed with the flour of other seeds to make it more palatable. Young leaves can be boiled or roasted between hot stones and eaten as green vegetables. The seeds were also used in poultices for wounds. However, one species, *D. pinnata*, is reported to be *poisonous* in large quantities, causing blindness and paralysis of the tongue (Kingsbury 1964).

Wallflower (*Erysimum*)
Description: Wallflowers are annual, biennial, or perennial herbs that are often taprooted. The leaves are usually narrow, and the flowers occur in dense racemes. The sepals are erect and narrow, with the two outer usually saccate at the base.

Quick Key to the Wallflowers

Plant is perennial; flowers are yellow to orange in color	Western Wallflower (*E. capitatum*)—This perennial species is common on dry slopes below 12,000 feet. It blooms from May to August.
Plant is annual; flowers are light yellow	Wormseed Wallflower (*E. cheiranthoides*)—This annual species is rather uncommon in disturbed areas below 8,000 feet. It flowers from June to August.

Interesting Facts: Wallflowers were once used as a poultice. *Erysio* means to draw out, as in drawing out pain or causing blisters. Additionally, an infusion of dried, pulverized *E. capitatum* was rubbed on the head and face to prevent sunburn or to alleviate heat exposure. A pneumonia victim was cured by having his back massaged with chewed wallflower root (Strike 1994).

Watercress, Yellowcress (*Rorippa*)
Description: These are taprooted annuals or rhizomatous perennials with simple or pinnately divided leaves. The flowers are yellow or white; the pods are elliptical to linear and three-sided to slightly compressed. The five species in the Sierra Nevada occur in moist, wet, or aquatic habitats up into the middle elevations.

Watercress (*Rorippa* spp.)

Plants annual
    Stems erect and branched above   Marsh Yellowcress (*R. palustris*)—This species occurs in wet places, often with roots immersed, below 8,000 feet. It blooms from April to June.

    Stems branching from base of plant
        Leaf segments are linear to oblong in shape, with an abrupt tip; fruit is obviously curved   Yellowcress (*R. curvipes*)—This species is occasionally found in wet places, with roots immersed, below 7,500 feet.

        Leaf segments are more rounded or may be lobed; fruit is not curved   Yellowcress (*R. sphaerocarpa*)—This species occurs in wet places between 5,000 and 8,000 feet. It flowers from June to September.

Plants perennial
    Plants terrestrial   Umbellate Yellowcress (*R. subumbellata*)—Look for this species in moist places between 6,000 and 7,000 feet. It flowers from June to July.

    Plants aquatic   Watercress (*R. nasturtium-aquaticum*)—This is a common species below 8,000 feet. It flowers from May to October.

Interesting Facts: The herbage of watercress is edible if the waters in which they grow are not polluted. However, finding unpolluted water may be difficult. One suggestion would be to soak the fresh greens in a disinfectant or treat the water with water purification tablets or a tablespoon of bleach in a quart of water. Then rinse the greens well in potable water to remove the chemicals. The peppery tasting plants were eaten raw or cooked as a potherb. A good source of vitamins, watercress is listed as efficient in preventing scurvy. Medicinally, the plant was used for freckles, pimples, liver, and kidney troubles.

Note: Bittercress (*Cardamine breweri*) looks similar to watercress. To quickly differentiate the two species, look at the fruits. Bittercress fruits are linear and narrow; watercress fruits are round or four-angled in cross section.

Western Bladder-pod (*Lesquerella occidentalis*)
Description: This perennial plant has stellate pubescent on the stems. It occurs on dry slopes below 8,000 feet from Placer County northward. It flowers from May to June. The genus is named after Leo Lesquereux, a late-nineteenth-century American paleobotanist. There are about forty species of annual to perennial densely hairy herbs that have small flowers.

Interesting Facts: A related species, Fendler's bladderpod, was used by the Navajo as a tea and used as a remedy for the bites of spiders (Elmore 1944).

Whitlow Grass, Draba (*Draba*)
Description: These are low, pubescent perennials with basal leaves in a tuft or cushiony rosette. The leaves are lanceolate, and the flowers occur in a terminal raceme. The small flowers are white or yellow that fade to white with age. The pods are egg shaped, elliptical, or club shaped and sometimes twisted. The various *Draba* species occur on shaded slopes and rocky areas between 7,000 and 11,000 feet. It flowers from July to August.

Interesting Facts: The species are mainly unpalatable. Whitlow grasses were formerly used for treating "whitlows," inflammations of the fingertip (Pojar and

MacKinnon 1994). Whitlow is a name applied loosely to any inflammation involving the pulp of the finger, attended by swelling and throbbing pain.

Wild Radish (*Raphanus sativus*)

Description: Wild radish is a branched herb up to three feet tall. The flowers are white with rose or purple veins, sometimes yellowish. The fruit is a rounded pod up to two inches long. Wild radish is a weed of waste places and fields at the lower elevations. The garden radish is a cultivated form of this species.

Interesting Facts: The leaves, flowers, and pods of wild radish are used as a food, rather than the root. The root of wild radish is a bit too woody to be eaten like garden radishes. The flowers can be tossed in a salad or eaten alone as a snack. The fruits can be used in a salad but must be collected before the seeds harden and the pods dry out. The taste of these pods resembles that of the garden radish.

Wintercress (*Barbarea orthoceras*)

Description: This herb has angled stems and pinnatifid leaves. The flowers are yellow, and the pods are linear and four-angled. The genus is named for Saint Barbara. The Latin name is derived from the fact that the young leaves can be eaten on Saint Barbara's Day in early December. Wintercress grows in moist places and along stream banks from 2,500 to 11,000 feet. It flowers from May to September.

Interesting Facts: The young stems and leaves of wintercress and *B. vulgaris* (garden yellowrocket) can be eaten raw in salads or prepared as a potherb. We like to boil the plants in at least two changes of water.

NETTLE FAMILY (Urticaceae)

There are about 45 genera and 550 species found in the nettle family. Six of the genera are native to the United States and are of little economic importance.

Stinging Nettle (*Urtica dioica*)

Description: This is an annual or perennial herb with stinging hairs. The flowers are numerous, small, and clustered on drooping branches at the base of the leaves. Stinging nettles can be found along roadsides and streams and in moist areas and waste places in the low to middle elevations. Stinging nettle is an indicator of good soil conditions. Many people consider stinging nettles, for obvious reasons, as obnoxious weeds. Recent taxonomic revisions have consolidated various *Urtica* species into a single species.

Interesting Facts: One of the first things a person learns about stinging nettles is their stinging effect. The intense burning and itching or stinging of the skin may persist for a length of time. If you look closely at the hairs, you will see a hypodermic mechanism consisting of a very fine capillary tube with a bladderlike base that is filled with chemical irritant. When brushed against, a minute spherical tip breaks off, uncovering a very sharp-pointed tip that easily penetrates the skin. The chemical is forced into the skin through the tube as the hair bends and constricts the bladderlike base. Therefore, stinging nettles should be collected with gloves.

Stinging Nettle (*Urtica dioica*)

The young stems and leaves of stinging nettle are edible after boiling or steaming and are very delicious as a spinach substitute. Boiling the leaves destroys the formic acid found in the hairs. The leaves are high in vitamins A, C, and D, the latter of which is rare in plants. The roots are also edible after they have been roasted. A tea made from the leaves is said to have astringent and diuretic qualities and has been used for internal bleeding and nosebleeds. Native Americans learned of stinging nettle as a food from early European travelers and settlers and possibly from Chinese immigrants.

The older stems become fibrous, which reduces their edible qualities but allows them to be used to produce strong cordage. The older leaves also contain cystoliths that can irritate the kidneys. A yellow dye may be obtained by boiling the roots.

A tea brewed from *Urtica* was said to relieve chest colds and internal pains. As a poultice, *Urtica* was used for headaches.

NIGHTSHADE FAMILY (Solanaceae)

There are about eighty-five genera and twenty-three hundred species in the nightshade family. About thirteen genera are indigenous to the United States. The family is noted for its edible, poisonous, and medicinal plants. Some of the economically important plants include tomato, potato, chili, tobacco, and eggplant. Medicinally, two alkaloids, atropine (belladonna) and scopolamine, which can be deadly in large amounts, are obtained from this family. Eye doctors today use atropine in very exact small doses as an effective dilator when examining eyes.

Boxthorn (*Lycium andersonii*)

Description: This thorny shrub may grow to nine feet tall. The leaves are clustered, drought deciduous, and fleshy. The flowers are lavender in color and arise from between the forks of branches. This species grows in woodlands and scrub habitat up to 6,000 feet.

Interesting Facts: The plants produce large crops of berries that were an important food item for California Natives (Sweet 1976). They were eaten raw or dried in the sun like raisins. The dried berries can be boiled into mush or ground into flour and mixed with water. The roots can also be dried, pounded, baked, and eaten.

Coyote Tobacco (*Nicotiana attenuata*)

Description: This annual plant grows up to forty inches tall and has glabrous to glandular pubescent stems. The leaves are large, ovate to ovate lanceolate in shape, and the upper leaves are narrowed. Flowers occur in a terminal raceme, and the calyx is five cleft. The corolla is white and tubular or funnel shaped. The flowers close in the sun. Coyote tobacco grows in disturbed places below 8,500 feet. It flowers from May to October.

Interesting Facts: All species of *Nicotiana* contain the highly toxic alkaloid nicotine. An effective insecticide against aphids can be prepared by steeping tobacco leaves in water and spraying the solution on affected parts of the plant.

Dwarf Chamaesaracha (*Chamaesaracha nana*)

Description: This plant with entire, white-pubescent leaves has flowers with five basal green spots. The berry is dull white to yellowish in color. Dwarf chamaesaracha occurs on sandy flats between 5,000 and 9,000 feet from Mono County northward.

Interesting Facts: The berries of a similar species are edible raw, cooked, or dried. However, it is unknown if this species can be used in the same way. Therefore, you are cautioned against using it.

Jimson Weed (*Datura wrightii*)

Description: This is a coarse, erect annual or perennial plant, growing up to forty inches tall and having a strong odor and narcotic qualities. Leaves are alternate, ovate, and unequal at the base, with wavy margins and petioles. Its flowers are large, solitary, and in the forks of branching stems. The calyx is tubular shaped and five toothed. The corolla is funnel shaped, white, and purple to violet. Its flowers are pleated in the bud and open in the evening. The fruit is a prickly, globose capsule with spines. Jimson weed grows in sandy or gravelly open areas, generally below 6,000 feet. It flowers from April to October.

Jimson Weed
(*Datura wrightii*)

Interesting Facts: Extracts from these plants are narcotic and, when improperly prepared, are very lethal. *Datura* is also known as "hell's bells" among drug users because of the popular misconception that the plants provide worthwhile hallucinogenic experiences. Unfortunately, many of those who have experimented with *Datura* as a drug have died as a result. Those who did survive realized that it was not worth the try. Its narcotic properties have been known since before recorded human history, and they once figured importantly in the religious ceremonies of many Native Americans (Avery, Satina, and Rietsema 1959).

Although potentially deadly, the plants do have valuable medicinal properties. There are several alkaloids including atropine and hyoscyamine that have been used in scientifically refined antispasmodic drugs. *Datura* also contains scopolamine, an antivertigo compound commonly used to treat motion sickness and other conditions involving disequilibrium. The concentrations of these compounds vary widely among plant parts, species, and localities, making *Datura* highly unreliable for internal use, even in trained hands (Balls 1970).

Warning: These plants should never be taken internally for any reason. They contain dangerous alkaloid compounds that could easily kill a person.

Purple Nightshade (*Solanum xantii*)

Description: This perennial herb has stems that grow up to sixteen inches tall. The stems are densely and gray hairy; and they may form mats on the ground. The leaves are ovate in shape. The corolla is five angled and deep violet, and the anthers are fused around the style. The fruit is a greenish berry. Purple nightshade grows on dry places from 5,000 to 9,000 feet. It flowers from May to September. This is a highly diverse genus comprising more than a thousand species worldwide. Solanum probably comes from the Latin *solamen,* which means quieting, referring to the sedative properties of some species.

Interesting Facts: Members of the genus *Solanum* are pollinated when an insect such as a bumblebee or hoverfly uses its wings and thorax to set up a vibration in and around the anthers. This causes the pollen to stream out onto the insect. This is called "vibrating pollen collection."

The most *poisonous* part of the plant is the unripe fruit, but the stems, leaves, and roots are also dangerous. In fact, the alkaloid content in the plants decreases in this order: unripe fruit, leaves, stems, then ripe fruit. Solanine is the predominant glyco-alkaloid, but others may be present. Solanine is highly toxic and can cause death, but the degree of toxicity varies among and within species. It is reported that cooking destroys the solanine, making the ripe fruit edible. The berries of *S. nigrum* (black nightshade) were formerly used as a diuretic (Foster and Duke 1990). Native Americans of the Southwest used the crushed berries to curdle milk for making cheese, and they have also been used in various preparations for sore throat and toothaches.

Solanum dulcamara (climbing nightshade), when used correctly and in appropriate dosages, is said to be useful in the treatment of skin disorders, rheumatism, and bronchitis. Recent studies have shown that this plant possesses anticancer qualities (Duke 1992a).

Caution: There are references listing the berries of some species of *Solanum* as edible. However, it is our recommendation that none of the plants that occur in southern California be consumed in any manner.

OLEASTER FAMILY (Elaeagnaceae)
Plants in this family are shrubs or trees with alternate or opposite and silvery-gray leaves. The fruits are drupe- or berrylike. There are three genera and approximately forty-five species in this family distributed in North America, south Europe, Asia, and eastern Australia.

Buffaloberry, Soapberry (*Shepherdia argentea*)
Description: Buffaloberry is a low, spreading shrub with opposite leaves, each with a dark-green upper surface and lighter underside that is covered with tiny brown scales. The inconspicuous flowers are yellow-green; the fruits range in color from yellow to bright red. Buffaloberry grows along streams from 3,500 to 6,500 feet. It blooms from April to May. It is also found in recently burned areas. The genus is named after John Shepherd (1764–1836), curator of the Liverpool Botanic Garden.

Interesting Facts: Another common name for this species is soapberry. This is because the berries contain a significant amount of saponin, which not only gives the plant its bitter taste but also whips up into a frothy mass called "Indian ice cream."

Native Americans used the berries of these plants extensively, both fresh and dried. The berries of this species are at first pleasant, then the soaplike bitterness prevails. We enjoy cooking them with sweeter-tasting berries such as thimbleberries and serviceberries and a large amount of sugar. We found the berries to be somewhat unattractive for general use but a valuable consideration in emergencies. The berries taste better after a few good frosts during the fall. They can also be used in the making of pemmican or jelly. Dried into cakes, the berries can be stored for winter.

Infusions of the stems and leaves were drunk as a tonic beverage. The berries can be crushed and made into a tea for use as a liquid soap. Native Americans used the tea to relieve constipation. Hart (1996) indicates that the Flathead and Kootenai Indians made solutions from the bark of buffaloberry for eye troubles.

Buffaloberry, Soapberry
(*Shepherdia argentea*)

"INDIAN ICE CREAM"
To make Indian ice cream, Native Americans placed a small number of berries into a bowl with a little water, then used a special stick with some grass tied on one end to beat the fruit. The result was a foamy concoction. In recent times, sugar was added to improve the taste. Care must also be taken in picking and preparing the berries so that they do not come in contact with oil or grease of any kind, or they will not whip. Indian ice cream is still served in many households, especially at parties and family gatherings in the Pacific Northwest.

OLIVE FAMILY (Oleaceae)
There are approximately twenty-nine genera and six hundred species in this family. Members are cosmopolitan in distribution, particularly well represented in temperate and tropical Asia. Five genera occur in the United States, and the family is of considerable economic importance because of olives, timber, and ornamentals. The family is one of the few with flowers having four sepals and petals and two stamens.

Oregon Ash (*Fraxinus latifolia*)
Description: This tree has oblong- to oval-shaped leaves. Fruits are winged nutlets (samaras), much like those of maple (*Acer*). It grows in canyons and near streams throughout the Sierra Nevada. It flowers from April to May.

Interesting Facts: Related species have edible inner bark. The wood is useful in making bows. In late spring, *Fraxinus* trees are heavily infested with army worms (*Homoncocnemis fortis*), which were gathered by Natives, then parched and eaten (Strike 1994).

A cold infusion of the twigs was used for fevers, whereas an infusion of the bark was taken for worms. The roots are used to make baskets.

PEA OR LEGUME FAMILY (Fabaceae; formerly Leguminosae)

The pea family is one of the largest plant families in the world. There are approximately six hundred genera and thirteen thousand species worldwide. Next to the grass family, which produces all our grains and cereals, the pea family is the second most economically important group of plants in the world. The beans and peas that we eat for dinner, as well as the traditional peanuts at baseball games, are found in this family. But before taking a bite of the next legume you see, be aware that the family also contains a number of highly toxic members. The various species of locoweeds and milkvetches (*Oxytropis* and *Astragalus*) have caused much loss of livestock.

The pea flower is referred to as "papilionaceous" and means butterfly-like. The flowers are bilaterally symmetrical, consisting of five petals. The largest upper petal is called the banner, the two lateral ones are the wings, and the two lowest ones are fused at the lower margins to form a boatlike structure called the keel.

The ripe seeds of various members of this family are good protein sources, especially when mixed with cereals (Poaceae). The amino acids of their respective proteins combine, which increases their nutritional efficiency markedly. Legumes are also rich in carbohydrates.

Pea Family (Fabaceae)
Flower (expanded)

Alfalfa (*Medicago sativa*)

Description: In general, this is described as a hairless, branching perennial or annual herb with leaves divided into three leaflets. The terminal leaflet is evidently longer than the other two. The pods are twisted. It is usually found in disturbed areas at the lower elevations. Burclover (*M. lupulina*) occurs too.

Interesting Facts: Alfalfa can cause bloat in livestock when it constitutes a high percentage of their diet, especially when it's young, before flowering. Saponins found in the leaves may contribute to the problem. Humans should, therefore, use this plant in moderation. The dried and powdered young leaves and flower heads of alfalfa are nutritious and can be steeped in hot water to make a bland tea. The tender leaves can also be added to salads and are rich in vitamins A, D, and K. Alfalfa also supplies calcium, magnesium, and phosphorus (C. E. Smith 1973). Alfalfa sprouts are a popular salad addition, and the seeds are available from various health stores. Nectar from the flowers produces a good honey. In addition to uses as food and medicine, alfalfa seeds contain an oil for use in paints and varnishes. Papermakers have used the stem fibers in their craft, and wool dyers extract a yellow dye from the seeds. The seeds of another species, *M. lupulina* (black medic), can be parched and eaten or ground into flour.

American Vetch (*Vicia americana*)

Description: Vetch are annual or perennial herbs with trailing to climbing stems. Leaves are pinnately divided with tendrils in place of terminal leaflets. It is found in waste places at lower elevations. *Vicia* closely resembles *Lathyrus* (sweetpea) and requires close examination of the stipules. The stipules of *Vicia* are usually cut into narrow lobes, whereas the stipules of *Lathyrus* are entire to dentate. This trailing, climbing perennial has sparsely pubescent stems and grows in open places below 5,000 feet. It flowers from April to June.

Interesting Facts: Many species contain toxic compounds and therefore should be considered *poisonous*. However, Kirk (1975) and Craighead, Craighead, and Davis (1963) state that the young stems and seeds can be boiled or baked. The seeds of some species contain compounds producing toxic levels of cyanide when digested. Because of the poisonous compounds found in vetch, it is not recommended for eating.

Clover (*Trifolium*)

Description: There are many species of clover in the Sierra Nevada. In general, they are annual and perennial plants from rhizomes with leaves that are divided into three or more leaflets. The flower colors range from white to pink, yellow, red, or purple, and the seedpods are round to elongated. They are found in various habitats at all elevations. The genus name refers to the three leaflets.

1. Anderson's Clover (*T. andersonii*)—This species occurs in many habitats up to 7,000 feet.
2. Beckwith's Clover (*T. beckwithii*)—This species may be found in moist meadows between 4,000 and 7,000 feet. It flowers from May to August.
3. Bolander's Clover (*T. bolanderi*)—This clover occurs in wet meadows to 7,000 feet. It flowers from June to July.
4. Brewer's Clover (*T. breweri*)—This species is found on wooded slopes below 7,000 feet from Madera County north. It flowers from May to August.
5. Tree Clover (*T. ciliolatum*)—This is a common plant on open and grassy slopes below 5,000 feet on the west side of the Sierra Nevada. It flowers from April to June.
6. Bowl Clover (*T. cyathiferum*)—This annual is found in moist places below 8,000 feet.
7. Clover (*T. eriocephalum*)—This hairy perennial is found in moist meadows to dry, open slopes between 900 and 4,500 feet in the northern Sierra Nevada.
8. Clover (*T. gracilentum*)—This is a common plant in open, grassy places below 5,000 feet throughout the western side of the Sierra Nevada.
9. Alsike Clover (*T. hybridum*)—Look for this species in damp places below 6,000 feet from Mariposa County north.
10. Shasta Clover (*T. kingii*)—This clover occurs in moist wooded places between 3,800 and 8,000 feet. It flowers from June to August.
11. Lemmon's Clover (*T. lemmonii*)—Look for this species on slopes and in valleys between 5,000 and 7,000 feet.
12. Long-stalked Clover (*T. longipes*)—This species occurs in moist places below 9,000 feet.
13. Dedecker's Clover (*T. macilentum*)—This is a rare plant in alpine habitats between 7,000 and 11,500 feet in the southern Sierra Nevada.
14. Large-headed Clover (*T. macrocephalum*)—This clover is found on rocky slopes and ridges below 8,000 feet.
15. Small-headed Clover (*T. microcephalum*)—This species occurs in open, grassy places below 8,500 feet.
16. Carpet Clover (*T. monanthum*)—Look for this species in wet places between 5,000 and 11,500 feet throughout the western side of the Sierra Nevada.
17. Clammy Clover (*T. obtusiflorum*)—This clover is found in moist places below 5,000 feet on the western slope of the Sierra Nevada.
18. White Clover (*T. repens*)—This is a common plant in disturbed habitats below 7,000 feet.
19. Tomcat Clover (*T. wildenovii*)—This is a common plant in grassy places below 5,000 feet throughout the western side of the Sierra Nevada.
20. Mountain Clover (*T. wormskioldii*)—This species is found in wet places below 10,000 feet throughout the Sierra Nevada. It flowers from May to October.

Mary Dedecker (1909–2000) lived in Independence, California, for more than sixty years and extensively explored Inyo and Mono Counties. She was a self-taught botanist who discovered several new species and one entirely new genus near Death Valley, which was named after her (*Dedeckera eurekensis*). She founded the Bristlecone Chapter of the California Native Plant Society and wrote *Flora of the Eastern Mojave*.

Interesting Facts: All species are nutritious and high in protein, but the flower heads and tender young leaves are hard to digest raw and may cause bloating. To improve digestibility of the plants, soak them in saltwater for several hours or overnight. Leaves prepared this way may be dried and stored for future use. The dried flower heads and seeds can be ground into a flour substitute or extender.

Trifolium was an important food source for many Native Americans. In the spring, explorers and settlers saw them in the meadows picking and eating large quantities of clover. This was an annual event for the Natives, who relished the greens of the spring season. Unfortunately, the nonnatives in their ignorance compared the Natives to grazing animals. This is just one of many disparaging comments made concerning misunderstood Native American behavior (Erichsen-Brown 1979; Murphey 1990).

A tonic tea can be made from the dried flowers. Made strong, the tea can be used as a gargle for sore mouths and throats and as a mild sedative. The tea can also be used as a wash for skin ailments. The dried leaves can be smoked.

False Lupine (*Thermopsis macrophylla*)

Description: This is a stout perennial plant growing twelve to thirty-two inches tall. The stem and leaves are densely covered with long, white, silky hairs, and the palmately three-foliate leaves occur on petioles that are one and one-half inches long. The yellow flowers are in a terminal raceme that is six to ten inches long. The fruit is a pod and is densely covered with short, appressed hairs. False lupine grows in open places (below 4,500 feet). It flowers from April to June.

Interesting Facts: A cold decoction of the leaves was used by Pomo Natives as a wash for sore eyes. A tea made from the leaves, roots, or bark was used by Kashaya women to slow their menstrual flow (Moerman 1986).

Locoweed, Oxytrope (*Oxytropis*)

Description: In general, these are perennial herbs that are stemless or have a short, leafy stem. The flowers are white to reddish or purple. The keel is prolonged into a point or tooth or a straight to curved beak. The name is from the Greek *oxys,* meaning sharp, and *tropis,* meaning keel, in reference to the beaked keel. It can be found in a variety of habitats from the lower elevations to subalpine.

Quick Key to the Oxytropes	
Plants silvery and having silky hairs	Parry's Oxytrope (*O. parryi*)—This species occurs on dry knolls and rocky ridges near timberline and above (11,000 to 12,000 feet) on the eastern side of the range in Inyo and Mono Counties. It flowers from June to July.
Plants green and sticky	Sticky Oxytrope (*O. borealis*)—Look for this species on bare crests and talus slopes between 11,200 and 12,200 feet in Inyo County. It flowers from July to August.

Interesting Facts: Neither species is known to be poisonous (loco producing) but probably best if avoided and treated as potentially toxic. However, these are some of the famous "locoweeds." The plants are poisonous to livestock. In order for the plants to be lethal to livestock, the animals must eat large amounts over a long period of time. Extensive grazing of a related species, *O. sericea* (white flowered loco), induces a chronic poisoning called locoism. It is strongly recommended that these plants be avoided for the purposes of human consumption.

LOCOWEEDS (*ASTRAGALUS* AND *OXYTROPIS*)
There are as many as three hundred species of "locoweeds" in North America, although not all are toxic. The two genera include *Astragalus* and *Oxytropis,* and identification requires a trained specialist. Locoweeds contain one of three toxic fractions: miserotoxin, swainsonine, or selenium. The fresh plants are toxic, and the toxicity is gradually lost as the plant dries, except that of selenium accumulators, which are not affected by drying. Swainsonine poisoning occurs after about two weeks or more of ingestion. Cattle and horses may become habituated to locoweed and seek it out, even when good forages are available. Swainsonine may be passed in the milk.

Lotus (*Lotus*)

Description: The thirteen species of *Lotus* in the Sierra Nevada are annual or perennial herbs with pinnately compound leaves. The flowers are pea shaped, yellow or white, often tinged reddish or purple. They occur in many habitats at various elevations.

Interesting Facts: Many of the species are presumed to be *poisonous* and should be avoided.

Lupine (*Lupinus*)

Description: The many species of lupine are showy perennial or annual herbs with palmately compound leaves. Its flowers are blue, violet, rarely white, and rose in elongated, narrow inflorescence. The pods are flattened and usually hairy. They are found on open slopes and meadows up into the alpine zone. From the Latin name *Lupinus* comes *lupus,* meaning wolf, alluding to the belief that this plant wolfed nutrients and caused poor soil conditions. To the contrary, lupines are nitrogen fixers that greatly improve soil conditions.

Lupine (*Lupinus* spp.)

1. Silky Lupine (*L. adsurgens*)—This species occurs in dry places below 12,000 feet.
2. Lupine (*L. albicaulis*)—This species grows on dry slopes and openings between 2,000 and 8,500 feet from Fresno County northward. It flowers from May to September.
3. Silver Lupine (*L. albifrons*)—Look for this species in sandy to rocky places throughout the Sierra Nevada. It flowers from May to September.
4. Christine's Lupine (*L. angustiflorus*)—This species is common in Lassen National Park and vicinity.
5. Crest Lupine (*L. arbustus*)—This species is found on dry slopes and meadows between 5,600 and 9,600 feet throughout the Sierra Nevada. It flowers from May to September.
6. Tahoe Lupine (*L. argenteus*)—Look for this species in dry, open places between 5,000 and 8,000 feet from Madera and Mono Counties north to Lassen County. It flowers from July to August.
7. Lupine (*L. bicolor*)—This erect, long, hairy annual is common in open, sandy, and gravelly areas below 5,000 feet. It flowers from March to May.
8. Brewer's Lupine (*L. breweri*)—This is a low, matted perennial with a woody base that grows on dry, stony slopes between 4,000 and 11,000 feet in the mountains. It flowers from June to August.
9. Orange-flowered Lupine (*L. citrinus*)—Look for this species on rocky hills between 4,000 and 5,300 feet. It flowers from April to June.

Lupine (*Lupinus*) leaf

10. Coville's Lupine (*L. covillei*)—This species occurs in rocky places between 8,500 and 10,000 feet from Tulare to Tuolumne Counties. It flowers from June to September.
11. Quincy Lupine (*L. dalesiae*)—This is a rare lupine that grows in pine forests between 3,500 and 8,000 feet in the northern Sierra Nevada.
12. California Green-stipuled Lupine (*L. fulcratus*)—This occurs in dry habitats between 6,000 and 10,000 feet from Tulare to Nevada Counties.
13. Slender Lupine (*L. gracilentus*)—This species occurs between 8,500 and 10,500 feet in Rock Creek Lake Basin to Yosemite National Park. It flowers from July to August.
14. Gray Lupine (*L. grayii*)—This lupine occurs in dry habitats below 7,800 feet from Plumas County south.
15. Broad-leaved Lupine (*L. latifolius*)—This erect, herbaceous perennial is common in the open woods below 7,000 feet. It flowers from April to July.
16. Dwarf Lupine (*L. lepidus*)—Look for this species on open slopes above 5,000 feet.
17. Obtuse-lobed Lupine (*L. obtusilobus*)—This species grows on gravelly summits between 5,000 and 10,000 feet. It flowers from June to September.
18. Father Crowley's Lupine (*L. padre-crowleyi*)—This is a rare lupine growing in rocky soils above 8,000 feet on the eastern side of the Sierra Nevada.

19. Plumas Lupine (*L. onustus*)—This species is found occasionally on dry slopes between 3,000 and 5,500 feet in Plumas County northward. It flowers from April to September.

20. Many-leaved Lupine (*L. polyphyllus*)—This stout, erect perennial grows in wet places from 6,500 to 8,500 feet. It flowers from June to August.

21. Inyo Meadow Lupine (*L. pratensis*)—Look for this species in moist places between 4,000 and 10,500 feet in Tulare, Fresno, Mono, and Inyo Counties.

22. Harlequin Lupine (*L. stiversii*)—This species grows in sandy or gravelly habitats below 4,600 feet. It flowers from May to July.

Interesting Facts: The pealike seeds have been wrongly recommended by some authors of edible-plant books as a substitute for peas. Lupines possess many complex alkaloids and should be considered *poisonous*. Records do indicate, however, that some species have been safely consumed. For example, Weedon (1996) and Scully (1970) indicate that the young leaves and unopened flowers were steamed and eaten with soup by some Native American tribes. But because of hybridization, the edible species can concentrate toxic alkaloids that could result in an unhealthy game of "lupine roulette."

It does, however, appear that some of the alkaloids found in lupines are removed by cooking and that toxins intensify with age. The toxic principle of lupines is excreted by the kidneys, and the poisoning is not cumulative. That is, a lethal dose must be eaten at one time to cause death. The poisonous effects produced by lupines are referred to as lupinosis, with nervousness, labored breathing, convulsions, and frothing at the mouth being the obvious signs (Muenscher 1962). *But until documentation on Sierra Nevada species is established, lupines as a food source are not recommended.* Many people use the larger, hairy-leaved species as an excellent toilet paper substitute.

Milkvetch (*Astragalus*)

Description: There are at least nine species of milkvetch in the Sierra Nevada. They are perennial herbs with odd-pinnate leaves that have leafy stipules. This is a difficult genus of perhaps sixteen hundred species, making it the largest genus in the pea family. The name comes from the ancient Greek name for a plant in the pea family.

Interesting Facts: Although the roots, pods, and peas of some species were reported to be eaten by Native Americans, this genus is *not recommended* for consumption. All milkvetches either produce a toxic alkaloid substance or accumulate selenium from the soil or both. Selenium poisoning of livestock has the following characteristics: lethargy, diarrhea, loss of hair, breakage at the base of the hoof, excessive urination, difficulty breathing, rapid and weak pulse, and coma. Death usually results from the failure of lungs and heart (Turner and Szczawinski 1991).

An interesting side benefit has developed from the discovery of plants that grow only in selenium soils. Scientists can use the plants to map areas high in selenium for the purpose of mining the valuable element. Some *Astragalus* species also provide good indicators of uranium ore and copper-molybdenum deposits.

Redbud (*Cercis occidentalis*)

Description: This is a deciduous tree or shrub with simple, alternate, and entire leaves. The flowers are lavender in color and occur in clusters or racemes. The species occurs below 5,000 feet in the Sierra Nevada.

Interesting Facts: The pods and seeds were supposedly eaten by the Maidu and Wintum. The Mendocino treated chills and fevers using redbud bark. This was one of the most important basketry plants and was used regularly in making twined and coiled baskets. Some Native Americans made an astringent from the bark and used it as a remedy for diarrhea and dysentery (Strike 1994).

Round-leaved Psoralea (*Hoita orbicularis*)

Description: This species was formerly known as *Psoralea orbicularis*. The stems of this plant are prostrate, and the herbage is heavy scented and glandular. The flowers are in dense racemes, and the petals are reddish purple in color, with the banner often having a white spot on each side. The pods are hairy. This species occurs in moist places below 4,000 feet from Mariposa County north. It flowers from May to July.

Interesting Facts: The genus name *Psoralea* is from the Greek *psora* for warty or scruffy and refers to the glandular, dotted leaves of some species. The foliage was used to make a tea, and the tubers of several species are edible raw or cooked. The Luiseno and Maidu ate the leaves of the plant, and the Costanoan used the plant as a tonic to reduce fever. A yellow dye was obtained from the roots and used to dye basketry material.

A related species, *H. macrostachys* (leatherroot), was also used by the Luiseno and Maidu to heal skin sores and irritation. The Pomo and Maidu used the inner bark and roots of leatherroot to make cordage.

Sweetclover (*Melilotus*)

Description: Sweetclover are strongly taprooted perennial or annual herbs. The leaves are divided into three fine-toothed wedge-shaped leaflets. The white or yellow flowers are loosely arranged in an inflorescence, and the pods are thickly spindle shaped. It is usually found in disturbed habitats at the lower elevations. The genus name is from the Greek *mel*, meaning honey, and lotus flower.

1. Yellow Sweet Clover (*M. indica*)—This glabrous plant with yellow flowers is common in waste places in the lower elevations. It flowers from April to October.
2. White Sweet Clover (*M. albus*)—This tall, glabrous plant with white flowers is a common weed of waste places, particularly where damp. It flowers from May to September.

Interesting Facts: The young leaves (before the flowers appear) of *M. albus* (yellow sweetclover) may be eaten raw or boiled. The fruit may be used as seasoning for soups. The older leaves are toxic and should be avoided. The dried flowering plant of *M. officinalis* was used in teas for neuralgic headaches, nervous stomach, diarrhea, and aching muscles. Elias and Dykeman (1982) indicate that improperly dried yellow sweetclover will easily mold and in the process produce dicoumaral, an anticoagulant that can cause severe bleeding and death. Dicoumarol is used in rat poisons. Molding yellow sweetclover mixed in hay has killed many cattle.

The plants are sweet scented due to coumarin and become more pleasant when dried. They have been used to scent clothes and protect them from moths as an alternative to mothballs. It has also been a traditional flavoring additive in smoking tobacco and snuff.

Sweetpea (*Lathyrus*)

Description: Sweetpeas are vines, climbing or supporting themselves on other vegetation. The plants have tendrils at the ends of their leaves. Sweetpeas are found in a variety of habitats from the foothills up to the subalpine.

Quick Key to the Sweetpeas

Flowers two to eight in number
    Leaves are ovate in shape; calyx lobes are about equal — Sierra Nevada Pea (*L. nevadensis*)—This species occurs on dry slopes below 7,000 feet from Fresno County northward. It flowers from April to June.

    Leaves are oblong to linear in shape; calyx lobes are unequal
        The banner is reflexed at about 90 degrees; stipules are toothed — Brush Pea (*L. brownii*)—This glabrous perennial grows on dry slopes from 4,000 to 6,000 feet in the Tehachapi Mountains northward. It flowers from April to June.

        Banner is reflexed at more than 90 degrees; stipules are entire — Nevada Pea (*L. lanszwertii*)—Look for this species on dry slopes between 4,000 and 6,500 feet from El Dorado County northward. It flowers from May to July.

Flowers ten to twenty in number
    Stems are winged; herbage is slightly pubescent — Jepson's Pea (*L. jepsonii*)—This species occurs along watercourses and on sandy slopes throughout the Sierra Nevada. It flowers from May to June.

    Stems are not winged, may be angled; herbage is not hairy — Snub Pea (*L. sulphureus*)—This species grows on dry slopes below 8,000 feet throughout the Sierra Nevada. It flowers from April to July.

Interesting Facts: Some species of *Lathyrus* have a history of poisoning humans. Kirk (1975) indicates that an exclusive diet of some species from ten to thirty days can bring on partial or total paralysis, and Willard (1992) suggests avoiding these plants entirely. Strike (1994) says that the greens and raw seeds of *Lathyrus* in California were eaten by Native Americans. Some of the seeds were parched and made into pinole and were stored for winter use. Weedon (1996) says that the fruits of many species are edible in small amounts but may cause paralysis and several secondary disorders if eaten in large quantities over time.

Caution: It is best to assume that these plants are *poisonous* if ingested.

Wild Licorice (*Glycyrrhiza lepidota*)

Description: This is a perennial herb growing up to three feet tall. The flowers are greenish white in dense racemes. Mature fruits are a conspicuous pod up to one-half inch long and densely covered with hooked spines. It is usually found in moist, sandy soils, along river banks at the lower elevations.

Interesting Facts: The plant contains glycyrrhizin, sugar, and other chemicals used in medicine as a mild laxative, a demulcent, and a flavoring to mask the taste of other drugs. It is also used in confections, root beer, and chewing tobacco. Licorice root has been used in the treatment of asthma, stomach ulcers, bronchitis, and urinary tract disorders. The plants were chewed by Natives and used as a flavoring.

Currently, researchers using highly refined licorice extract suggest that chemicals in glycyrrhizin called triterpenoids may be effective against cancer. They

GLYCYRRHIZIN
Licorice root contains glycyrrhizin, the source of most of the pharmacological effects of licorice and rhizome. Glycyrrhizin is about fifty times sweeter than sugar and has a cortisone-like effect that may result in minor "poisoning" if consumed in very large amounts. Glycyrrhizin increases extracellular fluid and plasma volume and induces sodium retention and loss of potassium, which often leads to edema or water retention. Too much licorice can cause cardiac depression and edema.

block the production of prostaglandin (a hormonelike fatty acid that may be responsible for stimulating the growth of cancer cells) and help get rid of cancer-causing invaders. Triterpenoids have been shown in test tubes to stunt the growth of rapidly multiplying cells, like cancer cells, and they may even help precancerous cells return to normal (Miller 1973).

Warning: Continual use of this plant in large doses may cause water retention and elevated blood pressure.

PEONY FAMILY (Peoniaceae)
The family has only one genus with approximately thirty-three species.

Wild Peony (*Paeonia brownii*)
Description: This spring annual has palmately divided leaves. The flowers are large and showy, and the petals are red-brown. The flowers are often subtended by reduced leaves. This plant occurs in woodsy, shaded areas in the lower elevations. The genus name is from the Greek for Paeon, the physician of the gods who supposedly used the plant medicinally.

Interesting Facts: The leaves are edible when cooked as greens. We have found it best to boil the leaves in several changes of water until the bitterness was removed.

Another method utilized by Natives was to pick the young leaves before the blossoms appeared in the spring, boil them, and then place them in a cloth sack and weigh the sack down in the river with a stone. By allowing the water to run through the sack overnight, the bitterness was removed. Medicinally, an infusion of sliced, oven-baked roots was taken for indigestion by the Diegueno. The genus is also important in Chinese medicine.

PHLOX FAMILY (Polemoniaceae)
There are about 18 genera and 320 species in this family, found chiefly in North America and particularly in the western United States. The family is a source of a few ornamentals.

Collomia (*Collomia*)
Description: These are annual or perennial herbs with simple or branched stems. There are approximately fifteen species in western North America and South America. The flowers are funnel shaped or tubular with throats that abruptly flare into an expanded limb. The genus name is from the Greek *kolla*, meaning glue, because of the mucilaginous layer on the seeds of most species.

1. Large-flowered Collomia (*C. grandiflora*)—This erect annual has salmon or almost-white-colored flowers and grows in dry, open, and wooded areas below 8,000 feet. It flowers from April to July.
2. Talus Collomia (*C. larsenii*)—This perennial plant occurs in loose volcanic soils at about 10,400 feet.
3. Narrow-leaved Collomia (*C. linearis*)—This is an erect annual that has pink- to purplish-colored flowers and grows in dry places from 3,000 to 10,500 feet. It flowers from May to August.
4. Flaming Trumpet (*C. rawsoniana*)—Look for this perennial plant along streams between 3,700 and 7,000 feet.
5. Yellow-staining Collomia (*C. tinctoria*)—This annual plant is found in dry, open, disturbed places below 9,000 feet from Mariposa County north. The crushed leaves or stems of yellow-staining collomia will stain one's hands a yellowish brown.

Interesting Facts: From the roots of large-flowered collomia, an infusion was made for high fevers. Additionally, an infusion of the leaves and stalks was taken for constipation and to "clean out the system."

Narrow-leaved collomia was used as a dermatological aid by the Gosiute. They made a poultice of the mashed plant and applied it to wounds and bruises (Chamberlain 1911).

The seed coat becomes mucilaginous when wet. This "glue" helps keep the germinated seeds from drying out. This is a mechanism that helps store water between the first autumn rains and those that may not come for several weeks.

Few-flowered Eriastrum (*Eriastrum sparsiflorum*)

Description: These are annual or perennial herbs with about sixty-four species occurring in the western United States. These are late-blooming plants that generally occur in dry areas.

Interesting Facts: The genus name is from the Greek *erion*, meaning wool, and *astrum*, for star plants with starlike flowers. A decoction of the stalks was taken as an emetic for stomach problems.

Gilia (*Gilia*)

Description: There are many species of *Gilia*, and they are characterized as annual, biennial, or perennial plants. The leaves are mostly alternate, lobed, or dissected with the tips acute. The seeds are sticky when wet. They are found in a variety of habitats at various elevations.

1. Canescent Gilia (*G. cana*)—This species grows on dry, gravelly areas between 6,000 and 10,000 feet.
2. Smooth-leaved Gilia (*G. capillaris*)—Look for this species on sandy slopes and flats between 2,500 and 10,500 feet.
3. Blue Field Gilia (*G. capitata*)—This tall, slender annual is fairly common in open, sandy, or gravelly and well-drained slopes and flats below 6,000 feet. It flowers from April to May.
4. Bridge's Gilia (*G. leptalea*)—This species is found in open woods between 4,500 and 9,000 feet.
5. Broad-flowered Gilia (*G. leptantha*)—This erect annual grows in sandy and gravelly spots from 5,000 to 7,700 feet. It flowers from June to August.

Interesting Facts: Strike (1994) indicates that *Gilia* seeds were eaten by many California Natives.

Granite Gilia (*Leptodactylon pungens*)

Description: This low-branching shrub is glandular villous throughout. There are many leaves, and they are palmately cleft into five to seven lobes. Its flowers tend to open in the evenings. Granite gilia occurs in dry, rocky, and sandy habitats between 5,000 and 12,000 feet.

Interesting Facts: This species was used as a decoction to bathe swellings, sore eyes, and scorpion stings.

Jacob's Ladder (*Polemonium*)

Description: These are perennials from woody rhizomes. The flowers occur in terminal or axillary cymes. The genus name is thought to honor Polemon, an early Greek philosopher. The common name of Jacob's ladder is based on the arrangement of the leaflets, and comes from the story in Genesis 28:12 of Jacob's dream of a ladder connecting heaven to Earth.

1. Low Polemonium (*P. californicum*)—Look for this species in moist, shaded places between 6,000 and 10,000 feet from Fresno County north. It flowers from June to August.

2. Sky Pilot (*P. eximium*)—This species is found on dry, rocky ridges and slopes between 10,000 and 14,000 feet from Tuolumne to Tulare and Inyo Counties. It flowers from July to August.

3. Western Polemonium (*P. occidentale*)—This species occurs in moist, shaded places between 6,000 and 10,000 feet from Fresno County north. It flowers from June to August.

4. Showy Polemonium (*P. pulcherrimum*)—Look for this species on dry, rocky, often volcanic slopes between 8,000 and 11,000 feet from Mono and Mariposa Counties north. It flowers from June to August.

Interesting Facts: Sky pilot grows at the highest elevations of any plant in the Sierra Nevada. The common name is a nickname for a person who leads others to heaven, because the plant only grows on lofty peaks. As such, there is an old custom by mountaineers of picking a sprig of sky pilot when one first climbs a peak. However, the continued practice of this tradition should be discontinued as it is illegal in some localities, and it could severely impact the species. Medicinally, a decoction of showy polemonium was used as a wash for the head and hair.

Linanthus (*Linanthus*)

Description: In general, they are low annuals with opposite leaves that are palmately parted into slender segments or reduced to linear blades. They are found on dry, open slopes.

1. Whisker Brush (*L. ciliatus*)—This is a stiff, erect annual that grows in dry, open places, mostly below 8,000 feet. It flowers from April to July.

2. Harkness' Linanthus (*L. harknessii*)—This plant is found in open, sandy, and gravelly places between 3,000 and 10,400 feet from Fresno County north.

3. Mustang-clover (*L. montanus*)—Mustang-clover occurs in dry, gravelly places below 6,700 feet from Nevada County southward.

4. Tehachapi Linanthus (*L. nudatus*)—Look for this species on open slopes between 2,000 and 7,000 feet.

5. Tulare Linanthus (*L. oblanceolatus*)—This species occurs near the edges of meadows between 7,500 and 11,000 feet.

6. Bush Gilia (*L. pachyphyllus*)—Look for bush gilia in dry, woody places between 4,000 and 12,000 feet.

Interesting Facts: An infusion was made from *L. ciliatus* by the Maidu and Pomo Indians to treat children's coughs and colds (Barrett 1952). An unheated decoction was drunk instead of water to purify blood.

Navarretia (*Navarretia*)

Description: These are low annual plants with alternate, pinnatifid, pungent-smelling leaves. The corolla is salverform in shape, and the stamens are equally inserted on the corolla. The stigma is two or three lobed. Seeds are usually mucilaginous when wetted. The genus name honors the Spanish physician F. Navarrete.

1. Brewer's Navarretia (*N. breweri*)—This species occurs in damp to dry flats and valleys between 4,000 and 11,000 feet throughout the Sierra Nevada. It flowers from June to August.

2. Mountain Navarretia (*N. divaricata*)—Look for mountain navarretia in dry, open places and at the edges of meadows throughout the Sierra Nevada. It flowers from June to August.

3. Needle-leaved Navarretia (*N. intertexta*)—This species is found in moist places below 7,000 feet on the western side of the Sierra Nevada. It flowers from May to July.

4. Navarretia (*N. leucocephala*)—This species has about five subspecies in California. In general, it occurs in vernal pools below 6,000 feet in the northern Sierra Nevada.

5. Least Navarretia (*N. minima*)—This species may be found in moist to dry places below 7,000 feet from Placer County northward. It flowers from June to August.

6. Bur Navarretia (*N. prolifera*)—This species occurs on dry slopes on the west side of the Sierra Nevada in Tulare, El Dorado, Amador, and Lassen Counties. It flowers from May to June.

7. Skunkweed (*N. squarrosa*)—This species occurs in open, wet, gravelly flats and slopes below 2,000 feet in the northern Sierra Nevada.

Interesting Facts: The seeds of skunkweed were gathered and dried in the sun and then stored. When needed, the seeds were parched and then eaten dry. Medicinally, skunkweed was used as a tonic, fever reducer, laxative, and dye. The seeds of *N. leucocephala* were also eaten, and a decoction of the plant was used to reduce swellings.

Phlox (*Phlox* spp.)

Phlox (*Phlox*)

Description: The plants in this genus are either low shrubs, perennials, or annuals with opposite leaves. The flowers are salverform in shape. The many species can be found in various habitats at all elevations.

1. Cushion Phlox (*P. condensata*)—This phlox is found on dry slopes and benches from 6,000 to 11,000 feet on the eastern side of the Sierra Nevada, mainly in Mono and Inyo Counties.

2. Spreading Phlox (*P. diffusa*)—This low, matted perennial plant has stems three to nine feet long. Spreading phlox grows on dry slopes and flats from 3,000 to 11,000 feet.

3. Matted Phlox (*P. dispersa*)—Look for this plant in loose disintegrated granite from 11,000 to 12,500 feet in Tulare and Inyo Counties.

4. Slender Phlox (*P. gracilis*)—This is a common species in open habitats below 10,000 feet.

5. Clustered Phlox (*P. pulvinata*)—This plant is found in dry, stony places between 10,000 and 13,000 feet, mainly in Inyo and Mono Counties.

6. Western Showy Phlox (*P. speciosa*)—This species occurs on rocky hillsides and on wooded slopes below 7,000 feet.

Interesting Facts: Slender phlox was eaten by the Miwok in California as greens (Barrett and Gifford 1933). The Maidu also used slender phlox as a poultice on bruises and wounds. A decoction from the roots of a related species, *P. longifolia* (longleaf phlox), and other species were used by some Native Americans as an eyewash for sore eyes. The scraped roots were soaked in water or steeped or boiled to make the wash.

Scarlet Gilia
(*Ipomopsis aggregata*)

Scarlet Gilia (*Ipomopsis*)

Description: These are annual and perennial herbs with basal rosettes of pinnately cut leaves and tubular flowers. The genus name is from the Greek *ipo* (to impress) and *opsis* (appearance), referring to the showy flowers.

1. Scarlet Gilia (*I. aggregata*)—This species grows in open places like sandy flats between 3,500 and 10,300 feet.

2. Many-flowered Gilia (*I. congesta*)—This species grows between 7,000 and 12,000 feet

Interesting Facts: *I. aggregata* was used by Native Americans as a tea to treat colds, to make glue, and to treat blood troubles. In Nevada, the principal use of this plant was for the treatment of venereal diseases. The whole plant was boiled for the purpose, and a solution was taken as a tea or used as a wash. The whole plant was also boiled by the Ute Indians in Utah to make glue. A blue dye can be extracted from the roots (Gifford 1967).

PINK FAMILY (Caryophyllaceae)

There are approximately eighty genera and two thousand species in this family found in the north temperate zone. About twenty genera are native to the United States. In general, they are annual or perennial herbs with opposite, simple leaves. The stems are often swollen at the joints. The flowers are five-merous, and the calyx is tubular or has distinct sepals. Petals are often deeply notched, appearing like ten petals. Some species are cultivated as a source of ornamentals, and several genera are regarded as weedy.

Baby's-breath (*Gypsophila elegans*)

Description: This is a nonnative, perennial species with much-branched stems. The genus contains approximately 125 species of annual and perennial herbs with simple, linear, and opposite leaves. Flowers have five sepals, five petals, and ten stamens. This species is found in open forests at about 6,500 feet.

Interesting Facts: Several species are grown as ornamentals, and some are reported to have medicinal qualities. Baby's-breath is often used by commercial florists in dried flower arrangements.

Bouncing-bet, Soapwort (*Saponaria officinalis*)

Description: This is an erect perennial herb with sessile or nearly sessile leaves. The flowers are showy, usually pale pink. Soapwort can be found along roadsides and in disturbed areas and waste places at the lower elevations. The plant has escaped from cultivation. The genus name is Latin for soap, because the juice of the plant lathers with water. It flowers from June to September.

Interesting Facts: The plant contains saponins and will irritate the digestive tract if eaten. The crushed green plant and roots can be used as a soap substitute.

Catchfly, Campion (*Silene*)

Description: There are many species of *Silene* in the Sierra Nevada. They are annual, biennial, or perennial herbs with opposite leaves. The sepals are united and often inflated into a five-lobed tube. The petals are lobed at the tip and have appendages at the point where the broader upper portion (blade) joins the narrower lower segment (claw). The various species are found in a variety of habitats. Most species of *Silene* have many gland-tipped hairs that tend to entrap insects, hence the common name of catchfly.

Interesting Facts: The young shoots of a related species, *S. acaulis* (bladder campion), can be used as potherbs. The sap of *S. antirrhina* was used by Miwok Indians in California to paint designs on the faces of young girls (Barrett and Gifford 1933). The designs were cosmetic, not ritualistic.

Mouse-ear Chickweed (*Cerastium*)

Description: Species in this genus are annual or perennial herbs with entire-margined leaves opposite on the stem. The herbage is usually hairy and often

POLLINATION ECOLOGY 101
Successful reproduction is defined by an individual's passing on its genes to the next generation. The showy wild-flowers of grasslands and meadows are insect- or hummingbird pollinated. The array of shapes, colors, and fragrances of these flowers is for the sole purpose of attracting pollinators and maintaining species' isolation. For the service of pollination, the pollinator receives nectar, pollen, or oil. This relationship between plant and pollinator is the result of a long and intimate coevolution. The animal pollinators have distinct color, shape, and food-type preferences in the flowers they visit and have specific anatomical features that maximize their abilities to collect pollen or nectar. For example, beetles have brush-like mouthparts for collecting pollen; bees have hollows or pollen baskets on their hind legs for storing pollen; hummingbirds have long, narrow bills and tongues for gathering nectar; and butterflies have tubular, bristled proboscises for nectar drinking. Additionally, the patches, streaks, and spots on flowers are called nectar guides and often direct the pollinator to the nectar glands. These guides are situated so that the pollinator must brush against the sex organs of the flowers on its way in. Many of these guides are conspicuous, such as the lines on the interior of some penstemon flowers; others are produced by specialized plant tissues that strongly reflect ultraviolet light and are seen only by the pollinators.

sticky. The few to several (rarely solitary) flowers are borne in an open inflorescence. There are five petals and five sepals, and the petals are white and deeply lobed at the tip. The genus name is from the Greek *keras,* meaning horn, referring to the tapered capsule, which in some species is bent slightly like a cow's horn.

Quick Key to the Mouse-ear Chickweeds

The inflorescence has bracts that are green, or only the upper portions are narrowly green; plants are usually found above 8,500 feet	Alpine Cerastium (*C. beeringianum*)—This species is occasionally found near snowbanks between 8,500 and 12,200 feet in Tuolumne, Mono, and Alpine Counties. It flowers from July to August.
The inflorescence has bracts with broad, membranous margins; plants are found between 5,000 and 8,000 feet	
The petals are about as long as the sepals	Common Mouse-ear Chickweed (*C. vulgatum*)—Look for this species in meadows throughout the Sierra Nevada. It flowers from May to August.
The petals are about two to three times as long as the sepals	Meadow Cerastium (*C. arvense*)—This species occurs in moist, rocky, or grassy banks. It flowers from May to August.

Interesting Facts: *Cerastium* is frequently confused with *Stellaria media* (chickweed), but to the general forager there is no danger. The tender leaves and stems of most *Cerastium* can be added to a salad, but we found they are better if boiled first and served as greens.

Pink (*Dianthus*)

Description: This is an introduced genus with about three hundred species of annual and perennial herbs with opposite leaves and fragrant flowers. Both of the following species are found at the lower elevations.

1. Grass Pink (*D. armeria*)—This species occurs in disturbed habitats usually below 4,500 feet.
2. Meadow Pink (*D. deltoides*)—This pink is found in wet, disturbed sites below 8,000 feet and can be found almost anywhere throughout the Sierra Nevada.

Interesting Facts: Most of the cultivated forms are doubles and bear many petals. Several species have given rise to some of the most important garden and cultivated flowers. In fact, there are more than thirty thousand cultivars recorded in the International Dianthus Register and supplements. The name *carnation* is applied to cultivars, whereas "pink" is loosely applied to any member of this genus.

Sandwort (*Arenaria*)

Description: The species in this genus are generally annual or perennial herbs with opposite leaves. The white flowers are borne in open to congested, flat-topped inflorescences. The species in the Sierra Nevada are found in various habitats from the foothills to above timberline in dry, rocky, and open areas, as well as moist, open forests. The genus name is from the Latin *arena,* referring to sand, the habitat of many species.

This and the related genus *Minuartia* are recognized by their linear leaves, undivided petals, and often matted growth habits.

Interesting Facts: The sandworts are known for their medicinal uses by several

Sandwort (*Arenaria* spp.)

SAPONINS AND SOAP
Many plants contain saponin, which, once extracted, can be used as soap. Saponins can be found in soapwort (*Saponaria* spp.), clematis (*Clematis* spp.), snowberry (*Symphoricarpus* spp.), elderberry (*Sambucus* spp.), and other species. Cuts, wounds, and rashes need to be cleansed of bacteria to prevent infections; when hunting, by eliminating the human scent we can avoid alerting game animals of our presence. Food and cooking gear also need to be kept clean for health reasons.

On the other hand, soaps can be detrimental to some organisms living at the surface of water in ponds or streams. In water, soaps break down the surface tension, making life more difficult for organisms who are dependent on it.

Native American tribes. The roots of *A. aculeata* (prickly sandwort) were used as a decoction by the Shoshone as an eyewash, whereas a poultice of steeped leaves of *A. congesta* (capitate sandwort) was applied to swellings (Strike 1994).

Sandwort (*Minuartia*)

Description: These are low, mat-forming perennials with glandular hairs. The flowers occur in open, few-flowered inflorescences. The genus name honors J. Minuart, an eighteenth-century Spanish botanist and pharmacist.

1.   Douglas' Sandwort (*M. douglasii* = *Arenaria douglasii*)—This delicate annual occurs in dry areas below 7,000 feet. It flowers from April to June.
2.   Nuttall's Sandwort (*M. nuttalli* = *Arenaria nuttallii*)—This species occurs in dry granitic areas between 6,500 and 11,000. It flowers from July to August.
3.   Alpine Sandwort (*M. obtusiloba*)—This species occurs in dry, rocky places between 10,500 and 12,500 feet in Fresno and Inyo Counties. It flowers from July to August.
4.   Dwarf Sandwort (*M. pusilla*)—This species is occasionally encountered in dry woods and on open slopes below 6,800 feet from Tuolumne County northward. It flowers from May to July.
5.   Red Sandwort (*M. rubella* = *Arenaria rubella*)—This species occurs in dry, rocky places above 11,000 feet in the southern Sierra Nevada. It flowers from July to August.
6.   Sandwort (*M. stricta*)—This species occurs on high plateaus between 12,000 and 13,200 feet from Mono County southward. It flowers in August.

Interesting Facts: Some species were used medicinally by some Native American tribes.

Starwort, Chickweeds (*Stellaria*)

Description: These are mostly low annual or perennial herbs with flowers in an open inflorescence in the leaf axils or at the ends of stems. The five sepals are separate to the base. Petals are white and deeply lobed or lacking.

Quick Key to the Chickweeds

Plants annual	Common Chickweed (*S. media*)—This species may occur in gardens and lawns in the resort areas. It blooms from February to September.
Plants perennial Petals equal to or longer than sepals	Starwort (*S. longipes*)—This tufted perennial grows up to ten inches tall. The leaves are opposite, lanceolate, and rigid. The flowers are terminal, solitary, or in few-flowered clusters. Starwort is fairly common in moist places from 4,500 to 10,500 feet. It flowers from May to August.

Petals much shorter than the sepals or none
    Margins of leaves fringed with hairs
        Mature sepals one-eighth inch or less long; fruiting pedicels ascending
	Chickweed (*S. calycantha*)—This prostrate to erect perennial occurs on mossy banks, bogs, dry creeks, wet meadows, and shaded areas from 5,000 to 11,000 feet.

Mature sepals greater than one-eighth inch long; fruiting pedicels are reflexed	Northern Starwort (*S. borealis*)—This species occurs in wet places below 6,000 feet from Mariposa County northward. It flowers from June to July.

Margins of leaves smooth		
Sepals usually four; leaf margins flat	Obtuse Stellaria (*S. obtusa*)—This is an uncommon species found in moist habitats below 7,000 feet in Mariposa County northward.	
Sepals five		
Leaf margins warped or curved; inflorescence with leafy bracts	Starwort (*S. crispa*)—This species occurs in moist places and meadows between 7,000 and 9,200 feet. It flowers from May to August.	
Leaf margins plane; inflorescence with small scarious bracts	Umbellate Chickweed (*S. umbellata*)—This species is infrequently encountered in damp, shaded places between 6,000 and 11,500 feet throughout the Sierra Nevada. It flowers from July to August.	

Interesting Facts: Although the uses of other starworts are unknown, the young shoots of *Stellaria media* have been used as salad herbs or potherbs if cooked like spinach. Although it is edible raw, we prefer to boil it for a few minutes before eating. Because the plants are usually quite small and only the youngest parts are good, chickweed can be tedious to collect. The greens are low in calories and packed with copper, iron, phosphorus, calcium, potassium, and vitamin C—valued in the prevention and treatment of scurvy.

Medicinally, *S. media* can be used as a tonic, in large quantities a laxative, and diuretic. For itchy skin, make a strong tea and wash the area. A poultice of the plant has been used to treat skin sores, ulcers, and infections as well as eye infections and hemorrhoids.

Sticky Starwort (*Pseudostellaria jamesiana*)

Description: This is a weak-stemmed, glandular perennial with lanceolate leaves and many few-flowered cymes of white flowers that have slightly two-lobed petals. This species is found about meadows and damp places between 4,000 and 8,500 feet throughout the Sierra Nevada. It flowers from May to July.

Interesting Facts: The tuberlike swellings can be eaten raw or dried in the sun. They have a thin, light-brown rind and a tender rather mealy texture inside, similar to a potato.

PITCHER-PLANT FAMILY (Sarraceniaceae)

Members of this family are perennial herbs with basal leaves that are tubular in form. The family is restricted to North and South America, chiefly the eastern United States.

California Pitcher-plant (*Darlingtonia californica*)

Description: This is an uncommon insectivorous perennial plant that grows up to three feet tall. The flowers are nodding with five yellow-green sepals that surround five purple petals. The leaf has two purple-green appendages. The plant is endemic to serpentine soils, and the plants trap insects to compensate for nutrient deficiencies in these soils. Pitcher plant grows below 6,000 feet in Plumas and Nevada Counties.

Interesting Facts: John Muir, in his book *Steep Trails* (1918), writes about pitcher plants as follows:

This is one of the few places in California where the charming linnaea is found, though it is common to the northward through Oregon and Washington. Here, too, you may find the curious but unlovable darlingtonia, a carnivorous plant that devours bumblebees, grasshoppers, ants, moths, and other insects, with insatiable appetite. In approaching it, its suspicious looking yellow spotted hood and watchful attitude will be likely to make you go cautiously through the bog where it stands, as if you were approaching a dangerous snake.

The leaf is tubular and has evolved to capture flying insects; the hood has translucent areas that suggest an escape route, but inside the leaf there are slippery surfaces and stiff, reflexed hairs that make escape impossible. Moisture at the base of the leaves contains enzymes that digest the insects.

PLANTAIN FAMILY (Plantanaceae)

Three genera and 270 species of this family are found worldwide. *Plantago* is widespread in the United States. In general, the family is of little economic importance, but several species of *Plantago* are weeds, and one (*P. psyllium*) is the source of seeds used to make a commercial laxative.

Plantain (*Plantago*)

Description: These are characterized as short-stemmed annual or perennial herbs with basal leaves. The flowers are greenish or purplish. Many of the species are introduced from Europe and can be found at the lower elevations, particularly in fields and waste places. *Plantago* means sole of foot and refers to the sole-shaped leaves of plantain that lie close to the ground as though stepped on.

Quick Key to the Plantains

Leaves broadly ovate; three to six inches long	Common Plantain (*P. major*)—This is a perennial plant with basal leaves. The leaves are ovate and are longitudinally ribbed with conspicuous nerves that merge at the base and apex. It is a common plant of damp places. It blooms from April to September.
Leaves narrower Leaves lanceolate	English Plantain (*P. lanceolata*)—This perennial plant with basal leaves is a common weed in moist places. It blooms from April to August.
Leaves linear; herbage silky pubescent	California Plantain (*P. erecta*)—This low, villous annual is a common plant of dry slopes. It flowers from March to May.

Interesting Facts: Because of their reputation as being weeds, plantains are a forgotten edible to many people. In fact, a lot of effort is spent trying to get rid of the plants from gardens. *P. major* and *P. lanceolata* were brought over by European settlers for use as potherbs and medicine. The Native Americans called the plants "white man's foot" because they followed the settlers west. The native species of plantain are uncommon in comparison, and it is suggested that they be used only when large populations are found.

As a food, the young leaves of common plantain and narrowleaf plantain were used fresh or cooked. They contain calcium and other minerals. One hundred grams of plantain is said to furnish as much vitamin A as a large carrot. The older leaves may be too fibrous and bitter for use, but they are usable if one is able to remove the fibers. Seeds are tedious to collect in quantity but can be ground and used as a flour substitute or extender (Doebley 1984).

The leaves and seeds of many species were used medicinally. The foliage contains tannins and iridoid glycosides, notably aucubin, which stimulates uric acid secretion from the kidneys. The crushed leaves of common plantain provide an astringent juice that can be used to soothe wounds, sores, insect bites, and the rash of poison oak. The plantain juice is a traditional treatment for earaches. The seeds contain up to 30 percent mucilage, which swells in the gut to act as a bulk laxative and soothes irritated membranes (Vestal 1952). Rubbing the leaves on one's skin works as a natural, moderately effective insect repellant.

POPPY FAMILY (Papaveraceae)

Most plants in this family are annual or perennial herbs and sometimes shrubs. The sap is often milky or colored. There are twenty-six genera and two hundred species in this family distributed in the subtropical and temperate areas of the Northern Hemisphere, particularly in western North America. Thirteen genera are native to the United States. The family is of little economic importance except for *Papaver somniferum,* which yields opium and its many derivatives, including morphine and heroin. A few species are cultivated as ornamentals.

Bleeding Heart (*Dicentra*)

Description: These are perennials with basal and sometimes stem leaves. The flowers occur in panicles or racemes and are heart shaped. There are four petals, and the outer two are spurred. The genus name is from the Greek *dis* (two) and *kentron* (spurred), referring to the shape of the flowers.

1. Golden Ear-drops (*D. chrysantha*)—This species is common on burns and in disturbed places below 5,000 feet from Calaveras County southward.
2. Bleeding Heart (*D. formosa*)—Look for this species in damp and shaded places below 7,000 feet.
3. Few-flowered Bleeding Heart (*D. pauciflora*)—This species grows in gravelly habitats between 6,000 and 10,000 feet in Tulare and Tuolumne Counties.
4. Steer's Head (*D. uniflora*)—This species occurs in gravelly and rocky places between 5,400 to 12,000 feet from Fresno County north.

Interesting Facts: The plants are considered to be *poisonous* and contain several different alkaloids. These alkaloids are found throughout the plant and can cause trembling, staggering, convulsions, and labored breathing. Large quantities can be fatal. A poultice from *D. cucullaria* (Dutchman's breeches) was apparently made to treat skin diseases.

Bush Poppy (*Dendromecon rigida*)

Description: This is a shrub with entire, lanceolate leaves. The yellow flowers are showy, and the sepals fall off early. This is a common species on chaparral slopes below 6,000 feet. It flowers from April to June.

Interesting Facts: Apparently, the Kawaiisu used the seeds for food. Unfortunately, we are not aware of how they prepared them. Otherwise, when preparing tobacco (*Nicotiana*), the Kawaiisu added a leaf or two of bush poppy to enhance the strength of the tobacco (Zigmond 1981).

California Poppy (*Eschscholzia californica*)

Description: This glaucous herb has leaves that are several times dissected into linear segments. The flowers are bright yellow to orange and showy. This is a common plant on grassy slopes at the lower elevations. Some plants have escaped gardens and can be found growing wild. It flowers from March to June.

Interesting Facts: The flowers and leaves were eaten by some Native Americans. The foliage was eaten by gathering the plants before the plants bloomed, leaching them in running water, and then cooking them.

Sap from the fresh root is mildly narcotic. It was apparently used by the Cahuilla Indians in California as a sedative for babies. Additionally, a piece of the root was placed in a tooth cavity to stop a toothache. A root extract was used as a wash or liniment for headaches and open sores. If taken internally, the root extract caused vomiting (Barrows 1967).

Warning: All parts of the plant should be considered toxic.

Cream Cups (*Platystemon californicus*)

Description: This low annual with long, white, hairy stems has opposite leaves growing mostly on the lower part of the plant. The flowers are yellow. Cream cups grows at lower elevations in the mountains. It flowers from March to May.

Interesting Facts: The leaves can be eaten as greens.

Prickly Poppy (*Argemone corymbosa*)

Description: This prickly stemmed annual or perennial herb has yellow sap and stands twenty-four to sixty inches tall. The pale green leaves are lanceolate to ovate and deeply lobed and are prickly on the veins. The upper leaves clasp the stem. The flowers are large and showy, up to five inches across. The petals are white, and the fruit is a lanceolate capsule. Prickly poppy grows in dry areas below 6,000 feet. It flowers from July to August.

Interesting Facts: The seeds of *Argemone* have been used in the past as food, but they are so difficult to extract that it hardly seems worth it. Medicinally, the ripe seeds of *A. munita* were roasted, mashed, and applied as a salve on burns and abrasions by some Native Americans (Sweet 1976). The seeds were also pounded and used as a poultice on open sores and as a hemorrhoid remedy. The juice of *Argemone* has a rubifacient and somewhat caustic effect and was used for burning off warts. A tea made from the plant is an analgesic topically and can be applied to sunburns and abrasions to relieve pain and swelling.

Warning: The plants contain toxic alkaloids.

Prickly Poppy
(*Argemone corymbosa*)

Sierra Corydalis (*Corydalis caseana*)

Description: This is an herb with dissected leaves. The flowers are yellow or white to pinkish, with two sepals. One of the outer petals is spurred at the base. The genus name is from the Greek *korudallis,* the ancient name of the crested lark. It can be found in moist, open areas and along streams.

Interesting Facts: The plant is considered to be poisonous and contains several different alkaloids. Native Americans apparently used a tea made from a related species, *C. aurea,* for painful backaches, diarrhea, menstruation, bronchitis, sore throats, and stomachaches and inhaled the fumes of burning roots for headaches (Foster and Duke 1990).

Primrose Family (Primulaceae)

Worldwide, there are approximately twenty-eight genera and eight hundred species found in this family. Eleven of the genera are native to the United States, mostly to the eastern part of the country. They are of minor economic importance as a source of ornamentals.

Androsace (*Androsace*)

Description: The genus name is from the Greek *andros* (make) and *sakus* (a buckle), alluding to the shape of the anther. The genus contains about one hundred species of alpine, annual, and perennial herbs with simple, often tuft-forming leaves and clusters of showy flowers. The petals are fused at the base. The genus differs from *Primula* in that the petal tube is shorter than the sepals and somewhat constricted at the mouth.

1. Western Androsace (*A. occidentalis*)—This species occurs at about 5,500 feet at Emigrant Gap in Alpine County.
2. Northern Androsace (*A. septentrionalis*)—This species occurs in dry, rocky places between 10,000 and 13,000 feet; scattered throughout the Sierra Nevada.

Interesting Facts: A compound decoction was made from the whole plant of western androsace and used for postpartum bleeding. A cold infusion of northern androsace was taken for internal pain. The plant was also used as a lotion to give protection from witches.

Scarlet Pimpernel (*Anagallis arvensis*)

Description: This is a small annual with opposite leaves that clasp the stem and are oval shaped. The small scarlet-colored flowers are on solitary stalks. This is an introduced plant from Europe, and in some places of the United States it is quite common and weedy. Look for it in waste places at the lower elevations. The genus name is from the Greek *anagelao*, which means to laugh. Dioscorides, physician to the Roman army, was said to give the plant to the men to relieve the depression that accompanies disorders of the liver. Another common name applied to this plant is "poor man's weather glass" because of its habit of closing before a rain.

Interesting Facts: Scarlet pimpernel is said to be an effective diuretic that helps eliminate gravel from the kidney and is used in dyspepsia. As a poultice, it was applied to the skin to relieve the itch and sting of insects (Callegari and Durand 1977).

Shooting Star (*Dodecatheon*)

Description: All leaves are basal and form a loose rosette. The flowers are located at the end of a stalk with narrow, reflexed, rose-colored petals. The species habitats range from grassland to shrubland, meadows, and riparian habitats up to the alpine zone.

Shooting Star
(*Dodecatheon* spp.)

Petals usually four; anthers four
   Herbage glandular-pubescent; leaves
   oblanceolate in shape

Jeffrey's Shooting Star (*D. jeffreyi*)—This species occurs in wet places between 2,300 and 10,000 feet.

   Herbage not hairy; leaves linear in shape

Alpine Shooting Star (*D. alpinum*)—This glabrous perennial with leafless, flowering stems grows from 4,000 to 11,000 feet. It flowers from May to August.

Petals five; anthers five
   Plants heavily glandular-pubescent

Mountaineer Shooting Star (*D. redolens*)— This glandular, pubescent perennial has leafless stems and grows in moist places from 8,000 to 11,500 feet. It flowers from July and August.

Plants not as above
   Herbage and roots reddish in color; plants found above 7,000 feet

Sierra Shooting Star (*D. subalpinum*)—Look for this species in moist shaded places from Tulare to Tuolumne Counties. It flowers from May to June.

   Herbage green; plants found below 7,000 feet

Henderson's Shooting Star (*D. hendersonii*)— This species is found in shaded places throughout the Sierra Nevada. It flowers from March to May.

Interesting Facts: Because none of the species is listed anywhere as poisonous, it is likely that all the species are edible. It is usually the texture that discourages people from using the plants. We have found that the leaves of many species have a good flavor when eaten raw. Weedon (1996) and Strike (1994) also indicate that the roots and leaves of *D. hendersonii* are edible after roasting or boiling. Scully (1970) believes that at least five species of shooting star in the Rocky Mountains were used by American Indians and that they ate the green leaves and roasted the roots. Thompson and Thompson provide some additional insight into their preparation of *D. jeffreyi:* "[W]e tried eating the leaves of Shootingstar raw, but decided that their texture made them unappealing to chew on. At least they do not seem to become bitter, even after the flowers are blooming. When boiling for about 15 minutes and seasoned with butter and salt, they make a satisfactory but bland green vegetable" (1972, 120).

Very little information regarding the medicinal uses of shooting stars could be found. However, the Native Americans of the Northwest used a leaf tea as a treatment for cold sores.

Sierra Primrose (*Primula suffrutescens*)

Description: The flowers of this plant are magenta in color with a yellow throat. It is found growing under overhanging rocks and about cliffs between 8,000 and 13,500 feet throughout the Sierra Nevada.

Interesting Facts: This plant was first collected in Yosemite in 1872 by the botanist Asa Gray. His guides were naturalist John Muir and mountaineer Galen Clark. The leaves of related species were edible raw or when used as potherbs.

Star-flower (*Trientalis latifolia*)

Description: This is a low, perennial herb with tubers and leaves crowded near the apex. The flowers are white or pinkish. The genus name is Latin, meaning having one-third of a foot, referring to the plant's height.

Interesting Facts: The sap was mixed with water and used as an eyewash.

Tufted Loosestrife (*Lysimachia thyrsiflora*)

Description: This erect perennial has opposite leaves that are sessile and gland dotted. The yellow flowers occur in spikelike racemes and are deeply five-parted into linear segments. The plant is occasionally found in wet places between 3,500 and 5,000 feet in Plumas and Shasta Counties. It flowers from June to August.

Interesting Facts: The genus name is from the Greek *lysis* (releasing) and *mache* (strife). King Lysimachos of Thrace (ca. 360–281 BC) was said to have pacified a bull with loosestrife. Some species are grown as ornamentals; others have local medicinal uses. For example, the Iroquois made a compound decoction of the plant and used it as a wash and applied it as a poultice to stop milk flow (Herrick 1977).

PURSLANE FAMILY (Portulacaceae)

Nineteen genera and six hundred species occur worldwide, of which nine genera are native to the United States. The family is particularly well represented along the Pacific Coast. It is of little economic importance but includes several ornamentals.

Bitterroot, Lewisia (*Lewisia*)

Description: Lewisias are indigenous to the western part of the United States. They can be found clinging precariously to rocky ledges among boulders, on rock-strewn slopes, damp gravelly places, alpine meadows, and in near-desert conditions where rainfall is seasonal and unpredictable. There are about eighteen species, many evergreen, but other are bulblike in that they are belowground for part of the year. Several species have large, showy flowers.

1. Columbia Lewisia (*L. congdonii*)—This species can be found in rocky places.
2. Yosemite Bitterroot (*L. disepala*)—This is found in rocky places between 6,500 and 8,500 feet in Tulare and Mariposa Counties. It flowers from May to June.
3. Kellogg's Lewisia (*L. kelloggii*)—Look for this species in sandy places on ridges between 4,500 and 7,700 feet from Mariposa to Plumas Counties.
4. Lee's Lewisia (*L. leana*)—This plant grows on cliffs and rocks between 9,000 and 10,000 feet in Fresno County.
5. Nevada Lewisia (*L. nevadensis*)—This perennial with a fleshy root and several stems grows on wet banks and in moist meadows from 4,500 to 12,000 feet. It flowers from May to July.
6. Alpine Lewisia (*L. pygmaea*)—This species occurs in damp gravel between 5,000 and 8,000 feet. It flowers from July to September.
7. Bitterroot (*L. rediviva*)—This perennial with a fleshy taproot grows from 6,000 to 9,000 feet. It flowers in May and June.
8. Three-leaved Lewisia (*L. triphylla*)—Look for this species in damp, gravelly places between 5,000 and 11,200 feet throughout the Sierra Nevada.

Interesting Facts: Although all species may be edible, *L. rediviva* is the species that has been used extensively. These plants were an important food item for many Natives. The root is remarkably large and thick for a small plant and contains nutritious farinaceous matter that is much prized. The roots are dug up in spring before flowering. Once dug, the root is peeled promptly, and the small red "heart" (embryo of next year's growth) is removed to reduce the root's bitter flavor. It is then steamed, boiled, or pit-cooked and eaten. The root can also be dried and will keep for a long time. The bitterness of the root varies, and cooking is said to improve the flavor. The root boiled to a jellylike consistency will be pink in color. The pounded root was chewed for a sore throat (Hart 1996).

Note: Though some still collect it today, bitterroot is considered a rare plant in many areas. There is little evidence, however, that harvesting by Native Americans has contributed to its rare status. Overgrazing and trampling by range livestock and habitat destruction from agricultural encroachment seem to have been a major impact on lewisia populations. Remember, digging the roots destroys the plant. Programs to maintain and enhance habitat for the plant are recommended.

Calyptridium, Pussy-paws (*Calyptridium*)

Description: The leaves of these plants are alternate or basal and spatulate in shape. The small flowers have sepals that are membranous or membranous-margined, and there are two or four petals.

1. Calyptridium (*C. monospermum*)—This annual species is found in sandy, open flats below 10,000 feet.
2. Parry's Calyptridium (*C. parryi*)—This annual from a taproot is found in open areas below 11,000 feet.
3. Dwarf Calyptridium (*C. pygmaeum*)—This is a rare species that occurs in sandy or gravelly places between 7,500 and 11,500 feet in Inyo County.
4. Rosy Calyptridium (*C. roseum*)—Look for this species in moist, often alkaline places from 5,000 to 10,000 feet in Lassen, Mono, Sierra, and Inyo Counties. It flowers from June to August.
5. Pussy-paws (*C. umbellatum*)—This species grows in loose sandy or gravelly places between 2,500 and 12,000 feet.

Interesting Facts: The stems of pussy-paws thermoregulate; that is, respond to temperature. When it is cool (evening and morning), the stems lay flat on the ground, protected from the chilling winds. As the day warms, the flower stems rise, absorbing the midday warmth but moving away from the scorching earth. The tiny black seeds are an important food for rodents.

Montia (*Montia*)

Description: The genus is composed of slightly succulent annual and perennial herbs. The flowers have two persistent sepals and five white or pinkish petals. Most *Montia* species grow in moist or seasonally wet areas that are partially to fully shaded.

1. Toad-lily (*M. chamissoi*)—This perennial with creeping or floating stems and erect branches grows in wet places, as in meadows from 4,000 to 11,000 feet. It flowers from June to August.
2. Water Chickweed (*M. fontana*)—This species is common in streams and pools below 10,000 feet throughout the Sierra Nevada.
3. Linear-leaved Montia (*M. linearis*)—Look for this plant in moist habitats below 7,500 feet.
4. Small-leaved Montia (*M. parvifolia*)—This species occurs in moist, rocky places below 8,500 feet from Tuolumne County northward.

Interesting Facts: All species of *Montia* have stems and leaves that can be eaten raw or boiled like spinach. The roots are also edible raw or boiled.

Purslane (*Portulaca oleracea*)

Description: This is a small, succulent annual herb found at the lower elevations. It has been used as a food for more than two thousand years in India and Persia. In Europe, it is grown as a garden vegetable. The genus name may be derived from *portula,* meaning little gate, referring to the lid on the capsule.

Interesting Facts: The stems and leaves of purslane have a tart taste. The entire aboveground part of the plant can be boiled, steamed, fried, or pickled. The mucilaginous juice of the stems makes a good thickener for soups. Because the plant tends to hold a lot of dirt and grit, you may want to wash it thoroughly. Besides the good flavor, purslane also provides vitamins A and C, iron, and calcium. The tiny black seeds are also nutritious. They can be ground and mixed with other flours.

Red Maids (*Calandrinia ciliata*)

Description: *Calandrinia ciliata* is an annual with linear to oblanceolate leaves and red flowers. The plant is found in open, grassy areas at the lower elevations. The genus is named for J. L. Calandrini, born in Switzerland in 1703.

Interesting Facts: The seeds of *C. ciliata* were prized by a great many Native Americans (Kirk 1975). After gathering and winnowing, the seeds were parched with coals, pulverized, and then pressed into balls and cakes for eating. The roots were also eaten, and the young leaves and stems were eaten raw or cooked.

Spring Beauty (*Claytonia*)

Description: The fifteen to twenty-five species in this genus are succulent herbs with simple leaves and racemes of white flowers. The genus is named for Dr. John Clayton, an American botanist and notable plant collector of colonial days.

1. Truncate-leaved Claytonia (*C. cordifolia*)—This species grows in seeps and wet meadows between 4,000 and 8,000 feet.
2. Western Spring Beauty (*C. lanceolata*)—Look for this plant in moist woods and along streams between 4,500 and 8,800 feet from El Dorado County northward. It flowers from May to July.
3. Fell-fields Claytonia (*C. megarhiza*)—This is a rather rare plant that grows on talus and loose rock or gravel between 9,000 and 11,000 feet around Yosemite National Park.
4. Sierra Claytonia (*C. nevadensis*)—This plant is infrequently found in gravelly, wet places between 5,000 and 12,000 feet in the Lassen Peak area and from Alpine County south. It flowers from July to August.
5. Marsh Claytonia (*C. palustris*)—This rare species is found below 8,000 feet.
6. Miner's Lettuce (*C. perfoliata*)—Look for miner's lettuce in moist areas below 7,500 feet throughout the Sierra Nevada. It flowers from April to June.

Spring Beauty
(*Claytonia lanceolata*)

Interesting Facts: Often called Indian potato, wild potato, or mountain potato, the small corms can be eaten raw, boiled, or roasted. For many Native Americans, spring beauty was an important "root vegetable." When collecting, keep only the largest corms and replant the others. At first, many find the corms distasteful, as they do take a little getting used to. The corms are high in starch and, when cooked, taste like potatoes. Boil or bake the corm for thirty minutes. Most species are not plentiful, so be conservative in your endeavor. They can also be dried on strings for long-term storage.

The rosettes can also be eaten raw or cooked and are high in vitamins A and C. They are better when mixed with other salad plants. The leaves of *C. sibirica* (Siberian springbeauty) were soaked and applied to the head as a remedy for a headache (Schofield 1989).

In California, some Native Americans picked miner's lettuce and placed it near the nests of red ants. The ants were allowed to crawl over the leaves and were then shaken off. The residue left on the leaves by the ants had an acerbic flavor. A tea from the leaves was used as a laxative. A poultice made from the plant was used for rheumatic pains and to stimulate a poor appetite. Miner's lettuce is one of the few native plants of the United States that has been cultivated elsewhere. Introduced into Europe, miner's lettuce is now cultivated and used for salads and as a potherb.

ROCK ROSE FAMILY (Cistaceae)
This family is mostly tropical and subtropical. In California there are two species of *Helianthemum* (rock-rose). These plants are often cultivated.

Peak Rush-rose (*Helianthemum scoparium*)
Description: The genus name is from the Greek *helios* (sun) and *anthemon* (flower), the flowers tending to open only in bright sunshine. It occurs in dry, sandy, or rocky soil on hills and ridges below 5,000 feet in the lower and northern Sierra Nevada.

Interesting Facts: The Kumeyaay made a tea from boiled flowers and gave it to pregnant women who were experiencing problems during labor (Strike 1994).

ROSE FAMILY (Rosaceae)
The rose family consists of approximately one hundred genera and three thousand species worldwide, with the family being particularly common in Europe, Asia, and North America. About fifty genera occur in the United States. The family is of considerable economic importance because of the edible fruits (apples, pears, cherries, plums, peaches, apricots, blackberries, raspberries, and strawberries are among the important fruits) and many ornamentals.

Antelope Bush, Bitterbrush (*Purshia tridentata*)
Description: This fragrant shrub grows up to eight feet tall. Leaves are deeply three cleft into linear lobes, glandular above and tomentose below. The leaf margins are rolled inward. Its flowers are pale yellow to white, and the fruit is a pubescent oblong achene. Bitterbrush grows on dry slopes and canyons of the desert mountains between 3,000 and 9,000 feet. It flowers from April to June. It is named for F. Pursh, author of an early book on the plants of North America.

Interesting Facts: The leaves and inner bark of this species were used to produce a tea that was used as an emetic or strong laxative. The tea was also used as an analgesic and to relieve menstrual cramps. The ripe seed coat produces a violet dye. Old *Purshia* stumps produced shredding bark that women would peel off, work with their hands to soften, and use as baby diapers. These were sometimes combined with juniper (*Juniperus*) bark.

Bitterbrush
(*Purshia tridentata*)

Avens (*Geum*)
Description: The two species in the Sierra Nevada are perennial herbs that are found in wet, open areas from low elevations to above timberline. Most of the

leaves are basal and pinnately divided, whereas the stem leaves are small and less divided. The bell- or saucer-shaped flowers are solitary or borne in an open, few-flowered inflorescence.

Quick Key to the Avens	
Stem leaves, at least lower ones, are not much smaller than basal leaves	Large-leaved Avens (*G. macrophyllum*)—This species occurs in moist places between 3,500 and 10,500 feet throughout the Sierra Nevada.
Stem leaves few and greatly reduced	Old Man's Whiskers (*G. triflorum*)—Old man's whiskers is found on dry to moist slopes and flats between 8,500 and 11,000 feet from Alpine County northward.

Interesting Facts: These species were used medicinally by a variety of Native Americans. A decoction of the root was taken for stomach pain, and a poultice of chewed or bruised leaves was applied to boils and wounds. To others, it was a panacea, where the leaves were chewed as a universal remedy—"good for everything."

Burnet (*Sanguisorba*)

Description: The two species in California include *S. minor* (small burnet) and *S. occidentalis* (western burnet), which are annual herbs with pinnately divided leaves. The flowers are clustered in heads and have no petals. The species are found in waste places or moist soils at the lower elevations. The generic name comes from the Latin *sanguis,* meaning blood, and *sorbeo,* meaning to staunch, referring to the herb's ability to stop bleeding.

Interesting Facts: The young leaves make a good salad plant, tasting somewhat like cucumbers. The leaves can be chopped and blended or mixed with other herbs as a seasoning. The dried flowers and leaves can be prepared as a tea. The roots are very astringent, and a decoction was used in the treatment of internal and external bleeding and dysentery. The brew can also be used as a mouthwash for gum problems.

Chamise (*Adenostoma fasciculatum*)

Description: The genus is composed of two species native to California and Baja California. These are shrubs with evergreen leaves that are clustered, needlelike or linear, and alternately arranged. This shrub is a common chaparral plant and occurs at lower elevations in the mountains. It blooms in the late spring, May to June. The flowers are small, white, saucerlike, and massed in a showy terminal inflorescence.

Interesting Facts: The seeds are edible but are very tedious to collect. An infusion of chamise was used to cure skin infections. Twigs were ground into a powder, mixed with animal fat, and used as a salve on sores. Chamise leaves were brewed into a tea that cured ulcers, colds, and chest ailments.

Chamise, a prominent plant in the chaparral, burns readily. Natives often tightly bound the branches and used them as a torch. Both species created coals that were useful in cooking and roasting food. Gum from a scale insect on chamise was used as an adhesive to bind arrow points to shafts, baskets to bed mortars, and so on.

Cinquefoil (*Potentilla*)

Description: The many species of cinquefoil include perennial, biennial, or annual herbs and one shrub, with alternate, mostly compound leaves that have a membranous appendage (stipules) at the base of the petioles. The flowers are yellow, white, or, in one case, purple. Fruit is a cluster of achenes borne on the convex receptacle. They can be encountered in various habitat types at all elevations.

1. Silverweed (*P. anserina*)—This low, flat-lying perennial with pinnate compound leaves grows in moist, alkaline areas between 4,000 and 8,000 feet. It flowers from May to October.
2. Diverse-leaved Cinquefoil (*P. diversifolia*)—This slender-stemmed plant is found in moist, rocky places between 8,000 and 11,600 feet throughout the Sierra Nevada.
3. Drummond's Cinquefoil (*P. drummondii*)—This species is found growing in moist habitats between 6,000 and 13,000 feet. It flowers from July to August.
4. Mount Rainier Cinquefoil (*P. flabellifolia*)—This plant grows in moist places between 5,800 to 12,000 feet throughout the Sierra Nevada. It flowers from June to September.
5. Bush Cinquefoil (*P. fruticosa*)—This much-branched shrub grows in moist places between 6,400 and 12,000 feet throughout the Sierra Nevada.
6. Sticky Cinquefoil (*P. glandulosa*)—This erect perennial with glandular, often reddish stems grows in dry or moist places up to 8,000 feet. It flowers from May to June.
7. Slender Cinquefoil (*P. gracilis*)—This plant with slender stems occurs in moist places between 2,500 and 11,000 feet. It flowers from June to August.
8. Many-leaved Cinquefoil (*P. millefolia*)—This slender, spreading to prostrate plant is found in damp grassy places between 2,500 and 6,000 feet from Nevada County north.
9. Rough Cinquefoil (*P. norvegica*)—This species grows in moist places between 4,500 and 7,500 feet and is scattered throughout the Sierra Nevada.
10. Purple Cinquefoil (*P. palustris*)—This plant with stout stems and wine-purple flowers grows in swamps and bogs below 8,000 feet from El Dorado County northward.
11. Pennsylvania Cinquefoil (*P. pennsylvanica*)—This densely puberulent plant occurs in moist habitats between 9,000 and 12,000 feet. It flowers from June to August.
12. Strigose Cinquefoil (*P. pseudosericea*)—Look for this plant in dry, rocky places between 10,500 and 13,000 feet from Inyo and Tuolumne Counties northward.
13. Wheeler's Cinquefoil (*P. wheeleri*)—This flat-lying or spreading perennial plant has silky, hairy stems. Wheeler's cinquefoil grows at the edge of open meadows from 6,500 to 11,500 feet. It flowers from June to August.

Interesting Facts: The large fleshy, older roots of *P. anserina* (= *Argentina anserina*) (silverweed) can be boiled or roasted and added to soups and stews. Prepared this way they are quite tasty and have a nutty or parsniplike texture, but more woody. They were a staple among many Native Americans. Today, they are seldom harvested but greatly enjoyed by those who still use them. Silverweed is high in tannins and can be used to tan leather. Other cinquefoils are considered astringent as well. For example, a tea can be made from the leaves of *P. fruticosa* (= *Pentaphylloides floribunda*) (bush cinquefoil). The whole plant or root of *P. arguta* (tall cinquefoil) in a tea or poultice stops bleeding and has been used on cuts and wounds and for diarrhea and dysentery. A strong tea may still be useful, although it is no longer used as a mouthwash and gargle for sore throats or tonsil inflammations to help reduce gum inflammation.

Caution: In ancient times, these plants were grown for food and medicine. Although there are no reports of toxic reactions from use of this genus, moderation is still advised.

Common Agrimony (*Agrimonia gryposepala*)

Description: This genus has about fifteen species of rhizomatous herbs. This glandular perennial has pinnately compound leaves divided into five to nine leaflets that are evenly serrate. The flowers are yellow. The genus name is possibly a misrendering of the name *Agremonia* used by the Greeks and derived from *argena*, a fleck in the eye. This plant is known to occur in Plumas County. It flowers from July to August.

Interesting Facts: The plant was used to treat eye disease.

Horkelia (*Horkelia*)

Description: These are perennial herbs with pinnate leaves. The white to pink flowers occur in cymose panicles, and there are usually ten stamens with dilated filaments. The genus name honors Johann Horkel (1769–1846), a German physiologist.

1. Dusky Horkelia (*H. fusca*)—This species is found in moist to dry habitats below 10,500 feet throughout the Sierra Nevada.
2. Three-toothed Horkelia (*H. tridentata*)—Look for this plant in woods between 2,000 and 6,500 feet throughout the Sierra Nevada.

Interesting Facts: The Kashaya drank a tea made from the roots of a related species, *H. californica*, to purify the blood (Goodrich and Lawson 1980).

Ivesia (*Ivesia*)

Description: These are perennials with thick rhizomes and numerous, mostly basal, pinnately compound leaves. Its flowers are yellow, white, or purple. About twenty-two species occur in western North America. The genus is named for Lt. Eli Ives, leader of a Pacific Railway survey.

1. Field Ivesia (*I. campestris*)—Look for this species in meadows between 6,500 and 11,000 feet in Tulare County.
2. Gordon's Ivesia (*I. gordonii*)—This plant is found in dry, rocky places between 7,500 and 13,000 feet in Tuolumne and Mono Counties.
3. Clubmoss Ivesia (*I. lycopodioides*)—This species is found in moist places between 7,500 and 12,500 feet from El Dorado County south.
4. Muir's Ivesia (*I. muirii*)—This slender-stemmed plant is found on gravelly slopes between 9,500 and 12,000 feet from Fresno to Tuolumne and Mono Counties. It flowers from July to August.
5. Dwarf Ivesia (*I. pygmaea*)—This densely glandular plant is found on rocky slopes from 9,500 to 13,000 feet.
6. Mousetail Ivesia (*I. santalinoides*)—This occurs on gravelly slopes and ridges, from 6,500 to 9,000 feet. It blooms from June to August.
7. Shockley's Ivesia (*I. shockleyi*)—Look for this species in gravelly and rocky places between 9,000 and 13,000 feet from Inyo to Placer Counties.
8. Yosemite Ivesia (*I. unguiculata*)—This ivesia is found on open slopes between 5,000 and 8,000 feet from Fresno to Mariposa Counties.

Interesting Facts: Gordon's ivesia was used by the Arapaho. They apparently made an infusion of the root for use as a tonic (Arnason, Hebda, and Johns 1981).

Mountain-ash (*Sorbus*)

Description: Two species of mountain-ash are found in the Sierra Nevada. They are trees or shrubs with deciduous, pinnately compound leaves. The flowers are white to cream colored and borne in densely branched, flat-topped inflorescences. The fruits are small and berrylike, ranging in color from orange to red. They can be found in moist meadows and forest openings at the middle elevations.

Mountain-ash (*Sorbus* spp.)

Quick Key to the Mountain-ashes	
Leaflets seven to eleven in number; inflorescence convex	California Mountain-ash (*S. californica*)— This many-stemmed shrub grows in moist, shady places between 5,000 and 11,000 feet throughout the Sierra Nevada.
Leaflets eleven to thirteen in number; inflorescence flat topped	Western Mountain-ash (*S. scopulina*)—This shrubby species is found in canyons and on wooded slopes between 4,000 and 9,000 feet.

Interesting Facts: The fruits may be eaten raw, cooked, or dried. They are high in vitamins A and C and carbohydrates. To make a "lemonade-like" drink, crush the fruit in water, strain, and sweeten as needed. However, unripe berries are very bitter and somewhat unpalatable. The fruits, which are pomes, are commonly processed into jams and jellies. They have a high pectin content and jell readily. As a coffee substitute, grind the dried, roasted seeds. The berry juice can be used as a gargle for a sore throat and as an antiseptic wash for cuts. Sorbitol, the sugar in the fruit of *Sorbus*, is used commercially for sweetening candies, toothpaste, and other products.

Mountain-mahogany (*Cercocarpus*)

Description: There are thirteen species growing in the western United States and Mexico. Five occur in California, and the following two are common. These are shrubs and small trees with hard wood. The flowers are not showy but instead are sweet with nectar. The fruit is quite characteristic and is an achene that ends with a long terminal style that is covered with shiny hairs at maturity. The shrubs glisten in the sun from the mass of silvery fruits, each one a "tailed fruit," as indicated by the generic name.

Mountain-mahogany (*Cercocarpus ledifolius*) fruits

1. Birch-leaf Mountain-mahogany (*C. betuloides*)—This evergreen shrub with obovate leaves toothed near the apex is common on chaparral slopes at lower elevations, generally below 6,000 feet. It flowers from March to May.
2. Curlleaf Mountain-mahogany (*C. ledifolius*)—This shrub or small tree with entire, lanceolate leaves is fairly common on dry mountain slopes between 4,000 and 10,000 feet.

Interesting Facts: The common name, mountain-mahogany, applied to this genus is somewhat misleading. These shrubby trees are not related to true mahogany (*Swietenia*), a valuable cabinet wood of tropical America. The dark reddish brown, mahogany-colored hardwood of *Cercocarpus* may have led to this name. Native Americans used the wood for spears, arrow shafts, and digging sticks. The inner brown bark produced a red-purple dye, as did the roots.

A tea to treat colds was prepared by peeling the bark, scraping out its inner layer, and drying and boiling it. The dried sap was pulverized and applied to the ears to treat an earache. A decoction of the bark and leaves was used for women's gynecological problems.

Caution: The leaves and seeds of curlleaf mountain-mahogany contain cyanogenic glycosides and should be considered toxic.

Mountain Misery (*Chamaebatia foliosa*)

Description: This evergreen shrub has sticky leaves and white flowers. Mountain misery grows in open forests between 2,000 and 7,000 feet throughout the Sierra Nevada. The name comes from the plant's sticky black gum that gets on clothing. The resinous sap makes it a fire hazard.

Interesting Facts: A hot tea was made from the leaves and drunk for rheumatism and for skin diseases. For coughs and colds, a decoction of leaves was sometimes mixed with other herbs and taken internally.

Oceanspray, Holodiscus (*Holodiscus*)

Description: Two species may be found in the Sierra Nevada. The egg-shaped leaves are shallowly lobed with toothed margins. The creamy white–colored flowers are small and borne in a diffusely branched inflorescence. Oceanspray is common on rocky slopes and in open forests in the middle elevations.

Quick Key to the Holodiscuses	
Leaves toothed to below the middle; inflorescence one to eight inches long; petioles one-eighth to three-quarters inch long	Oceanspray (*H. discolor*)—This shrub grows at lower elevations in the woods or fairly moist areas below 4,500 feet. It flowers from May to August.
Leaves toothed near apex only; inflorescence one to one and a half inches long; petioles one-sixteenth inch long	Holodiscus (*H. microphyllus*)—This bushy shrub grows in dry, rocky places from 5,000 to 11,000 feet. It flowers from June to August.

Interesting Facts: The small, dry, one-seeded fruits of oceanspray can be eaten raw or cooked. This plant is also used as an astringent, diuretic, tonic, and emetic. The stem bark may be decocted and drunk to treat upset stomach, diarrhea, colds, and influenza. As a tonic, it is said to give athletes endurance. Pojar and MacKinnon (1994) also indicate that some aboriginal peoples steeped the brownish fruiting clusters in boiling water to make an infusion that was drunk for diarrhea. The hardwood can be used for digging sticks.

Oso Berry (*Oemleria cerasiformis*)

Description: This shrub or tree may grow up to thirteen feet tall. It has alternate, thin, deciduous leaves. It produces two kinds of flowers (seed bearing and pollen bearing) that are found on separate plants. The flowers are white and fragrant. The fruits consist of one to five drupes per flower. It is found in canyons below 5,000 feet.

Interesting Facts: The berry is edible raw or cooked, although the taste is not very appealing. A poultice of the chewed, burned plant mixed with oil was applied to sore places. The bark was used a mild laxative.

Pacific Ninebark (*Physocarpus capitatus*)

Description: This is a deciduous, erect, or spreading shrub, with exfoliating bark. The leaves are three to five lobed, with the middle lobe conspicuously longer and with stellate-pubescent hairs underneath. Look for this species on moist banks and north slopes below 4,500 feet throughout the Sierra Nevada. The genus name is from the Greek *physa* (bladder) and *karpon* (fruit), in reference to the inflated fruits.

Interesting Facts: Although most Native Americans considered Pacific ninebark as highly *poisonous,* a tea was made from a stick with the outer bark peeled off. It was used as an emetic and purgative (Pojar and MacKinnon 1994). The Miwok apparently ate ninebark pods raw (Strike 1994). The branches of these plants were straightened and used to make arrow shafts. The stems were also peeled and pounded to obtain fiber for cordage making.

Plum, Stone Fruits (*Prunus*)

Description: This genus has about four hundred species. They are shrubs or trees with simple leaves. Many species have a pair of warty glands present at the tops of the petioles or at the bases of the leaf blades. Its flowers are pink to white and rather showy. Fruits are drupes with one stone and typically embedded in the fleshy pulp. Seeds and leaves are toxic, as they contain hydrocyanic acid.

Quick Key to the Plums

Plant with spinose branchlets	Sierra Plum (*P. subcordata*)—Sierra plum is found on dry, rocky, or moist slopes below 6,000 feet throughout most of the range. It flowers from April to May.
Plant without spinose branchlets Leaves three-quarters to two inches long; young twigs with shiny red color	Bitter Cherry (*P. emarginata*)—This deciduous shrub or tree grows in canyons and moist slopes below 9,000 feet in the mountains. It flowers from April to May.
Leaves one to three and a half inches long; young twigs gray-brown in color	Western Choke Cherry (*P. virginiana*)—This deciduous shrub or small tree grows in damp places in the woods and mountains below 8,000 feet. It flowers from May to June.

Interesting Facts: In general, the fruits of these species are sour or bitter when raw, but after cooking the sourness disappears. Native Americans dried the berries whole or in cakes for use in winter. When needed, the dried fruits were soaked in water and then eaten. Lewis and Clark's expedition members ate western choke cherry when other foods were scarce. It seems that after drying, the fruits lose some of their bitterness, resulting in an almost sweet taste.

To make cakes, the ripe fruits are usually ground up, pits and all, and dried in the sun. When needed, the cakes, or portions thereof, can be soaked in water, mixed with flour and sugar, and made into a sauce or gravy. This sauce was eagerly traded among some Native Americans such as the Navajo, Shoshone, Arapaho, and Utes (Bailey 1940). The only difficulty we've found in preparing cakes in this manner is that the pits do not grind down nicely into a fine material, leaving larger chunks that could have resulted in broken teeth.

Other uses of the berries were their incorporation into pemmican. They can also be used in making jelly, but because choke cherries are low in natural pectin, it is advisable to add pectin.

The leaves of these species contain toxic amounts of cyanide, as do the seeds (pits). Cyanide is highly volatile, and the pits can be rendered safe by long-term drying, by boiling in several changes of water, or by dry roasting. Do not eat them in significant amounts, and even then mix them with larger quantities of other foods. *Prunus* shoots, peeled and split, were used in basketry. The wood was used for various implements, such as digging sticks, arrows, and arrow foreshafts.

Warning: The leaves, bark, and seeds of all *Prunus* contain cyanide-producing glycosides. Therefore, eating large quantities of ripe berries with their pits could cause nausea and vomiting. In some instances, it could be fatal. Cooking and drying the seeds appears to dispel most of the glycosides; then, the seeds in dried, mashed choke cherries are not as significant a problem. To be safe, it is best to discard the seeds before eating the fruits.

Raspberry, Blackberry, Thimbleberry (*Rubus*)

Description: Our species are deciduous shrubs with arching or trailing stems covered with bristles and prickles. The flowers have white petals, and the fruit is a coherent cluster of small, one-seeded drupes (raspberries, blackberries, dewberries, cloudberries, and marionberries). The genus name is derived from the Latin *ruber,* meaning red, in reference to the color of the fruit. This is a large and complicated group taxonomically.

Quick Key to the Raspberries and Thimbleberries	
Plants without prickles; leaves palmately lobed	Thimbleberry (*R. parviflorus*)—This shrub with compound leaves grows in open woods below 8,000 feet. It flowers from March to August.
Plants with prickles; leaves pinnately compound	
Sepals are deflexed when flowering	Western Raspberry (*R. leucodermis*)—This straggly shrub with simple leaves has trailing, prickly stems and grows on dry slopes from 4,000 to 7,000 feet. It flowers from May to June.
Sepals are not deflexed	Glaucous-leaved Raspberry (*R. glaucifolius*)—This plant with erect flowering shoots and white petals grows on dry slopes between 3,000 and 7,400 feet throughout the Sierra Nevada.

Interesting Facts: All species produce edible berries. Fossil evidence shows that *Rubus* species have formed part of the human diet from very early times (C. E. Smith 1973).

Flowers of all species can be added to salads and can be nibbled upon when hiking. The fresh or dried leaves can be steeped for a tea, alone or in herbal blends. Do not use the wilted or molded foliage, as it may be toxic. The young shoots cut just above ground can be peeled and eaten raw or cooked. A tea from the roots was used to dry runny noses, and a tea from the bark was used to stop dysentery. The plants can also provide a uterine astringent, diuretic, laxative, and mild sedative.

DEWBERRY JUICE
Cover the raw berries with water, crush them, and strain through cheesecloth. Add sweetening to taste. Uncooked juice should be refrigerated and consumed within a day or two.

Rockmat, Rock-spiraea (*Petrophytum caespitosum*)

Description: This evergreen subshrub has intricately branched, prostrate stems, covered in old leaf bases. It tends to form dense mats up to two feet across. The tufted basal leaves have short petioles and narrowly spoon-shaped, bluish green blades. The small flowers have a cone-shaped calyx with five triangular lobes and five creamy white petals, about as long as the sepals: The genus name is from the Greek *petros* (rock) and *phytum* (plant), referring to the normal habitat of these plants.

Interesting Facts: The plants are often used as ornamentals. However, a poultice of boiled roots was applied to burns as a salve. The plants were also utilized as a narcotic.

Rose (*Rosa*)

Description: These are shrubs with prickles and leaves that are pinnately divided into three to eleven leaflets. The large, red to pink flowers are borne singly or a few together. The fruits, called hips, are orange, red, or purplish and urn shaped. This is a taxonomically difficult genus with many variable species that tend to hybridize.

Wild Rose (*Rosa* spp.)

Quick Key to the Roses	
Leaves are once-serrate, and the teeth are not gland tipped	
Stems armed with stout, flattened, recurved prickles	California Rose (*R. californica*)—This shrub is found on dry slopes from 5,000 to 9,000 feet. It flowers from June to August.
Stems armed with straight and slender prickles	Mountain Rose (*R. woodsii*)—This shrub grows in moist places below 6,000 feet. It flowers from May to August.
Leaves are usually doubly serrate and gland tipped	
Prickles are stout	Ground Rose (*R. spithamea*)—This species of rose is found in open woods below 5,000 feet from Tulare to Yuba Counties.
Prickles are slender	
Sepals and styles deciduous	Wood Rose (*R. gymnocarpa*)—This species occurs in shaded woods below 6,000 feet. It flowers from May to July.
Sepals and styles persistent	Pine Rose (*R. pinetorum*)—This rose grows in open woods between 2,000 and 6,500 feet throughout the Sierra Nevada.

WILD ROSE SYRUP
Gather rose hips in late summer or autumn. Snip off both ends, then cut the hips in half, put them into a pot, and barely cover with water. Bring to a boil, cover, and simmer until the hips are tender. Strain off and retain the liquid, then cover the hips with more water and boil for fifteen minutes. Strain off and retain the liquid and add it to the first batch. Measure the liquid, then add half as much sugar. Bring to a boil and simmer until the syrup thickens. Pour into sterilized bottles. Serve over pancakes.

Interesting Facts: The hips are edible raw, stewed, candied, or made into preserves. They are high in vitamin C and also contain vitamins E, B, and K and beta-carotene, calcium, iron, and phosphorus. There are many other edible parts, besides the fruit. Young *Rosa* shoots in spring make an excellent potherb, and the roots and stems can be used to make a tea. The petals may be used in salads. The peeled spring shoots can also be nibbled upon. Almost all parts of the plant have been made into a wash or dressing for cuts or sores to coagulate blood. One of the more common methods is to sprinkle fine shavings of debarked stems into a washed wound. The petals can be used as a dressing. A poultice of leaves can be used to relieve insect stings. In addition, the young leaves can be washed, cut into small pieces, and dried for a hot tea.

Serviceberry (*Amelanchier*)

Description: These are shrubs or small trees with simple leaves that are serrate on the terminal half. The white flowers have five petals, five reflexed sepals, and many stamens. The ovary is inferior, and the fruit a pome. The genus name is the French Savoy word for the medlar (*Mespilus germanica*), a species that has similar fruits.

The leaves and petioles are glabrous	Smooth Serviceberry (*A. alnifolia*)—This species produces dark purple fruits and is found in damp places between 4,500 and 8,000 feet from El Dorado to Nevada Counties. It flowers from May to June.
The leaves and petioles are somewhat pubescent	Utah Serviceberry (*A. utahensis*)—The fruits of this species are purplish black. It is found growing on rocky slopes and forests up to 11,000 feet throughout most of the Sierra Nevada. It flowers from April to June.

Serviceberry
(*Amelanchier* spp.)

Interesting Facts: All species within the genus *Amelanchier* produce edible pomes that ripen in late spring and the summer. They were a considered to be a major food for many Native peoples. In fact, some Native Americans intentionally moved their camps to locations where they could be more easily harvested. The pomes may be eaten raw, cooked, or dried. After drying, the pomes can be pounded into loaves or cakes. These in turn may be eaten after softening a piece in water or placing them in soups or stews. Prepared this way, the pomes could be kept for several years. Additionally, the dried pomes could be incorporated into pemmican.

The boiled inner bark was a Native American remedy for treating snowblindness. One drop of strained fluid would be placed in an afflicted eye three times daily. It was also used for eardrops and to stop vaginal bleeding. These applications are probably due to the astringency of the plant's tannic acid content.

The wood can be used for arrows, digging sticks, and other useful items. The berry juice makes a purple dye.

Spiraea (*Spiraea*)

Description: Two species of *Spiraea* can be found in the Sierra Nevada. They are small shrubs with deciduous leaves with white to pink flowers that are densely clustered in showy, flat-topped to spikelike inflorescences. The species can be found in brushy, open slopes and in moist habitats up to timberline.

PEMMICAN
Pemmican is a concentrated food used by many early peoples. It is extremely nourishing and does not spoil. To make pemmican, cut meat into strips and dry it until it completely dries and crumbles. It is then ground as finely as possible, usually by pounding. Melted suet (fat) is then poured over the meat, and salt is added for taste, as are fresh berries such as currants and serviceberries. The mixture is then kneaded into a paste and packed into containers. In the "old days," the containers were intestines, but you can use plasticware. The finished product can be eaten raw, boiled in stews and soups, or fried like sausages.

Inflorescence flat topped or nearly so	Mountain Spiraea (*S. densiflora*)—This species occurs in moist, rocky habitats between 5,000 and 11,000 feet.
Inflorescence elongated or spikelike	Douglas' Spiraea (*S. douglasii*)—This erect shrub grows in moist habitats below 6,400 feet in Butte and Plumas Counties northward. It flowers from June to September.

Interesting Facts: Spiraea is a source of methyl salicylate, similar to the active ingredient in aspirin. Native Americans brewed a tea from the stem, leaves, and flowers of some species to use as a pain reliever (Craighead, Craighead, and Davis 1963). The plants are astringent, and a poultice made from the leaves and bark was used to treat ulcers, burns, and tumors. The roots were also peeled and boiled until soft, mashed, and used as a poultice for burns. The wiry, branching twigs can be used to make broomlike implements for collecting tubers.

Strawberry (*Fragaria*)

Description: These white-flowered perennial herbs are produced from root-stocks and long runners that root at the nodes. Its flowers are white to pinkish, and borne in cymes. The leaves are clustered at the base of the stem and are divided into three egg-shaped, coarsely toothed leaflets. Woodland strawberry is found in moist, humus-rich, well-drained soils of open forest and forest margins up the subalpine zone (4,500 to 7,500 feet). The genus name comes from the Latin *fraga,* the classical name used for the strawberry fruit and referring to its fragrance. The common name, strawberry, comes from the Anglo-Saxon *streawberige* and refers to the berries "strewing" their runners out over the ground.

Quick Key to the Strawberries	
Leaves yellow-green, upper surface bulged between the veins; apical tooth of the leaflets is greater than those on either side	California Strawberry (*F. vesca*)—This strawberry grows in shady, damp places below 7,000 feet throughout most of the Sierra Nevada. It flowers from April to June.
Leaves blue-green; apical tooth of leaflets is smaller than those on either side	Broad-petaled Strawberry (*F. virginiana*)— This species is found in woods between 4,000 and 10,000 feet throughout most of the range. It flowers from May to July.

Interesting Facts: Strawberries do not keep well and should be dried for future use if not eaten soon after being picked. Tea made from the green or dried leaves is said to tone up one's appetite. It may also be a nerve tonic and was used for bladder and kidney ailments, jaundice, scurvy, diarrhea, and stomachaches. Externally, the leaf tea can also be used as an antiseptic wash for eczema and wounds and as a gargle for sore throats and mouth ulcers. The plants do contain substantial amounts of vitamins A and C, sulfur, calcium, potassium, and iron. To remove tartar, rub the berries on your teeth and let the juice sit for a few minutes. Afterwards, brush your teeth thoroughly with baking soda and water.

Toyon (*Heteromeles arbutifolia*)

Description: This is a small tree or shrub that can grow up to fifteen feet tall. The leaves are evergreen, leathery, and glossy. The white flowers occur in flat-topped clusters, and the fruits (pomes) are red or sometimes yellow. Toyon occurs at elevations below 5,000 feet in foothill, chaparral, and woodland habitats.

Interesting Facts: The fruit, though rather dry and mealy, is edible raw, roasted, or dried. After boiling, the berries can be baked in earthen ovens for a couple of days. The Maidu used toyon berries to make a cider.

Medicinal uses include a tea made from the bark and leaves to relieve stomach discomfort and for other aches and pains. A dye was also extracted from the red berries by simmering them gently in water and crushing them as they simmered.

The wood was used to make arrows, cooking implements, and tools (for instance, scrapers, awls, and the like).

Warning: The leaves, berries, and seeds contain cyanogenic glycosides and should be considered toxic.

SANDALWOOD FAMILY (Santalaceae)

The sandalwood family is composed of partially parasitic herbs. There are about thirty genera and four hundred species distributed in the Tropics and temperate parts of the world. The most common representative in North America is the genus *Comandra*.

Toolmaking is not unique to humans. Chimpanzees are known to make a variety of implements, from probes to extract termites to sponges to collect water. There are even some birds that occasionally make simple tools to get at food. But in the line leading to our species, toolmaking seems to have snowballed through a form of evolutionary feedback, with each advance opening a way for the next. At first, this feedback worked slowly, and innovations were few and far between. Our immediate ancestors (*Homo erectus*) were proficient toolmakers but not great inventors. We do not know what they made from bone and wood or what they did with the finished products. However, progress in using stone for tools was very slow. In one area of East Africa, there appears to be only four new designs that appeared in a span of about one million years.

Bastard Toadflax (*Comandra umbellata*)

Description: Bastard toadflax is a partially parasitic perennial herb with a waxy surface and a rather woody base. The leaves are linear, and the flowers are bell shaped. The fruit is a one-seeded, berrylike drupe. Bastard toadflax is common and widespread in shrublands up to the subalpine zone. *Comandra* comes from the Greek *kome*, meaning tufts of hairs, and *aner*, which means man, in reference to the stamens. The roots are blue when cut.

Interesting Facts: The mature, brown, urn-shaped fruit of bastard toadflax may be eaten raw and is best when slightly green. They were popular with Native Americans because of their sweet taste. The berries, however, are rarely found in sufficient quantities for more than a pleasant tidbit. Consuming too many berries may cause nausea. Strike (1994) indicates that a root preparation was used to soothe sore, inflamed eyes.

SAXIFRAGE FAMILY (Saxifragaceae)

There are about 30 genera and 580 species in this family, found chiefly in cooler and temperate regions of the Northern Hemisphere. About 20 genera are native to the United States, with more species occurring in the western part of the country. Several species are cultivated as ornamentals. The family name is derived from the Latin *saxum* (rock) and *frango* (to break), as many members are found growing in rock crevices. All members of the saxifrage family are at least somewhat edible.

Alumroot (*Heuchera*)

Description: In general, *Heuchera* species are perennial herbs with basal leaves. Its flowers are small, saucer to bell shaped, and greenish, white, or pinkish in color. Their various habitats include moist soils and rocky areas up to the alpine zone. The genus is named in honor of Professor J. H. Heucher (1677–1747), custodian of the Botanic Garden in Germany. Many species of alumroot readily hybridize, making identification of some plants difficult.

Quick Key to the Alumroots	
Petals white; plants found below 7,000 feet	Small-flowered Heuchera (*H. micrantha*)—This heuchera grows on moist banks of humus and rocks.
Petals pink; plants found between 6,000 and 12,000 feet	Pink Heuchera (*H. rubescens*)—This species occurs in dry, rocky places between 6,000 and 12,000 feet from Plumas County southward.

Interesting Facts: The leaves of all species in the Sierra Nevada are edible, although they are not choice. They have a sour taste because of the high tannin content. Therefore, the leaves should be boiled or steamed. Because they are rather tough, we found them to be more palatable if chopped and added to soups or salads.

Heuchera is said to be one of the strongest astringents due to their high tannin content. Tilford (1997) indicates that the root of these plants may contain as much as 20 percent of its weight in tannins. Tannins tend to shrink swollen, moist tissues. Therefore, they are also gastrointestinal irritants and have been known to cause kidney and liver failure. Ingestion of the plant should be in moderation. Otherwise, the pounded, dried roots of many species have been used as a poultice that stops bleeding and promotes healing when applied to cuts and sores. The raw root, eaten in small amounts, has been used as a cure for diarrhea. A tea from the roots can also be used as a gargle for sore throats. The powdered roots have been used as an antiseptic.

Alumroots are also commonly used as mordants, substances that make natural dyes colorfast. The alumroot of choice, however, occurs in western deserts near sulfur springs, but satisfactory substitutions probably do exist in California.

Boykinia (*Boykinia*)

Description: These are glandular pubescent perennials with horizontal rhizomes and more or less leafy stems. Flowers occur in cymose panicles, with the lower half of the calyx adnate to the ovary. The petals are white, usually soon becoming deciduous. Stamens occur opposite the sepals, and the ovary is two celled. The genus is named for Dr. Samuel Boykin (1786–1846), an American botanist, naturalist, and physician.

1. Mountain Boykinia (*B. major*)—This species occurs in shaded, moist habitats below 5,000 feet from Amador County northward.

2. Brook Foam (*B. occidentalis*)—This species grows in rocky habitats below 7,500 feet from Madera County northward.

Interesting Facts: The raw leaves of mountain boykinia were eaten for tuberculosis, and the roots were used medicinally (in an unspecified manner) (Strike 1994). Native Americans also dried the leaves of it and wore them inside their basket caps and hats for the pleasant fragrance.

Fringe-cups (*Tellima grandiflora*)

Description: The flowers are unusual in that the sepals are greenish and form a rather broad, cuplike tube. The conspicuously fringed petals are greenish white to reddish and extend out and back over the top of the sepal tube. The resulting impression is a fringed cup. This species grows in moist woods and rocky places below 5,000 feet.

Interesting Facts: The roots were chewed by the Maidu to relieve colds and stomachaches. A decoction was used to stimulate a poor appetite (Strike 1994).

Grass-of-parnassus (*Parnassia fimbriata*)

Description: This familiar plant has sessile, spade-shaped leaves near the midlength of the stem. The basal leaves have petioles up to four inches long and broadly spade-shaped blades with prominent parallel nerves. Calyx lobes are elliptical with fringed tips, and the petals have five to seven prominent, parallel veins and long, filiform fringes on both sides near the base. The genus is named for Mount Parnassus in Greece.

Interesting Facts: An infusion of powdered leaves was given to babies for dullness or sickness of the stomach. A poultice of the plant was applied, or the plant was used as a wash for venereal disease.

Saxifrage (*Saxifraga*)

Description: Saxifrages are glabrous or, more often, glandular hairy perennials with simple to divided leaves that are basal or alternate (opposite in one species) and sometimes have small bulbs in the axils. Flowers (sometimes replaced by small bulbs) are solitary or borne on stalks in a simple to branched inflorescence. The calyx is five lobed and saucer- to bell shaped, and there are five petals. The generic name is from the Latin *saxum*, meaning rock, and *frangere*, meaning to break. It alludes to the species' rocky habitat. Herbalists once used some species in a treatment for "stones" in the urinary tract.

1. Sierra Saxifrage (*S. aprica*)—This species grows in moist, gravelly, and stony places between 5,500 and 12,000 feet.
2. Bud Saxifrage (*S. bryophora*)—This plant is found in moist, gravelly places between 7,000 and 11,200 feet from Tulare to Plumas Counties. It flowers from July to August.
3. Green's Saxifrage (*S. fallax*)—Look for this species in rocky places between 3,000 and 7,500 feet from Fresno to Plumas Counties.
4. Wood saxifrage (*S. mertensiana*)—This species grows in moist, rocky places.
5. Peak saxifrage (*S. nidifica*)—Peak saxifrage is found moist places between 2,500 and 11,000 feet.
6. Brook Saxifrage (*S. odontoloma*)—Look for this plant on moist stream banks between 6,500 and 11,200 feet throughout most of the range. It flowers from June to August.
7. Bog Saxifrage (*S. oregana*)—This plant grows in wet meadows and boggy places between 3,500 and 11,000 feet.
8. Lobed-leaved Saxifrage (*S. rivularis*)—This saxifrage occurs in damp places in the shade of overhanging rocks between 11,000 and 12,000 feet in Inyo, Tulare, Madera, and Tuolumne Counties. It flowers from July to August.
9. Alpine Saxifrage (*S. tolmiei*)—Look for alpine saxifrage in moist, rocky places between 8,500 and 11,800 feet.

Interesting Facts: The genus as a whole is regarded as a safe group of plants. The leaves can be used fresh or in stews and are high in vitamins A and C. Richard Scott (formerly of Central Wyoming College) shared with us this favorite back-country salad of his:

> When backpacking, we always carry a little bottle of vinegar, one of oil, and one of dried or powdered salad dressing. In fact, an excellent salad is based on saxifrage, particularly *S. arguta* (Brook Saxifrage) leaves with a few alpine sorrel leaves (*Oxyria digyna*), a couple of Indian Paintbrush inflorescences (*Castilleja miniata* or *C. rhexifolia*), and a couple Columbine flowers (*Aquillea caerulea*).

In China some species were used in the treatment of nausea and ear infections. In our area, there is little documentation regarding medicinal uses.

Woodland Star (*Lithophragma*)

Description: These perennial herbs have slender, bulbed rhizomes. There are approximately ten species in western North America. The genus name is from the Greek *lithos* (rock) and *phragma* (fence). Incidently, generic names that end in *phragma* are considered of neuter, not feminine, gender.

1. Sierra Star (*L. bolanderi*)—Look for this species in dry habitats below 7,000 feet.
2. Rock Star (*L. glabrum*)—This plant occurs in dry to moist open places.
3. Small-flowered Lithophragma (*L. parviflorum*)—This species occurs on open slopes between 2,000 and 6,000 feet.
4. Lithophragma (*L. tenellum*)—This perennial grows in dry areas from 5,000 to 7,500 feet. It flowers from May to June.

Interesting Facts: The root of *L. affinis* was chewed by the Maidu, Mendocino, and Yuki Indians to treat stomach ailments and colds (Chestnut 1902).

SILKTASSEL FAMILY (Garryaceae)

These are shrubs with four-angled stems and opposite, evergreen leaves. The flowers occur in catkinlike racemes, and the fruit is a berry. There is one genus (*Garrya*) with perhaps eighteen species in North America and the West Indies. Some species are cultivated as ornamentals.

Silktassel (*Garrya*)

Description: The genus consists of fourteen New World species ranging from the Pacific Northwest to Panama. *Garrya* is a highland genus occurring in chaparral and coniferous forests above the lowland deserts. First discovered by David Douglas in the Pacific Northwest in 1826, it was named in honor of Nicholas Garry, the first secretary of the Hudson's Bay Company.

Quick Key to the Silktassels

Leaves gray above, slightly hairy below	Silktassel Bush (*G. flavescens*)—This is an evergreen shrub that grows on dry slopes between 3,000 and 8,000 feet in the mountains.
Leaves glossy above, glabrous below	Fremont's Silktassel (*G. fremontii*)—This evergreen shrub grows on dry slopes below 7,000 feet and blooms from February to April.

Interesting Facts: These plants contain quinine and garryine and were used medicinally by several Native American groups. The inner bark was used to reduce fevers. The leaves were boiled, and the resulting infusion was used for stomach problems and diarrhea (Chatfield 1997). The fire-hardened wood can be used as a digging stick.

SNAPDRAGON FAMILY (Scrophulariaceae)

The snapdragon family has about 220 genera and 3,000 species distributed worldwide. About 40 of the genera are native to the United States. The family is of economic importance because of the cardiac glycosides derived from *Digitalis* (foxglove) and the many fine ornamentals.

Bird's-beak (*Cordylanthus*)

Description: These are annual herbs and may be partial root parasites. Leaves are narrow or cut into narrow divisions. The flowers are tubular and two lipped, with the upper lip incurved and enclosing the style and two or four of the stamens.

1. Stiffy-branched Bird's-beak (*C. rigidus*)—This plant with light-yellow flowers is found on dry, granitic slopes between 3,000 and 6,000 feet from Mariposa County southward.
2. Slender Bird's-beak (*C. tenuis*)—This glabrous or minutely pubescent plant grows on dry, open slopes between 4,500 and 8,500 feet from Fresno to Butte and Plumas Counties. It flowers from July to September.

Interesting Facts: Strike (1994) indicates that *Cordylanthus* was used by the Luiseno Indians as an emetic.

Blue-eyed Mary, Collinsia (*Collinsia*)

Description: These are annual plants that are often glandular and brown staining. The leaves are opposite and the flowers are five lobed, appearing pealike. The genus name honors Zaccheus Collins, a Philadelphia botanist who lived from 1764 to 1831.

1. Child's Blue-eyed Mary (*C. childii*)—This plant with white to pale-violet flowers is found in dry, shaded places between 3,000 and 7,000 feet from Mariposa County south. It flowers from May to June.
2. Small-flowered Collinsia (*C. parviflora*)—This species occurs in moist, shaded places between 2,500 and 11,000 feet throughout the Sierra Nevada.
3. Few-flowered Collinsia (*C. sparsiflora*)—This purple-flowered plant is found in grassy places and open woods below 5,000 feet from Fresno to El Dorado Counties.
4. Tincture Plant (*C. tinctoria*)—This plant with yellow to greenish white flowers with purple dots or lines is found in dry or moist, mainly stony habitats between 2,000 and 7,500 feet.
5. Blue-eyed Mary (*C. torreyi*)—Look for this species in damp or partially dry, sandy banks below 10,000 feet

Interesting Facts: Small-flowered collinsia was used to make a horse run fast and was used externally for sore flesh. The leaves of another species, *C. heterophylla* (Chinese houses), were used by the Maidu as a poultice for bites from insects or snakes.

California Figwort (*Scrophularia californica*)

Description: This coarse, tall perennial grows up to six feet tall and has four-angled stems. The leaves are opposite and ovate in shape. Its flowers are small and occur in an open inflorescence. The corolla is two lipped and maroon. Figwort is common in moist places in the woods. It flowers from February to July. There are four functional stamens, and the fifth one is reduced to a small knob on the corolla.

Interesting Facts: A strong tea made from the plant can be applied to fungal infections of the skin (for example, athlete's foot) and can help eczema, rashes, burns, and hemorrhoids.

Gaping Penstemon (*Keckellia breviflora*)

Description: This is a shrub with many slender, glaucous stems. The leaves are opposite and oblong. The white flowers are often tinged with pink, and the upper lip is arched and galeate (see *Castilleja*), over half of the corolla length, and the lower lip is reflexed. Gaping penstemon occurs on dry, rocky slopes below 8,000 feet throughout most of the Sierra Nevada. It flowers from May to August.

Interesting Facts: An infusion of the plant was taken for colds by the Miwok, whereas the Paiute ground the dry leaves and used them on running sores (Strike 1994).

Lousewort (*Pedicularis*)

Description: These are perennial, partially parasitic herbs with toothed or pinnately divided leaves. The flowers are subtended by leaflike bracts and occur in a spikelike inflorescence. The tubular flower is strongly two lipped at the mouth. They are found in open, dry, or moist habitats, including meadows up to the alpine zone. The genus name comes from the Latin *pediculus* (little louse), alluding to the superstition that livestock that ate the plants would suffer an infestation of lice. Although many members of this genus are attractive, they are partially parasitic on the roots of other plants; thus, none of them has been cultivated for garden use.

1. Little Elephant Heads (*P. attollens*)—This plant with lavender or pink flowers is common in meadows and moist places between 5,000 and 12,000 feet throughout the Sierra Nevada.

2. Indian Warrior (*P. densiflora*)—This perennial herb grows on dry slopes up to 6,000 feet elevation.

3. Elephant Heads (*P. groelandica*)—This species is found in meadows and wet places between 6,000 and 11,200 feet.

4. Leafy Lousewort (*P. racemosa*)—This species is found on dry slopes between 4,000 and 7,000 feet. It flowers from June to August.

5. Pine Lousewort (*P. semibarbata*)—This perennial herb is found in the pine forests between 5,000 and 11,000 feet.

Interesting Facts: *P. groelandica* and *P. attollens* often grow near each other, although the former appears to require a much wetter habitat. Both species are pollinated by the same bumblebee, but the shape of the two flowers is different enough so that the pollen adheres to a different part of the bee's body, and therefore no hybrids are possible.

Tilford (1997) indicates that as long as lousewort is not attached to an unpalatable host, the fleshy roots can be prepared and eaten in moderation. He does not, however, describe methods of preparation. He also states that the leaves and stems of some species may be steamed or boiled as potherbs, but this is not recommended.

Moore (1979) describes *Pedicularis* as an effective sedative for children and tranquilizer for adults. The whole stalk, when dried and prepared as a tea, acts as a mild relaxant, quieting anxiety and tension. The fresh or dried plant is a vulnerary for minor injuries, with mild astringent and antiseptic properties.

Warning: Because of the sedative nature of these plants, the potential toxic alkaloids, and the host species it may be attached to, ingestion of this plant is not recommended.

Monkeyflower (*Mimulus*)

Description: Many species of monkeyflower can be found in the Sierra Nevada. They are annual or perennial herbs with opposite leaves. The flowers flare at the mouth to form five lobes, two of which form the upper lip and three of which form the lower. The perennial species occur in permanently wet soil, whereas the annual are found in vernally moist habitats (Grant 1924). There are about twenty-three species of monkeyflowers in the Sierra Nevada.

Interesting Facts: The genus name of *Mimulus* is from the Latin word *minus* (comic actor), because the flowers have grinning "faces" like an early-day comic actor. In some species, the two-lipped stigma closes immediately when touched lightly. This is believed to be a mechanism to prevent self-pollination. When a bee enters the flower for nectar, the pollen of that flower cannot pollinate itself as the bee brushes by the closed stigma as it exits. Therefore, cross-pollination is ensured.

The young stems and leaves of *M. guttatus* (seep monkeyflower) have been used as salad greens. Sometimes, leaves were burned and the ash used a salt. Weedon (1996) indicates that the young herbage of *Mimulus* species may be eaten in salads and that they grow bitter with age but remain edible. For example, the leaves of *M. primuloides* (primrose monkeyflower) were eaten by the Maidu Indians, and the young plants of *M. moschatus* (musk monkeyflower) were boiled and eaten by other California Natives (Strike 1994).

There are also reports of some Native Americans making a tea from monkey-flower stems, leaves, and flowers as a treatment for kidney or urinary problems and to cure diarrhea (Moerman 1986). Monkeyflower leaves and stems were used externally to poultice wounds or internally to reduce fevers. *Mimulus* roots were used to treat fevers, dysentery, and diarrhea and to curtail hemorrhages. The raw leaves and stems can be applied to burns and wounds as a poultice, as well as insect bites and pain from contacting stinging nettle.

Mudwort (*Limosella*)

Description: These are acaulescent annual glabrous herbs with stolons and rosulate leaves. The flowers are solitary on a scapelike peduncle, and the calyx is five lobed and campanulate. The corolla is five cleft, campanulate, and nearly regular. There are four stamens. The genus name is Latin for mud and seat.

1. Southern Mudwort (*L. acaulis*)—This species occurs on muddy shores below 8,000 feet from Plumas County south.
2. Northern Mudwort (*L. aquatica*)—This plant is occasionally found below 10,000 feet throughout the Sierra Nevada. It flowers from June to September.

Interesting Facts: The Navajo used the leaves of northern mudwort ceremonially by rubbing them on the body for protection in hunting and from witches. A roll of washed leaves was used to plug a bullet or arrow wound (Elmore 1944).

Mullein (*Verbascum thapsus*)

Description: This is a coarse plant with stems up to six feet tall. The basal leaves are broadly lance shaped, entire margined, or shallowly toothed and up to sixteen inches long. The stem leaves are smaller, and the upper ones clasp the stems, forming low wings below the point of attachment. The whole plant is densely covered with long, soft hairs. Its flowers are borne in a densely congested inflorescence. The saucer-shaped flowers are white or yellow. The upper three stamens are densely hairy. The species can be found in disturbed places up to the subalpine zone.

Interesting Facts: The leaves of mullein are said to be edible when eaten in small quantities and cooked. Because of their woolly texture, however, we have found the plants to be undesirable.

Native Americans smoked the dried leaves of mullein for asthma and sore throats. A tea from the leaves was used to treat colds, and the flowers contain an oil that has been used for earaches. The flowers and first-year leaves make a soothing decoction for coughs and sore throats. Boil leaves in water for ten minutes, then strain the liquid through a cheesecloth to remove the tiny hairs. The leaves can also be used as poultices applied locally to hemorrhoids, sunburn, and inflammations. The dried stalks are ideal for use as hand drills to start fires. The flowers and leaves produce yellow dye; as a toilet paper substitute, the large fresh leaves are a good choice.

Caution: Mullein does contain coumarin and rotenone, two substances that may be toxic in large quantities if ingested. Also, the seeds are not recommended for consumption.

Owl's Clover (*Orthocarpus cuspidatus*)

Description: Members of this genus are semiparasitic slender plants with sessile flowers that are subtended by a large, colorful bract. The tubular flowers are two lipped with two upper lobes united to form a small hood-shaped galea. The three lower lobes are partly united and nearly as long as the galea. Owl's clovers resemble paintbrushes (*Castilleja*) but can be distinguished by the annual growth form and by having a lower corolla that nearly equals the galea.

Interesting Facts: The foliage of a related species, *O. luteus,* was used by some Natives to dye small skins and feathers a reddish tan color.

Paintbrush (*Castilleja*)

Description: This is a large genus found primarily in western North America that contains many species found in California. The genus is easily recognized, but many species are notoriously difficult to identify. They are perennials with deeply lobed to entire leaves. The flowers are subtended by colorful leaflike bracts. The calyx is composed of four sepals that are united below. The five petals are united below into a tube but separate above into a small, three-lobed lower lip and an elongated upper lip (galea) that encloses the four stamens. Some paintbrushes are partial root parasites. Their roots attach to those of neighboring plants from which they obtain supplemental water and nutrients. The genus name honors the Spanish botanist Domingo Castillejo.

1. Wavy-leaved Paintbrush (*C. applegatei*)—This is a common plant in dry habitats below 11,000 feet throughout the Sierra Nevada.
2. Field Owl's-clover (*C. campestris*)—This yellow-flowered plant occurs in moist habitats below 5,000 feet from Fresno County north.
3. Cut-leaved Owl's-clover (*C. lacerus*)—Look for this species in open habitats below 7,400 feet from Fresno County northward.
4. Lemmon's Paintbrush (*C. lemmonii*)—This paintbrush is found in moist meadows between 7,000 and 11,000 feet from Fresno and Inyo Counties north to Lassen Peak.
5. Great Red Paintbrush (*C. miniata*)—This species occurs below 11,000 feet throughout the Sierra Nevada.
6. Alpine Paintbrush (*C. nana*)—This paintbrush with yellow or purplish red bracts and sepals is found in rocky habitats between 6,400 and 12,000 feet.
7. Pierson's Paintbrush (*C. parviflora*)—Pierson's paintbrush grows in moist, rocky places between 7,800 and 11,000 feet from El Dorado County southward. It flowers from July to August.
8. Hairy Paintbrush (*C. pilosa*)—Look for this species in dry habitats between 5,000 and 10,000 feet mainly on the eastern side of the Sierra Nevada.
9. Paintbrush (*C. praeterita*)—Look for this species in alpine sagebrush fields between 7,500 and 12,000 feet in Inyo and Tulare Counties.
10. Gray Paintbrush (*C. pruinosa*)—This densely hairy plant occurs in rocky habitats below 8,000 feet from Tuolumne County north. It flowers from May to August.
11. Hairy Owl's-clover (*C. tenuis*)—This species is found in meadows between 3,000 and 8,000 feet from Tulare County north.

Interesting Facts: Many, if not all, of the species have flowers and bracts that can be eaten raw. The seeds of some species were gathered, winnowed, dried, and stored for winter use. In winter they were parched, pounded, and eaten dry. The plants, however, absorb selenium from the soil and so should be taken in moderation. Symptoms in humans of selenium poisoning will vary with the amount and form ingested but may include difficulty in breathing, excessive urine production, loss of appetite, mental depression, a weak and rapid pulse, blurry vision, digestive upset, and eventually coma and death (Kirk 1975).

Penstemon (*Penstemon*)

Description: There are many species of penstemon. In general, they are perennial herbs with opposite leaves. The flower is strongly to indistinctly two lipped at the mouth with a two-lobed upper lip and a lower lip with three lobes. There are four anther-bearing (fertile) stamens and a single sterile stamen that is often hairy at the tip. The fruit is a many-seeded capsule. Penstemons occur in dry or moist meadows or forest openings up into the alpine zone. The genus name is from the Greek *pete*, meaning five, and *stemon*, meaning thread, referring to the slender fifth stamen. Several species of penstemon are cultivated, and most are easily raised from seed. Approximately sixteen species of *Penstemon* occur in the Sierra Nevada in various habitats.

Interesting Facts: There are more than two hundred species of *Penstemon*, and most are very beautiful. One who carefully observes penstemon will soon realize that different species of bees and hummingbirds pollinate the plants. That is, each species has a specific pollinator. This is one way of ensuring hybridization between the species.

Native Americans had numerous uses for many of the penstemons, including an eyewash from a solution of soaked leaves, a tea for use as a laxative, powdered roots and leaves for treating sores, and the juice from leaves to treat venereal disease. As a topical astringent, pureed or juiced, the plants can be used as a general dressing for minor irritations of the skin (such as insect bites). Penstemon oil is a good addition to an all-purpose salve.

Speedwell, Brooklime (*Veronica*)

Description: These are annual or perennial herbs with only stem leaves that are opposite each other. The stalked flowers are borne on the end of the stem or on paired stalks that arise from the upper leaf axils. The saucer-shaped corollas are four lobed with the upper lobe the largest and the lower one the smallest. The underside of the petals is lighter in color, often nearly white. There are two stamens, and the seed capsule is somewhat compressed and usually notched or two lobed at the top. The mature fruit is often necessary for identification.

1. Brooklime (*V. americana*)—This species is found in wet places along streams below 10,000 feet throughout the Sierra Nevada.
2. Cusick's Speedwell (*V. cusickii*)—This is a rarely found species that grows in meadows and moist places between 8,200 to 9,500 feet.
3. Purslane Speedwell (*V. peregrina*)—Look for this species in moist places below 10,500 feet throughout the Sierra Nevada.
4. Marsh Speedwell (*V. scutellata*)—This species is found in wet places between 3,500 and 7,000 feet from Fresno County north.
5. Thyme-leaved Speedwell (*V. serpyllifolia*)—This plant with bright-blue to whitish flowers is common in moist places between 6,000 and 11,000 feet throughout the Sierra Nevada.
6. Alpine Speedwell (*V. wormskjoldii*)—This species grows in wet places between 7,000 and 11,500 feet throughout the Sierra Nevada.

Interesting Facts: The genus may be named after Saint Veronica, who was said to have wiped the face of Christ on route to his crucifixion. The name may also have arisen from its beginning as a medicinal herb, after a shepherd observed an injured deer heal its wounds by rolling in and eating the herb. The shepherd then reported this to his sick king. The king became well after trying it and showered the shepherd with riches.

The leaves and stems of all species, when collected during the spring and early summer, can be eaten like watercress, added to salads, or prepared as potherbs. The taste of the various species ranges from spicy to bitter to bland depending on personal taste. The plants also contain moderate amounts of vitamin C and were once used to prevent scurvy. The leaves and stems can also be steeped as a tea. Care should be taken to avoid plants growing in polluted waters. If desired, the addition of halazone tablets or chlorine bleach to the wash water may kill harmful microbes. After flowering, it is better to boil the plants to eliminate the bitterness.

Medicinally, the plants were used mainly as an expectorant for respiratory problems. Infusions were also used in hair-conditioning rinses and skin-cleaning herbal steams and as an ingredient in massage oils and ointments that were added to baths for a soothing soak. Pojar and MacKinnon (1994) indicate that *V. americana* has been used for centuries to treat urinary and kidney complaints and as a blood purifier. The leaf juice from *V. serpyllifolia* has been used for earaches, its leaves were poulticed for boils, and a tea was used for chills and coughs.

Spurge (*Euphorbia crenulata*)
Flower—cyathium

SPURGE FAMILY (Euphorbiaceae)

The spurge family has about 290 genera and 7,500 species distributed worldwide. Among the valuable products of the family are rubber, castor and tung oils, and tapioca. Most members are *poisonous* and have milky sap that will irritate eyes and mouth.

Chinese Caps, Spurge (*Euphorbia crenulata*)

Description: The plant grows to about eighteen inches tall, and the stem leaves are obovate to spatulate in shape. The floral leaves are opposite or occur in threes and often appear perfoliate. It is common in dry places below 5,000 feet. It flowers from May to August.

The flowers are borne in a complex structure called a cyathium. This cuplike structure contains several male flowers and a single female flower. The genus name is from the Greek *euphorbion,* a plant named after Euphorbos, a celebrated Greek physician of the first century BC.

Interesting Facts: *Euphorbia* contains toxic principles that will cause severe *poisoning* if ingested in quantity. Most species contain carcinogenic, highly irritant, diterpene esters and are strong purgatives. The white sap can cause skin irritations and blisters. The Cahuilla Indians in California used both the native and the introduced species as a medicine for reducing fever and as a cure for chicken pox and smallpox. The plant was boiled, and the afflicted person was bathed in the decoction (Bean and Saubel 1972).

Thyme-leaved Spurge (*Chamaesyce serpyllifolia*)

Description: The leaves of this plant are ovate to obovate in shape with distinct linear and entire stipules. It is common in dry, disturbed areas below 7,000 feet throughout the Sierra Nevada. It flowers from August to October.

Interesting Facts: The dried leaves were rubbed into abdominal scratches for children experiencing dysentery and bloating. A decoction of the plant was taken to encourage milk flow in nursing mothers. Plants were also used topically for poison oak.

Turkey Mullein (*Eremocarpus setigerus*)

Description: Turkey mullein is a grayish green annual with a musky smell. It is usually found in dry, often rocky areas at low elevations. The genus name is from the Greek *eremos*, for solitary, and *karpus*, for fruit.

Interesting Facts: The herbage of the plant is probably *poisonous*. Strike (1994) indicates that the Costanoan Indians used a decoction of the root for dysentery, whereas the Kawaiisu used the decoction, internally or externally or both, to relieve headaches and rheumatic pains.

Sweet (1976) says that the plant contains a narcotic, and the foliage was used by some aboriginal peoples to stupefy fish and poison their arrow points. It has been suggested that the stellate hairs get into the fish's gills and hold them open, so in time the fish would drown. Moore (1989), on the other hand, indicates that it may be diterpenes that are the main cause of the plant's effects on fish. Fresh leaves were bruised and applied as a counterirritant poultice for internal pain and asthma.

STAFF-TREE FAMILY (Celastraceae)

These are shrubs with small inconspicuous flowers borne in the axils of the leaves. The four to five sepals are united at the base, and the petals are separate. The fruit is a capsule. There are 60 genera and 850 species distributed in tropical and temperate regions of the world. Ten genera are native to the United States. Several species are cultivated as ornamentals.

Mountain Lover, Oregon Boxwood (*Paxistima myrsinites*)

Description: This is a low, dense, evergreen shrub. The thick, leathery leaves are opposite, oval to elliptic in shape, and the toothed margins are slightly rolled under. The small flowers are maroon in color. The plant is found in coniferous forest, rocky openings, and dry mountain slopes from low to midelevations. The genus is often spelled *Pachistima*.

Interesting Facts: The fruits were eaten by some Native Americans in California, whereas other sources indicate that the plant is inedible. Additionally, the Maidu used the boiled leaves to poultice pain or inflammation (Strike 1994).

ST. JOHN'S-WORT FAMILY (Hypericaceae)

There are forty genera and one thousand species worldwide. The family is of little economic importance in North America. A few species are used as ornamentals.

St. John's-wort (*Hypericum*)

Description: The two perennial species have yellow flowers and small, translucent glands on the leaves and petals. The species can be found in moist areas at various elevations.

FISH POISONS
Fish poisons (piscicides or ichthyotoxins) were used to stupefy or kill fish without making them toxic to humans. A number of plant species are known to possess chemicals toxic to fish. A general rule, however, is that fish poisons are effective only on relatively small fish. These poisons were used in small rivers or pools with slow-moving water. Their use can be tricky, because too much poison could kill all the fish in an area or make them too toxic to eat. The poisons are prepared in a number of ways. The simplest is to crush plants with a rock and throw them in the water.

Some plant poisons and their mode of action:

1. rotenone (Fabaceae)—interferes with respiration;
2. saponins (various families)—causes asphyxiation;
3. cardiac glycosides (Apocynaceae, Asclepidaceae)—affects central nervous system and nerve mechanism of heart;
4. alkaloids (Solanaceae, Ranunculaceae)—many effects;
5. tannins (Fagaceae)—slow acting;
6. ichthyoetherol (Asteraceae)—interferes with respiration; and
7. cyanogenic compounds (Rosaceae, Euphorbiaceae)—produces hydrogen cyanide.

Quick Key to the St. John's-worts

Plants usually growing in mats; stamens distinct	Tinker's Penny (*H. anagalloides*)—This straggly, often matted annual or perennial plant grows in wet places in the mountains from 4,000 to 10,000 feet. It flowers from June to August.
Plants growing erect; stamens occurring in groups	
Leaves are black dotted along the lower margin	St. John's-wort (*H. formosum*)—This slender, erect perennial grows on wet stream banks and in mountain meadows from 4,000 to 7,500 feet. It flowers from June to August.
Leaves with margins that are curled under	Klamath Weed (*H. perforatum*)—This weedy species occurs below 4,500 feet from Tuolumne County northward. It flowers from June to September.

St. John's-wort
(*Hypericum* spp.)

CAM PHOTOSYNTHESIS
This is one of three variations of the photosynthetic process known to occur in plants. Known as crassulacean acid metabolism, or CAM, it is named for the succulent plant family Crassulaceae. This photosynthetic process requires more energy than the other processes (for example, C_4 and C_3). The advantage of CAM photosynthesis for these plants is that it allows them to close their stomates during the day in order to reduce water loss. Therefore, the plants carry on photosynthesis using stored water and carbon dioxide. As such, the energy demand dictates that CAM plants have very slow growth rates.

Interesting Facts: The largest and most widespread of the species is *Hypericum perforatum*. Weedon (1996) indicates that the leaves may be eaten fresh or may be dried and ground to a flour that can be used like acorn meal. However, only a small amount of the herbage should be consumed.

Despite its reputation as a weed, St. John's-wort may have much to offer humans as a medicinal plant. A number of clinical studies strongly suggest the plant may be effective in treating depression. Other laboratory studies reveal that St. John's-wort has at least two compounds, hypericin and pseudohypericin, that are active against retroviruses. As such, it is being closely looked at in Acquired Immune Deficiency Syndrome (AIDS) research (Tilford 1997). The fresh flowers of St. John's-wort in tea, tincture, or olive oil were once a popular domestic medicine for the treatment of external ulcers, wounds, sores, cuts, and bruises. The ancient alleged magical properties of St. John's-wort were partly due to the fluorescent red pigment, hypericin, which oozes like blood from the crushed flowers. The red dye and extracts are used in cosmetics.

Caution: Craighead, Craighead, and Davis (1963) indicate that white-skinned animals feeding on these plants develop scabby sores and a skin itch. Apparently, the plants contain photosensitive toxins and alkaloids. Therefore, ingestion is not recommended in large quantities.

STONECROP FAMILY (Crassulaceae)

Members of this family are succulent herbs or shrubs. There are thirty-five genera and more than fifteen hundred species worldwide, of which nine genera are native to the United States. They are of no real economic importance except as ornamentals.

Quick Key to the Stonecrop Family

Petals one-half to one inch long and fused at the base	Live-forever (*Dudleya*)
Petals less than one inch long and distinct to base	Stonecrop (*Sedum*)

Hen and Chickens, Live-forever (*Dudleya cymosa*)

Description: This glabrous and pale perennial plant with bright yellow– to orange-colored flowers grows in dry, rocky places between 2,000 and 8,500 feet. The genus is named after W. R. Dudley, a western American botanist who lived from 1849 to 1911. It flowers from April to July.

Interesting Facts: The stems and laves of many *Dudleya* species were eaten. Some species were considered delicacies, as the fleshy leaves were eaten raw.

Heated *Dudleya* leaves were used as a poultice and to remove calluses. A leaf poultice of *D. cymosa* hastened the healing of sores or wounds. It was also used as a diuretic.

Stonecrop (*Sedum*)

Description: The Latin word *sedere* means to sit, possibly referring to the tendency of many species growing low to the ground. Stonecrops are well adapted to survival in shallow soil or on rocky outcroppings. The succulent leaves and stems have a waxy coating to help reduce water loss. The reddish color of the foliage in some species is enhanced by sunlight and occurs most often in plants in hot, exposed sites.

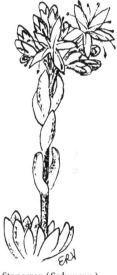

Stonecrop (*Sedum* spp.)

Quick Key to the Stonecrops

Flowers are purple and four-merous; leaves do not form rosettes	Rosy Sedum (*S. rosea*)—This species is found in moist, rocky places between 7,500 and 12,500 feet from Tulare to El Dorado Counties. It flowers from May to July.
Flowers are yellow and five-merous, leaves form rosettes	
Leaves of rosette are spatulate in shape and appear different from those on stem	
Petals are separate to the base	Pacific Sedum (*S. spathulifolium*)—Look for this species in rocky habitats. It flowers from June to July.
Petals are united for at least one-quarter of their length	Sierra Stonecrop (*S. obtusatum*)—This species occurs on rocky ridges and slopes from Tulare to Plumas Counties. It flowers from June to July.
Leaves of rosette are not differently shaped than those on the stem	
Leaves are linear shaped and rounded in cross section	Narrow-petaled Stonecrop (*S. lanceolatum*)—Look for this species in rocky places between 6,000 and 12,000 feet from Tulare to Alpine Counties. It flowers from June to July.
Leaves are lanceolate to elliptical in shape and flattened in cross section	Star-fruited Stonecrop (*S. stenopetalum*)—This species is found in rocky places between 4,500 and 5,500 feet in Lassen County.

Interesting Facts: The young leaves and stems of all species can be eaten as a salad or boiled as a potherb. We find them slightly tart and crisp—a wonderful addition to salads and trail snacks. However, some species have emetic and cathartic properties and can cause headaches. In an emergency, stonecrop can be eaten raw to allay hunger and thirst. The plants are best when collected before flowering because they tend to become bitter and fibrous in late summer. The green fleshy leaves are high in vitamins A and C. The tubers can also be boiled and eaten.

Sedum has also been reported as being slightly astringent and mucilaginous

(Willard 1992). It is valuable in the treatment of wounds, ulcers, lung disorders, and diarrhea. As a field remedy for minor burns, insect bites, and other skin irritations, just squeeze the juice onto the affected area. Decoctions of the plant were also used for sore throats and colds and as an eyewash.

STYRAX FAMILY (Styraceae)

This family occurs in the warmer parts of the Americas, in the Mediterranean, and in eastern Asia.

Snowdrop Bush (*Styrax officinalis*)

Description: This is a deciduous shrub with grayish twigs that grows to be three to twelve or more feet tall. The entire leaves are simple, alternate, round-ovate, and pubescent above and below. The white, showy flowers are fragrant and occur in drooping, terminal clusters. The corolla is variable in the number of lobes but is usually six. The flowers resemble orange blossoms. The fruit is globular and dry. Snowdrop bush grows on mountain slopes and in canyons at lower elevations. It flowers from April to May.

Interesting Facts: The Maidu used sap from this plant as an antiseptic and as an expectorant.

SUMAC OR CASHEW FAMILY (Anacardiaceae)

There are approximately seventy-nine genera and six hundred species in this family. Products originating from the sumac family include resins, oils, lacquers, edible fruits, ornamentals, and tannic acid. Although many members of this family produce edible fruits, the resinous oils can produce extreme dermatitis in sensitive individuals. The family contains the infamous poison ivy (*Toxicodendron rydbergii*), poison oak (*T. diversilobum*), and poison sumac (*T. vernix*).

Poison Oak (*Toxicodendron diversilobum*)

Description: Poison oak is a low shrub or woody vine found in waste places, hillsides, and rocky ravines in the lower elevations of the Sierra Nevada. The leaves are compound with three green, oval-shaped, pointed leaflets, which turn bright red or orange in the fall. The white flowers arise from the leaf axils, and the fruits are white berries.

Interesting Facts: The foliage of poison oak is *poisonous,* causing contact dermatitis. The actual skin irritant is found in the sap. The itchy or painful rash that develops from contact with the sap is greatest in spring and summer when the sap is abundant and the plant is easily bruised. Shortly after contact, the symptoms include itching, burning, and redness. Small blisters may appear after a few to several hours. Severe dermatitis, with large blisters and local swelling, can remain for several days and may require hospitalization. Secondary infections may occur when the blisters are broken. To help alleviate itching immediately, thorough washing with soap and water after contact is recommended.

Because droplets of the irritating chemical can be carried in smoke on dust particles and ash, do not burn poison oak. Smoke carries the oil and can produce a rash over the whole body. If inhaled, a rash can develop in the throat, bronchial tubes, and lungs. The oil can even spread through the body in the bloodstream. Although poison oak can cause havoc for humans, robins, cedar waxwings, flickers, woodpeckers, and other birds relish the berries (Muenscher 1962; Turner and Szczawinski 1991).

Poison Oak
(*Toxicodendron diversilobum*)

URUSHIOL
The heavy oil found in all parts of poison oak is called urushiol. It is incredibly potent, and even a pinhead amount of it can cause rashes and blisters to sensitive people. It is also long lasting, so much so that botanists handling hundred-year-old specimens have developed rashes. Urushiol is confined to the canals inside the plant's leaves, stems, and roots and oozes out readily when bruised or chewed upon. It only takes the oil ten minutes to penetrate the skin, so the adage of quickly washing with soap and water is the simplest field remedy.

Skunkbush (*Rhus trilobata*)

Description: Approximately sixty species of this genus occur worldwide. Skunkbush is a shrub that occurs on dry, sunny slopes and has compound leaves with three leaflets, of which the middle leaflet is largest. Skunkbush has small yellow-green flowers that bloom before the leaves come out. The red-orange fruits are sour tasting. The genus name comes from the Greek *rhous,* which is the name of a bushy sumac. This species is found in woodlands below 3,000 feet.

Interesting Facts: The berries of skunkbush can be eaten raw or soaked in cold water to make a refreshing drink. Malic acid (the cause of tartness in apples) flavors the sour fruits of the sumacs.

Tannic acid, which is present in all parts of the plants, can be used in tanning leather (hides). The leaves, branches, and fruits provide colorfast dyes for wool. The stem produces a yellow dye and the berries a tan or beige dye. Because tannic acid acts as a natural mordant (dye fixer), the fiber does not need to be treated with other chemicals (Bryan and Young 1940).

The slender, flexible branches of skunkbush can be used for weaving baskets, as they are somewhat vinelike (Barrett 1908). The branches can also be used as chew sticks to clean teeth and massage gums. Take a small stem several inches long, remove the outer bark, and chew on the tip to soften fibers. Because some people may have an allergic reaction to the oils of sumac, it is recommended that this be done sparingly.

Skunkbush (*Rhus trilobata*)

SUNDEW FAMILY (Droseraceae)

Members of the sundew family are insectivorous herbs growing in acidic bogs. The leaves are covered with sticky glandular hairs on which insects become trapped. There are approximately four genera and one hundred species, all of which grow in very nutrient-poor soil conditions. *Drosera,* a native of the United States, is cosmopolitan in its distribution, whereas the other genera are monotypic and restricted in their distribution. The family is of no economic importance, except that several species are cultivated as novelties because of their insectivorous habit.

Sundew (*Drosera*)

Description: These are perennials with basal rosettes of leaves covered with viscid, stalked glands that trap and digest small insects. Few to several short-stalked flowers are borne on one side at the end of an erect, naked stem. Each flower has five petals and sepals that are separate to the base, or nearly so. There are four to twenty stamens and three to five deeply divided styles. They are usually found in bogs in association with sphagnum moss.

SUMAC-ADE
The drink "sumac-ade" can be made from the juice of the fruit. To make this drink, simply collect several clusters of the ripened berries, clean off any excess herbage, then bruise the fruits slightly and extract the juice by soaking them in enough cool to warm water to cover the fruit fully. Because sumacs contain high levels of tannic acid, use cool water rather than hot water so little or no tannic acid will be extracted. Soak the berries for about twenty to thirty minutes, strain through a cloth, and drink.

Quick Key to the Sundews	
Leaves are spatulate in shape	English Sundew (*D. anglica*)—This species is infrequently found below 6,000 feet from Nevada County northward. It flowers from July to August.
Leaves are broader than long	Round-leaved Sundew (*D. rotundifolia*)— This species is found throughout the Sierra Nevada. It flowers from July to August.

Interesting Facts: The juice of some *Drosera* species has been used to curdle milk. In the Mediterranean region, sundews are mixed with brandy, raisins, and sugar and allowed to ferment into a drink called rossolis. However, the main use of *Drosera* has been medicinal.

The leaves of *D. rotundifolia* apparently have antispasmodic and expectorant properties. It has been used in the treatment of whooping cough, bronchitis, asthma, and other respiratory problems. To relieve a bad cough, the leaves were tinctured or made into a tea and sipped throughout the day. The plant also contains an antibiotic substance that, in pure form, is effective against streptococcus, staphylococcus, and pneumococcus (Lust 1987).

Schofield (1989) reports that sundew is helpful in a tinctured form for nausea associated with seasickness. Apparently, some drowsiness was associated with the tincture. Sundew juice has been used for removing warts and has been blended with milk to lighten freckles.

Caution: Although there are no reports of human poisoning from ingestion of sundew, it contains corrosive, irritating substances and should be used only in small doses. Cats have been poisoned from daily doses of the plants.

SUNFLOWER FAMILY (Asteraceae; formerly Compositae)

A very diverse family with more than twenty thousand species, the sunflower family is the second-largest plant family in the world. The sunflower family contains many economically important crop plants such as sunflowers, lettuce, and artichokes. Numerous edible and useful composites (members of the sunflower family) are found in California.

Although the family is considered by many botanists as a "difficult" group, composites, in general, are relatively easy to recognize. The small flowers are arranged in heads that at first appear to be an individual flower, although it may actually consist of several to hundreds of florets (little flowers). Each flower has an inferior ovary, five stamens fused at the anthers, and five fused petals. The flowers at the center of the head are disk flowers, whereas the peripheral ones are called ray. Surrounding the head are a series of bracts called the involucre. The calyx, if present (called the pappus and usually modified into thin hairs for dispersal), crowns the summit of the ovary in the form of awns, capillary bristles, scales, or teeth. Nearly all composites are herbs or shrubs. Table 3 summarizes the various tribes of the family (note the suffix *eae*). The pollen of many composites is allergenic. The colorful flowers of many species produce yellow and orange dyes.

Arnica (*Arnica*)

Description: These perennials arise from a rhizome or caudex, and the rootstalk produces a cluster of leaves the first year and a flowering stem the next. The leaves are simple and opposite. Flowering heads, one to several per stem, are radiate or discoid. The pappus consists of many capillary bristles. The name translates as lamb's skin and refers to the modified leaves (bracts) that are usually woolly.

1. Streambank Arnica (*A. amplexicaulis*)—This species is found in moist habitats from 7,000 to 10,000 feet throughout the Sierra Nevada. It blooms from July to August.
2. Meadow Arnica (*A. chamissonis*)—Look for this species in moist habitats between 5,000 and 10,000 feet. It flowers from July to August.
3. Heart-leaved Arnica (*A. cordifolia*)—Heart-leaved arnica is found in many dry to moist habitats between 3,000 and 10,000 feet throughout the Sierra Nevada. It flowers from May to August.

Wild Plants of the Sierra Nevada

A CARNIVOROUS PLANT
Found mostly in boggy places, sundews produces a rosette of small basal leaves and a short, naked stem bearing a more or less one-sided raceme of small, mostly white flowers. Each leaf consists of a slender petiole and a small blade, the upper surface of which is clothed with reddish bristles bearing, at their ends, knoblike glands. These glands secrete a sticky semifluid that glistens in the sun like dewdrops. When an insect touches any of the glands, it sticks fast. Other bristles of the leaf then bend toward the insect until many of the glands are in contact with it, suffocating it. Then they secrete protein-digesting enzymes until the digestible portions of the insect are made soluble and are absorbed by the leaf, after which the bristles resume their former position and are ready for the next victim.

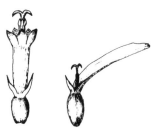

Sunflower Family (Asteraceae)
Flowers—Disk (tubular) and Ray (ligulate)

4. Rayless Arnica (*A. discoidea*)—Look for this species in open woods below 6,000 feet in the northern Sierra Nevada. It blooms from June to August.

5. Lawless Arnica (*A. diversifolia*)—This species is found in moist habitats from 7,000 to 11,000 feet from the central Sierra Nevada northward. It flowers from July to August.

6. Hillside Arnica (*A. fulgens*)—Hillside arnica occurs in moist, open places on the eastern slopes between 6,000 and 9,500 feet in the northern Sierra Nevada. It blooms from May to July.

7. Mountain Arnica (*A. latifolia*)—This rare species can be found in moist habitats in the northern Sierra Nevada between 5,500 and 7,000 feet. It blooms from July to August.

8. Seep-spring Arnica (*A. longifolia*)—This species is found in wet places between 5,000 and 11,000 feet throughout the Sierra Nevada. It flowers from July to August.

9. Cordilleran Arnica (*A. mollis*)—This species can be found in moist places from 6,800 to 11,500 feet throughout the Sierra Nevada. It flowers from July to September.

10. Sierra Arnica (*A. nevadensis*)—Look for this species in dry to moist habitats between 6,600 and 12,000 feet throughout the Sierra Nevada. It blooms from July to August.

11. Nodding Arnica (*A. parryi*)—This species is found in moist habitats between 6,400 and 11,000 feet in the central and northern Sierra Nevada. It flowers from July to August.

12. Recondite Arnica (*A. tomentella*)—This species occurs on open slopes and in woods between 5,000 and 7,000 feet in the central and southern Sierra Nevada. It flowers from July to August.

Interesting Facts: All the species are reported to be *poisonous* if taken internally. *Arnica* contains arnicin; choline; a volatile oil, arnidendiol; angelic and formic acid; and other unidentified substances that can alter cardiovascular activity. The Federal Drug Administration lists *Arnica* as "unsafe" and bans its use for human consumption. Moore (1979) states that *Arnica* is an *external* remedy only. The chopped plant is steeped in rubbing alcohol for about a week and squeezed through a cloth. The liniment is then used for joint inflammations, sprains, and

Arnica (*Arnica chamissonis*)

Table 3 Summary Table of Sunflower Family Tribes

Tribe	Brief Description	Representative Genera
Anthemideae	Aromatic plants with lacy leaves. Only disk flowers or disk and ray flowers. The flower head bracts have membranous edges at tips.	*Achillea, Anthemis, Artemisia, Chrysanthemum, Matricaria*
Asteraceae	These plants have only disk flowers or disk and ray flowers. If both flowers are present in the same head, disk flowers will usually be yellow and ray flowers a contrasting color. The flower head bracts are greenish and shingled in graduated lengths, making several rows.	*Aster, Chrysopsis, Conyza, Erigeron, Grindelia, Solidago*
Cichorieae	Ray flowers only. Stamens and pistils present in each flower. Milky sap.	*Agoseris, Cichorium, Lactuca, Microseris, Sonchus, Taraxacum, Tragopogon*
Cynareae	Disk flowers. Spiny plants that are thistle-like.	*Carduus, Cirsium, Silybum*
Eupatorieae	Disk flowers white, sometimes tinged with pink.	*Eupatorium, Brickellia*
Helenieae	Only disk flowers, or ray and disk flowers. No papery scales in any flowers.	*Helenium, Chaenactis, Eriophyllum*
Heliantheae	Only disk flowers, or ray and disk flowers. One papery scale below all disk flowers. Pappus not hairy.	*Helianthus*

sore muscles. It should not be used if the skin is broken because it is toxic if it enters the bloodstream. *Arnica* is useful as a topical preparation for bruises, sprains, and other closed injuries. When gathering, grasp the plant at the base of the stem just below the ground to leave the rhizome for continued growth. Wear gloves, as the volatile oils can be absorbed.

Arnica has also been used extensively in European folk medicine. Its flower and root have been used as a pain reliever, expectorant, and stimulant. Johann Wolfgang von Goethe (1749–1832), the German philosopher and poet, drank *Arnica* tea to ease his angina in old age (Coffey 1993).

Warning: All species of *Arnica* are reported to be *poisonous* if taken internally.

Aster (*Aster*)

Description: The many aster species in the Sierra Nevada are perennial herbs with alternate leaves. The ray flowers are pistillate containing female parts, ranging in color from blue, purple, and white to pink (never yellow). The central disk flowers are usually yellow. The involucral bracts are in many overlapping series, like shingles on a roof. Pappus is composed of capillary bristles. Asters usually flower from late summer into fall. A related genus, *Erigeron* (fleabane), is often confused with *Aster,* but fleabanes usually flower from late spring to midsummer and the bracts of the involucre are in one to two series. Asters are found in various habitats from low elevations into the alpine zone. The genus name is from the Greek and Latin meaning star, referring to the radiating ray flowers.

1. Aster (*A. alpigenus*)—This aster grows in moist or boggy meadows from 4,000 to 11,500 feet. It blooms from June to September.
2. Long-leaved Aster (*A. ascendens*)—This aster grows in moist or dry soil from 200 to 7,500 feet. It blooms from July to October.
3. Brewer's Aster (*A. breweri*)—Look for this species on open, rocky slopes and in forests between 4,500 and 10,900 feet throughout the Sierra Nevada.
4. Western Meadow Aster (*A. campestris*)—This species occurs on dry, open slopes between 6,000 and 8,000 feet in the northern Sierra Nevada and mainly on the east side.
5. Eaton's Aster (*A. eatoni*)—This species occurs in wet, cool habitats.
6. Alpine Leafy-bract Aster (*A. foliaceus*)—This species occurs in open woods to alpine meadows between 4,500 and 10,000 feet.
7. Marsh Aster (*A. fondosus*)—This species can be found below 6,000 feet in the southern Sierra Nevada.
8. Entire-leaved Aster (*A. integrifolius*)—This aster grows in dry habitats between 5,500 and 10,500 feet.
9. Western Mountain Aster (*A. occidentalis*)—This aster grows in moist meadows from 4,000 to 10,500 feet. It blooms from July to September.
10. Oregon White Aster (*A. oregonensis*)—Look for this species in dry woods from 3,500 to 6,500 feet.
11. Peirson's Aster (*A. peirsonii*)—This aster can be found in moist habitats between 11,000 and 12,500 feet in Tulare, Inyo, and Fresno Counties.
12. Broad-leaf Aster (*A. radulinus*)—This is found in dry forests below 5,000 feet.

SUN TRACKING
To survive the high elevation environment of the alpine zone, many plants have developed interesting adaptations. One such adaptation includes sun tracking. The flower head is rather large in comparison to the rest of the plant, and the flower faces the sun at all times of the day. This action enables direct rays of sunlight to warm reproductive parts, which accelerates their rates of development.

Interesting Facts: The leaves of several species were boiled and eaten by some Native American tribes. Some asters are known to absorb selenium. The Cheyenne Indians used a tea from *A. foliaceus* in the form of eardrops to relieve earaches. *A. laevis* (smooth aster) was burned to create smoke in a sweat bath, and the crushed foliage of *A. hesperius* (= *A. lanceolatus* ssp. *hesperius*) (Siskiyou aster) was sprinkled on live coals and inhaled to treat nosebleeds (Chamberlain 1901; Fernald and Kinsey 1958; Hart 1981).

Balsamroot (*Balsamorhiza*)

Description: These are low perennial herbs with thick rhizomes, and the leaves are mostly basal, large, and long petioled. The yellow flowering heads are large and showy, mostly on long peduncles. Balsamroot is often confused with *Wyethia* (mule-ears), which can be found in similar habitats. However, *Wyethia* leaves lack the fuzzy gray appearance seen on the balsamroot.

Quick Key to the Balsamroots

Plants densely hairy; leaves hairy beneath when young	Arrow-leaved Balsamroot (*B. sagittata*)—This perennial plant grows from 4,300 to 8,300 feet in Kern and Inyo Counties northward. It is common in sandy soil. It flowers from May to July.
Plants not densely hairy; leaves are green on both sides	Balsamroot (*B. deltoidea*)—This perennial herb can be found in open areas from 2,000 to 7,000 feet.

Arrowleaf Balsamroot
(*Balsamorhiza sagittata*)

Interesting Facts: Although *B. sagittata* is considered one of the most versatile sources of food, it is not necessarily palatable. The plants contain a bitter, strongly pine-scented sap. The large taproot, root crowns, young shoots, young leafstalks and leaves, flower budstalks, and the seeds were all eaten by various Native Americans. The larger mature leaves were often used in food preparation (that is, wrapping).

The woody taproot of perhaps all species is edible raw or cooked. The polysaccharide inulin is the major carbohydrate found within the root. The roots can be collected throughout the year but are very difficult to dig out. In some species, the taproot may be as large as one's forearm. Cooking the roots is yet another challenge. One method we have used involves peeling the roots by pounding them to remove the bark. These were then pit cooked for twenty-four or more hours. When properly cooked, the roots turn brownish and sweet tasting. Another way to prepare the roots is to pit steam large quantities for a day and then mash and shape them into cakes for storage. Cooked this way, the roots were called "pash" or "kayoum" (Hart 1996).

The young shoots are edible raw or pit cooked before they emerge in early spring. The young stems and leaves can also be eaten raw or boiled as greens. The older stems are fibrous and tough and will require some additional boiling.

The flower budstalks are collected while the buds are still tightly closed, then peeled and eaten raw or cooked as a green vegetable. They have a slightly nutty taste.

When harvested from dried heads, the seeds can be roasted and eaten or ground into flour. The chaff is usually removed by winnowing.

The roots are said to be antimicrobial and an expectorant, disinfectant, and immuno-stimulant. They can be mashed and applied to swellings and insect bites. Native Americans considered a boiled solution from the root of *B. hirsuta* (= *B. hookeri* var. *neglecta*) (neglected balsamroot) to be an excellent medicine for stomachaches and bladder troubles. The mashed roots of arrowleaf balsamroot were also used by Native Americans to treat swellings or insect bites (Tilford 1997).

Bidens (*Bidens*)

Description: These are annual plants usually found in moist areas. The stems are erect and have opposite, simple leaves. There are few or no yellow ray flowers, whereas there are many yellow disk flowers. There are two species in the Sierra Nevada.

Wild Plants of the Sierra Nevada

Quick Key to the Bidens	
Leaves sessile	Nodding Blue-marigold (*B. cernua*)—This species is found in elevations below 6,000 feet.
Leaves with petioles	Sticktight (*B. tripartita*)—Look for this species in elevations below 4,000 feet.

Interesting Facts: The genus name is the Latin *bidens,* referring to "with two prongs or points," in reference to the prominent prongs on the seeds. Once these seeds penetrate the clothing, they are very difficult to remove. The small seeds are edible.

Brickellbush (*Brickellia*)

Description: The three species of *Brickellia* that occur in the Sierra Nevada are perennial herbs with fibrous roots. The disk flowers are all tubular, white or creamy to pink-purple. They can be found in a variety of habitats at the lower elevations. This is a large and complex genus consisting mostly of shrubs.

Quick Key to the Brickellbushes	
Leaves opposite	Large-flowered Brickellbush (*B. grandiflora*)—This species is found on dry, rocky slopes between 4,500 and 8,000 feet in the central Sierra Nevada northward.
Leaves alternate	
Flowering heads in small clusters	California Brickellbush (*B. californica*)—California brickellbush is found in washes and on dry slopes of the foothills and chaparral, usually below 8,000 feet. It blooms from August to October.
Flowering heads solitary	Mountain Brickellbush (*B. greenei*)—This species occurs mainly in dry, open, rocky places between 2,700 and 8,000 feet in the northern Sierra Nevada northward.

Interesting Facts: Moore (1989) states that a tea or tincture from *B. grandiflora* has three distinct uses: lowering blood sugar in certain types of diabetes, stimulating hydrochloric acid secretions by the stomach, and stimulating bile synthesis and gallbladder evacuation. Others species were also probably used medicinally by Native Americans.

Broom Snakeweed (*Gutierrezia sarothrae*)

Description: This is a sticky, glandular perennial, appearing almost shrubby, and grows six to twenty-four inches high. The leaves are entire, linear-filiform, and less than one-eighth inch wide. The flowering heads are small, numerous, and with whitish, leathery, involucral bracts. Ray flowers are yellow, number three to eight per head, and are one-eighth inch long. There are three to eight disk flowers, and the pappus consists of two to eight stiff awns. Broom snakeweed grows on dry slopes up to 8,000 feet. It blooms from May to October. Another common name, matchweed, refers to the matchlike appearance of the flower heads.

TINCTURES
Here is one general procedure for making tinctures. First, place the fresh plant in a glass jar and cover with eighty-proof brandy or vodka. Keep the jar in a warm dark place for about two weeks, shaking it daily. After two weeks, strain the herbs through muslin or two layers of cheesecloth. Squeeze well to extract as much fluid as possible. Discard the herbs and bottle the tincture. The dosage varies depending on what the herb is to be used for. For sundew, three to six drops in a cup of water is recommended. Vinegar can be used in place of alcohol but has a shorter shelf life (one to two years) than alcohol tinctures (thirty to forty years). Vinegar tinctures are often used for babies, alcoholics, and persons with liver problems.

Interesting Facts: As with many aromatic plants, this species was used medicinally. The plant was boiled to make a tea for colds, coughs, and dizziness. The
tops of fresh, mature snakeweed were boiled until strong and dark. The liquid
could be drunk for lung trouble and colds or applied externally for skin ailments
such as heat rash, poisoning, and athlete's foot. For respiratory ailments, the root
was boiled in water and the steam inhaled (Moerman 1986).

California Helianthella (*Helianthella californica*)

Description: This plant with slender stems has herbage that is more or less glabrous. The leaves are lanceolate, often mostly basal. Flowering heads are solitary
on long peduncles with yellow disk and ray flowers. The pappus consists of two
short, marginal awns. The genus name is diminutive for *Helianthus*. It occurs on
grassy sites and in woodlands below 8,000 feet and flowers from May to September.

Interesting Facts: Flowers were cooked and eaten by several California Native
peoples (Strike 1994).

Canadian Horseweed (*Conyza canadensis*)

Description: This is an annual weed similar to *Erigeron* that grows to about two
feet tall with numerous narrow leaves. There are also numerous white flower
heads. Canadian horseweed is usually found growing in waste places at the lower
elevations, below 6,000 feet.

Interesting Facts: A native to North America, Canadian horseweed was introduced into Europe around the mid-seventeenth century, where it became widely
known for its tonic and astringent properties. A tea was made from the entire
dried plant and used for gravel dropsy, diarrhea, and scalding urine. Native
Americans used the plant in the form of a tea for leukorrhea and applied the
solution to external sores in cases of gonorrhea (Callegari and Durand 1977).
Foster and Duke (1990) also indicate that *E. canadensis* (= *C. canadensis*) was
used as a folk diuretic, astringent for diarrhea, and for kidney stones, nosebleeds,
fevers, and cough. The leaves and tops of Canadian horseweed can be pounded
and eaten uncooked.

Chamomile (*Anthemis*)

Description: Three species of *Anthemis* are found in California: *A. arvensis* (corn
chamomile), *A. cotula* (stinking chamomile), and *A. tinctoria* (golden chamomile). They are annual or short-lived perennial herbaceous plants with radiate
flowering heads composed of white or yellow ray flowers. Introduced European
weeds, they are usually found at the lower elevations as escapees from gardens
and cultivation.

Interesting Facts: A tea can be made from stinking chamomile to induce
sweating and vomiting. An astringent and diuretic, it has been used for ailments
such as fevers, colds, diarrhea, dropsy, rheumatism, and headaches (Foster and
Duke 1990). The leaves were rubbed on insect bites and stings. Golden chamomile was originally considered a noxious weed of clover fields but has since been
brought into cultivation for horticultural purposes.

Chicory (*Cichorium intybus*)

Description: This is a perennial herb that grows up to three feet tall with
dandelion-like leaves. The blue flower heads, which can be seen from spring to
fall, are composed of fifteen to twenty or more ray flowers. The sap is milky.

Chicory is a plant of waste places and is found at the lower elevations. Introduced from Europe, it now grows throughout the United States.

Interesting Facts: Although the roasted root is used for coffee, it is not considered a very satisfactory substitute by itself. Many coffee producers have used chicory as a coffee additive.

The young basal leaves and flower buds hidden at the base of the leaves are edible and best if collected from fall to spring. Because they are bitter, we found it necessary to boil them in at least one to three changes of water. When collected very young, the plants are milder when eaten raw. In some European countries, the buds are pickled and canned (Tull 1987).

CHICORY COFFEE ADDITIVE
To make a coffee additive, dig up chicory roots in the fall through spring, scrub them, and slice in half. Roast them in an oven at a low temperature (for example, 250 degrees Fahrenheit) for two to four hours or until they become dark brown and brittle. Break up and grind as you would coffee. One part chicory to four parts coffee is a common ratio when brewing.

Cocklebur *(Xanthium)*

Description: Two species, *X. spinosum* (spiny cocklebur) and *X. strumarium* (rough cocklebur), occur in California. These are coarse annual weeds of uncertain origin that have a cosmopolitan distribution. The stems of the plants are simple, and the leaves are alternate. The flower heads are solitary or clustered in the leaf axils. The bur (seed) has conspicuous, slender, hooked prickles. Cockleburs can be found in the lower elevations.

Interesting Facts: The uses of cocklebur are primarily medicinal. They have been used by many aboriginal people throughout North and South America and as a herbal medicine in China. The seeds were ground, mixed with cornmeal, made into cakes or balls, and steamed by the poorer class of Zuni Pueblo.

Historically, the roots of *X. strumarium* were used for scrofulous tumors (related to the lymph glands in the neck). Both species were also used for rabies, fevers, and malaria. They possess diuretic, fever-reducing, and sedative properties. Native Americans used a leaf tea for kidney disease, rheumatism, tuberculosis, and diarrhea and as a blood tonic. The seeds have germicidal qualities and were ground and applied to wounds. The seeds also contain an oil that can be used as lamp fuel.

Common Burdock *(Arctium minus)*

Description: This is a coarse biennial with large rounded to ovate leaves. There are several flowering heads composed of pink or purplish tubular flowers. This naturalized species from Europe is an occasional weed at the lower elevations. It flowers from June to August. *A. lappa* (greater burdock) also occurs in California and can be used in much the same way as common burdock.

Interesting Facts: Rich in vitamins and iron, the young leaves and shoots can be gathered for use as a potherb or eaten raw in salads. The plant has a strong rank taste and an objectionable odor. The inner pithlike material of the young stems can be eaten raw, but we find it better when boiled in one or two changes of water. The roots of young plants can be sliced and cooked, then eaten. The older roots can be roasted and ground for use as a tea or coffee substitute. Seeds can be dampened and grown as sprouts.

There is no clear evidence about exactly when our ancestors started to use fire. Among the earliest signs are layers of ash found in a cave (Zhoukoudian) in China. These date back to about half a million years, a time when the cave was inhabited by *Homo erectus*, the species that preceded humans. Making fire, as opposed to starting it from a natural blaze, is a much more recent skill. It is believed that this ability probably started less than twenty-five thousand years ago.

The medicinal uses of the plant predate its use as a food plant. The Chinese are said to have used the plants as a blood purifier for thousands of years. Current research has confirmed the usefulness of common burdock in the treatment of rheumatism, water retention, and high blood pressure (Moerman 1977). As a wash, it was used externally for hives, eczema, and skin problems. The crushed seeds were used as a poultice.

The tall rigid stems were used as drills for primitive fire-starting techniques. The burs can be used as a survival "Velcro" for holding clothes together.

Common Dandelion (*Taraxacum*)

Description: Dandelions need very little introduction. All species are taprooted perennials with milky juice and leaves that form a dense, basal rosette. The solitary flower head is composed of bright-yellow ray flowers. They are found in a variety of habitats up to the alpine zone. The genus name is from the Greek *tarrassein*, meaning to stir up. The plant was an ancient remedy for ailments ranging from spring doldrums to mononucleosis. The latexlike sap was a folk medicine for warts. Red-seeded dandelion (*T. laevigatum*), formerly considered a separate species, is now referred to as *T. officinale*.

Interesting Facts: Common dandelion was introduced into North America by European settlers as a food crop and a medicinal cure-all. Every part of common dandelion is edible. The young leaves may be eaten raw or cooked like spinach. The older leaves are also edible, but we find it is better to boil the older leaves in one or two changes of water to eliminate the bitterness that comes with age. The plants are high in vitamins A and C and a good source of B complex, iron, calcium, phosphorous, and potassium. The roots can also be eaten raw, boiled as a vegetable, baked as potatoes, or added to soups and stews. The roasted root can be used as a substitute for coffee, but it lacks the caffeine buzz. The flower buds can be pickled and added to meals such as omelettes. In general, common dandelion is good for blood circulation.

The less common native species of *Taraxacum* may also be edible, but no authenticated information has been located. Kirk (1975) also believes that the other native species have similar qualities. During World War II, a species of dandelion was cultivated by Russians as a commercial source of rubber. In spite of its many uses, many Americans consider the dandelion a pest and spend much time and money trying to eradicate it.

Coneflower (*Rudbeckia*)

Description: The flowering heads are solitary and terminal, large and showy. Both ray and disk flowers are present. The disk flowers are conical, greenish yellow, and elevated above ray flowers. The ray flowers are yellow and there are eight to twenty-one per head.

Coneflower (*Rudbeckia* spp.)

Quick Key to the Coneflowers	
No ray flowers in the flowering head	Blackheads (*R. occidentalis*)—Look for this species on wet ground in woods between 4,000 and 6,000 feet from Placer to Plumas and Butte Counties.
Ray flowers present in flowering head Pappus is present	California Coneflower (*R. californica*)—This species is occasionally found in moist meadows between 5,500 and 7,800 feet from El Dorado County southward.
No pappus is present	Black-eyed Susan (*R. hirta*)—This is an introduced species in the mid-altitude meadows of the Sierra Nevada from Mariposa to Amador Counties.

Interesting Facts: The genus name honors the Swedish father and son who were professors of botany and predecessors of Linnaeus—O. J. Rudbeck (1630–1702) and O. O. Rudbeck (1660–1740).

Several species are suspected of poisoning livestock when eaten in quantity. A root tea from *R. hirta* (black-eyed Susan) was used for worms and colds and as an external wash for sores, snakebites, and swelling. The root juice can also be used for earaches. A root tea from *R. lacinata* (cutleaf coneflower) was drunk for indigestion, a poultice made from the flowers was applied to burns, and the cooked spring greens were eaten for "good health." Some Native Americans cooked the entire flower bud of California coneflower for food and collected the ripe seeds in the fall to grind into a meal.

Corethrogyne (*Corethrogyne filaginifolia* = *Lessingia filaginifolia*)
Description: This is a perennial herb clothed with a soft white deciduous wool, resembling an aster. The genus name is from the Greek *korethron*, a brush for sweeping, and *gyne*, for style, from the brushlike-style appendages. There are about eleven varieties of this species. It flowers from June to October.

Interesting Facts: Native Americans in northern California made an infusion of leaves and twigs that then was burned in sweatlodges to induce sweating. Inhaling the herbal steam was said to relieve colds. Additionally, an infusion of boiled flowers was drunk to relieve chest pains.

Cudweed (*Gnaphalium*)
Description: These are woolly annual, biennial, or short-lived perennials herbs with alternate leaves. The plants are often confused with *Antennaria*, but cudweeds have both male and female flowers on the same plant and are taprooted. The disk flowers are yellow or whitish. The species are found from the low to midelevations in moist, open areas to well-drained soils.

1. Cudweed (*G. canescens*)—Look for this species in dry, open habitats below 8,700 feet. It flowers from July to October.
2. Lowland Cudweed (*G. palustre*)—This odorless annual is found below 9,500 feet. It flowers from May to October
3. Gnaphalium (*G. stramineum*)—This is an aromatic annual or biennial plant that grows in moist, open waste places below 6,000 feet. It flowers from June to October.

Interesting Facts: The bruised plant assists in healing wounds, and steeping the leaves in cold water increases perspiration. Some species contain pyrrolizidine alkaloids and should be regarded as potentially toxic.

PYRROLIZIDINE ALKALOIDS Most alkaloids are amino acid derivatives and have no certain role in plant metabolism except for serving to repel insects and herbivore predators due to their bitter taste. Pyrrolizidine alkaloids (PAS) are of special interest currently because several of them have been shown to cause toxic reactions in humans, primarily veno-occlusive liver disease, when ingested with foods or herbal medicines. Comfrey, a well-known medicinal herb characterized by U.S. Food and Drug Administration researchers as having been "one of the most popular herb teas in the world," contains PAS that are capable of causing liver damage. The PAS are primarily found in members of three plant families: Asteraceae, Boraginaceae, and Fabaceae.

Cushion Stenotus (*Stenotus acaulis*)
Description: Formerly known as *Haplopappus acaulis*, this plant with numerous stems arising from a woody base has leaves that are erect, entire, and spatulate or narrower in shape. The species occurs in dry, rocky habitats between 5,400 and 10,500 feet on the eastern side of the range. It flowers from May to August.

Interesting Facts: A related species (*S. lanuginosus*) was used by the Navajo for indigestion, nose troubles, sore gums, throat troubles, and as a toothache medicine (Zigmond 1981).

Fleabane (*Erigeron*)
Description: Many species of fleabane occur in California. They are characterized as annual, biennial, or perennial herbs with alternate or basal leaves. The flowering heads are radiate with narrow ray flowers that may be white, pink, blue, purple, or occasionally yellow. The numerous disk flowers are yellow, and the pappus is composed of capillary bristles. The various species bloom mostly in the spring and early summer, except at the higher elevations (see aster), and can be

found in a variety of habitats. Fleabanes resemble aster, but are distinguished by fewer rows of involucral bracts. Fleabanes also bloom earlier than asters. The genus name comes from the Greek *eri* (early) and *geron* (old man). The common name, fleabane, comes from the belief that these plants repelled fleas. This genus is rather difficult to work with in the field. At least eighteen species of *Erigeron* are known to occur in the Sierra Nevada.

Interesting Facts: Fleabanes, in general, are listed as astringents and diuretics. The disk flowers of a related species, *E. philadelphicus* (Philadelphia fleabane), were powdered to make a snuff to cause one to sneeze and break up a cold or catarrh (Arnason, Hebda, and Johns 1981). A tea from the entire plant of *E. annuus* (eastern daisy fleabane) was used to treat a sore mouth. The dried roots, stems, and flowers of *E. peregrinus* (subalpine fleabane) were steeped in hot water, and the patient would breathe the vapors. Fleabanes may cause dermatitis in some people.

Glyptopleura (*Glyptopleura marginata*)

Description: This tufted annual has milky sap. The stems are prostrate, and the leaves are lobed with toothed lobes. Flowering heads are solitary or few, and the flowers are pale yellow inside.

Interesting Facts: The leaves and stems were eaten raw by the Paiute (Merriam 1966; Fowler 1989).

Goldenbush (*Ericameria*)

Description: These are erect, resinous shrubs that bloom in the late summer and fall. Leaves are alternate. Many of the species below were in the genus *Haplopappus*.

1. Golden Fleece (*E. arborescens*)—This species is found in dry habitats mainly below 4,000 feet but sometimes up to 9,000 feet in the southern Sierra Nevada.
2. Bloomer's Goldenbush (*E. blommeri*)—Look for this species in dry habitats throughout the Sierra Nevada between 3,500 and 9,500 feet.
3. Wedgeleaf Goldenbush (*E. cuneatus*)—This species grows in cliffs and ridges below 9,000 feet from Plumas County southward.
4. Whitestem Goldenbush (*E. discoidea*)—This species occurs in rocky, mainly open slopes between 9,000 and 12,000 feet from Nevada County southward.
5. Green's Goldenbush (*E. greenei*)—This species occurs in rocky openings in forest situations in the northern Sierra Nevada below 6,500 feet.
6. Singlehead Goldenbush (*E. suffruticosus*)—This species is found on open, rocky slopes and ridges between 7,800 and 12,000 feet from Nevada County southward.

Interesting Facts: In general, the seeds and stems were eaten by many Native Americans. A leaf decoction of golden fleece was used by the Miwok to treat stomach ailments (Barrett and Gifford 1933). Women drank the tea to relieve cramps and pain associated with menstruation and childbirth. A poultice of bloomer's goldenbush leaves and twigs eased rheumatic pains. Additionally, the leaves were applied to boils to bring them to a head. The Miwok used a stem decoction of wedgeleaf goldenbush to treat colds (Barrett and Gifford 1933).

Goldenrod (*Solidago*)

Description: The various species of goldenrod are perennial herbs with fibrous roots. The leaves are alternate, simple, and either tooth or entire. The heads are made up of yellow ray flowers. Goldenrods may be found in dry to moist habitats from the foothills to timberline, often in dense patches. *Solidago* means to make whole in Greek.

Quick Key to the Goldenrods	
The leaves located in the middle of the stem are largest	Meadow Goldenrod (*S. canadensis*)—This species occurs in meadows and moist openings in woods. It flowers from May to September.
The leaves on the lower part of the stem are largest	
The inflorescence is flat topped or spherical in shape; there are ten to twenty flowering heads; plants occur mainly above 8,000 feet	Alpine Goldenrod (*S. multiradiata*)—Look for this species in sunny rocky or grassy places between 8,000 and 12,000 feet. It blooms from June to September.
The inflorescence is elongated and has many flowering heads; plants are found mainly below 8,000 feet	
Plants densely hairy and growing in dry habitats	California Goldenrod (*S. californica*)—This is a perennial plant that is common in dry or moist fields and in clearings or openings from 200 to 7,500 feet. It blooms from July to October.
Plants not hairy and growing in wet habitats	Showy Goldenrod (*S. spectabilis*)—This species is found in bog and alkaline meadows on the eastern side of the Sierra Nevada northward.

Interesting Facts: Goldenrods are a leading contestant for being our national flower, as there is hardly a part of this land where it is not found. Young leaves can be prepared as potherbs or added to soups. Depending on habitat, age, and personal preference, their palatability is quite variable. The dried leaves and dried, fully expanded flowers can be used to make a tea. The seeds can be used to thicken stews. Large amounts of the raw herbage should be avoided, as they may be toxic.

Medicinally, *Solidago* was employed for checking internal and external bleeding. An antiseptic lotion may be made by boiling the stems and leaves, or by using dry, powdered leaves. The powdered dry leaves were also sprinkled on cuts as a styptic. For insect bites and minor scrapes, apply fresh, crushed, or chewed leaves. A tea wash is said to be good for rheumatism, neuralgia, and headaches.

The fluffy down from the flower heads is a good additive for tinder bundles. All goldenrods contain small quantities of natural rubber; Thomas Edison at one time had hoped to be able to use them to make a rubber substitute.

Gray Tansy (*Sphaeromeria cana*)

Description: This is an aromatic perennial with several slender stems. Leaves are linear oblong to narrowly obovate with three linear oblong segments. Disk flowers are lemon-yellow. The species is found in dry, rocky places between 9,000 and 12,000 feet from the central Sierra Nevada southward. It flowers from July to August.

Interesting Facts: Young leaves can be used to flavor stews. The herbage is probably poisonous if eaten in quantity.

Groundsel (*Senecio*)

Description: These are annual, biennial, or perennial herbs with alternate or basal leaves. The flower heads are yellow, and the pappus is made up of hairlike bristles. Groundsels can be found in various habitats and elevations. *Senecio* is one of the largest genera of plants with nearly two to three thousand species distributed worldwide. Approximately one hundred species are found in the western United States. There are many species of *Senecio* in the Sierra Nevada.

Interesting Facts: Many species contain highly toxic alkaloids and should therefore be avoided. A related species, *S. douglasii* (Douglas' groundsel), was used medicinally by southwestern Native Americans as a laxative, although misuse could result in death. Strike (1994) indicates that young *Senecio* leaves were eaten by Maidu Indians as cooking herbs. Additionally, the seeds may have been eaten by the Chumash Indians. *Senecio* leaves were apparently used to line earthen ovens.

Gumweed (*Grindelia*)

Description: These species are biennials or short-lived perennials of waste places at low elevations. The leaves are alternate and have toothed margins. A sticky, resinous sap covers the leaves and bracts of the yellow flowers.

Interesting Facts: In general, gumweeds are considered toxic, and the toxicity appears to be dependent upon the soil in which they grow. However, many species have been used medicinally for hundreds of years.

The sticky flowers heads of *G. squarrosa* (curlycup gumweed) were used as a chewing gum substitute. The young leaves make an aromatic, bitter tea. The flower heads can be boiled in water and used as an external remedy for skin diseases, scabs, and sores. A hot poultice of the plant was used for swellings. A tea made from the plant was used for coughs, pneumonia, bronchitis, asthma, and colds.

Caution: *G. squarrosa* tends to concentrate selenium and is therefore considered *toxic*.

Hawksbeard (*Crepis*)

Description: In general, these are perennial, taprooted herbs with milky juice. The leaves are alternate or all basal, and the yellow flowers are all ray. The various species can be found in dry, open places at lower elevations to gravelly or rocky places in alpine or subalpine areas. This is a rather difficult genus to work with in the field. At least eight species are known to occur in the Sierra Nevada.

Interesting Facts: The stems and leaves of *Crepis* were eaten by Native Americans. The Karok Indians of northern California peeled the stems of *C. acuminata* before eating (Baker 1981).

The seeds or whole plant of *C. acuminata* was thoroughly crushed and applied as a poultice to breasts after childbirth to induce milk flow. The root of the plant was used to remove a foreign object from the eye. The root can also be ground into a smooth powder and sprinkled in the eye to treat eye problems. Several applications are necessary.

Hawkweed (*Hieracium*)

Description: Hawkweeds are fibrous-rooted perennial herbs with milky juice. Their flowers are all ray, yellow to sometimes orange or white in color. The name hawkweed comes from the belief by the ancient Greeks that hawks would tear apart a plant called the hieracion (from the Greek *hierax,* meaning hawk) and wet their eyes with the juice to clear their eyesight. Hawkweeds are found in a variety of habitats up to the subalpine.

Quick Key to the Hawkweeds	
Ray flowers white or pale yellow	White-flowered Hawkweed (*H. albiflorum*)—This is an erect perennial plant with white flowers that grows in dry, open, wooded places below 9,700 feet. It flowers from June to August.
Ray flowers yellow Leaves mainly basal; stems twelve to twenty-eight inches tall	Scouler's Hawkweed (*H. scouleri*)—This species is found in open woods and rocky places below 6,500 feet. It flowers from May to July.
Stems usually with leaves Herbage densely long-hairy	Hawkweed (*H. horridum*)—This perennial plant with yellow flowers is common on dry and rocky slopes from 5,000 to 11,000 feet. It blooms from July to August.
Herbage with some pubescence; leaves mainly basal	Alpine Hawkweed (*H. gracile*)—Look for this species in woods and rocky places between 8,000 and 11,000 feet. It flowers from July to August.

Interesting Facts: The green plant and juices of white-flowered hawkweed may be used as a substitute for chewing gum, although it is best when dried first. The plant was also used to ease toothaches, to cure warts, as an astringent in treating hemorrhages, and as a general tonic (Strike 1994).

Hazardia (*Hazardia whitneyi*)

Description: This perennial grows to about three feet tall. The leaves are oblong and serrated. There are no ray flowers, or there may be five to eight of them and fifteen to thirty disk flowers. It occurs on rocky, open slopes between 4,000 and 10,000 feet from Plumas County southward. It flowers from July to September.

Interesting Facts: The entire plant of a related species (*H. squarrosa*) was used by the Diegueno as a decoction for bathing aches and pains of the body.

Hoary-aster (*Macharenthera canescens*)

Description: This is a taprooted perennial from a woody crown called a caudex. The leaves are narrow and about one to two inches long, but they are reduced to mere bracts at the top of the branches. The ray flowers on the outside of the flower heads are bright bluish purple to pink. It flowers from July to October.

Interesting Facts: The genus name is from the Greek *machaira* (sword) and *anthera* (anther), referring to the shape of the anthers. The dried plant was pulverized and used as a snuff for nose troubles. An infusion of the whole plant taken and rubbed on the belly was an emetic. A poultice of mashed leaves was applied to a swollen jaw or neck glands, whereas a decoction of fresh or dried leaves was taken for headaches.

Indian Blanketflower (*Gaillardia aristata*)

Description: This is a perennial herb with a slender taproot. The leaves are linear to lance shaped. The flowering heads are solitary or few flowered with yellow ray flowers and purplish disk flowers. It is found in open places at low and middle elevations.

Interesting Facts: The blanketflower is so-called in allusions to the colorful Indian blankets of the Southwest. Several species actually do "blanket" vast fields in Texas. The Blackfoot Indians drank a tea made from the root for gastroenteritis, and the chewed powdered root was applied to skin disorders. They also bathed sore nipples of nursing mothers in a tea made from the plant or used the liquid as an eyewash or nosedrops.

Lessingia (*Lessingia leptoclada*)

Description: Lessingia grows in open areas, pine forests, and sandy soils up to 5,500 feet. It flowers from July to October.

Interesting Facts: This species was used medicinally by some Native Americans. For example, the bark provided a treatment for general aches and pains. In the late summer, the outer bark was peeled off the stems and rolled into small pellets. A few pellets were then placed on the body where there was pain and lit on fire. When these pellets were completely burned, they were replaced by others, and the process was repeated.

Milk Thistle (*Silybum marianum*)

Description: The name milk thistle refers to the white streaks along the leaf veins. In Germany, where the plant is often depicted as a religious symbol associated with the Virgin Mary, legends ascribe the white mottling to a drop of the Virgin Mary's milk. The species name (*marianum*) honors the symbolic association of the plant with the Virgin Mary.

Interesting Facts: This naturalized species from the Mediterranean was brought to America by early settlers, probably as food. The black shiny seeds are crowned with feathery tufts and have been roasted as a coffee substitute.

Medicinally, the plant has gained prominence based on research conducted in the past thirty years. However, it has been considered a "liver-protecting" herb since the first century (Callegari and Durand 1977).

Mountain Dandelion (*Agoseris*)

Description: These are annual or perennial, taprooted herbs with milky juice that resemble common dandelion (*Taraxacum*). The flowers are all ray, yellow, or occasionally orange in color. The pappus is white with hairlike bristles. The fruit (achene) is conspicuously ten-nerved. Mountain dandelions occur on moist to dry ground in meadows and open areas at various elevations. The genus name is from the Greek, meaning goat chicory. At least six species of mountain dandelion are known to occur in the Sierra Nevada.

Flowers are orange, drying to a purplish color	Orange-flowered Agoseris (*A. aurantiaca*)—This species is found in moist, mainly grassy places between 6,000 and 11,500 feet. It flowers from July to August.
Flowers are yellow and often dry a pinkish color	
Beak of achene less than two times as long as the fruit; plants found usually above 5,000 feet	
Beak of achene about half as long as fruit; flowering stalk four to twelve inches tall	Short-beaked Agoseris (*A. glauca*)—Look for this species in dry habitats between 5,000 and 10,500 feet throughout the Sierra Nevada. It flowers from July to August.
Beak of achene about as long as fruit; flowering stalk twelve to twenty-four inches tall	Tall Agoseris (*A. elata*)—This is a rather uncommon species, usually found in moist habitats between 5,000 and 10,500 feet. It blooms from July to September.
Beak of achene about two to four times as long as the fruit; plants found below 8,000 feet	
Plant annual	Annual Agoseris (*A. heterophylla*)—This agoseris grows on open, grassy slopes and flats below 7,500 feet. It blooms from April to July.
Plant perennial	
Pappus nine- to eleven-sixteenths inches long; leaf segments pinnatifid into regular lobes that point backward	Agoseris (*A. retrorsa*)—This agoseris grows on dry ridges and slopes from 2,500 to 8,000 feet. It blooms from May to August.
Pappus five-sixteenths to one-half inch long; leaves irregularly lobed	Large-flowered Agoseris (*A. grandiflora*)—This agoseris grows in dry or moist areas below 6,200 feet and blooms from May to July.

Interesting Facts: The leaves and roots of some species are edible when cooked but are bitter, especially in late season. Strike (1994) indicates that the seeds were eaten by the Chumash Indians in southern California. The sap from the leaves of some species, when hardened, can be used as chewing gum. Because the sap from some species is very thick and insoluble, it may be useful for waterproofing containers (for example, coiled baskets) and footwear.

Nodding Silverpuffs (*Microseris nutans*)

Description: This annual grows 4 to 24 inches tall and has mostly basal leaves and a leafless stem. Leaves are 1½ to 10 inches long, linear to narrow, elliptic in shape, and entire or with a few teeth or pinnatifid. The flowering heads are terminal on the stems. Involucres are ⅝ to 1½ inches long. The heads are composed of all ray flowers, and the ligules are yellow. Achenes are not beaked, and the pappus is of five silvery, papery scales that are tipped with an awn or bristle. *Microseris* grows in grassy places or in open woods below 6,000 feet. It flowers from April to June.

Interesting Facts: The slender roots of nodding silverpuffs are apparently edible raw.

Orange Sneezeweed (*Dugaldia hoopsesii*)

Description: This plant is recognized by its mound of disk flowers, drooping, narrow yellow-orange ray flowers, and the deeply veined, almost white rib in the leaf. The genus name honors Dugald Stewart, an eighteenth-century Scottish philosopher. The species name is for Thomas Hoopes (1834–1925), a prospector and seed collector in Colorado.

Interesting Facts: The whole plant can be used as a tincture as a counterirritant liniment. A snuff was made from the crushed blossoms, and a string of plant leaves was inhaled for headaches and hay fever. The Navajo used the plant to inhibit vomiting, whereas the roots were used as chewing gum (Zigmond 1981). The crushed flowers boiled in juniper ash were used as a yellow dye.

Ox-eye Daisy (*Leucanthemum vulgare*)
Description: The genus name is from the Greek *leukos* (white) and *anthemon* (flower). This is a Eurasian species now naturalized in North America.

Interesting Facts: The species is a popular ornamental and is used in home remedies for catarrh and is more or less edible. An infusion of flowers and roots was used as an eyewash, and the flowers were used to make a tonic.

Pericome (*Pericome caudata*)
Description: This aromatic perennial has many stems arising from the base. The leaves are simple, deltoid-lanceolate, and usually opposite. It is found in dry habitats mainly below 7,500 feet on the eastern side of the range from Mono County southward. It flowers from July to October.

Interesting Facts: A decoction of the root was taken for general body pain and for a cough. A cold infusion of leaves was taken for fever and influenza, whereas a poultice of heated root was applied for toothaches.

Pincushion, Chaenactis (*Chaenactis*)
Description: These are biennial or perennial herbs from a taproot. The leaves are pinnately dissected, and the flowering heads are composed of disk flowers that are white to pink to rose in color. The four species in the Sierra Nevada can be found in open, dry, and rocky habitats from the lower elevations into the alpine. The genus is endemic to the western United States.

Quick Key to the Pincushions

Stems matted and less than four inches tall	
Leaves with few lobes, less than one-half inch wide	Southern Sierra Chaenactis (*C. alpigena*)—This species is occasionally found in sandy or gravelly soils between 8,000 and 12,500 feet. It flowers from July to August.
Leaves bipinnate, one-half to one inch wide	Northern Sierra Chaenactis (*C. nevadensis*)—This species occurs in gravelly soils or talus between 8,000 and 10,900 feet. It flowers from July to August.
Stems usually erect, four to sixteen inches tall	
Plants annual; leaves entire to once pinnate	Chaenactis (*C. xantiana*)—This stout annual is common on slopes of desert mountains from 1,400 to 7,000 feet. It flowers from April to June.
Plant perennial; leaves two to three times pinnately lobed	Hoary Chaenactis (*C. douglasii*)—Look for this species in dry, rocky habitats between 4,000 to 7,000 feet throughout the Sierra Nevada. It flowers from May to July.

Interesting Facts: Leaves of *C. douglasii* were mashed and used to poultice sprains and swellings. A decoction of the plant was used for indigestion, coughs, and colds. Mashed leaves were used on rattlesnake bites (Strike 1994).

Plumeless Thistle (*Carduus*)

Description: Three species are reported to occur in California: *C. nutans* (nodding plumeless thistle), *C. pycnocephalus* (Italian plumeless thistle), and *C. tenuiflorus*. *Carduus* is distinguished from *Cirsium* in that the pappus of *Carduus* is simple and smooth, not a plume. These are weedy species that may occasionally be found along roadsides and other waste places at the lower elevations.

Interesting Facts: Kirk (1975) indicates that the pith of four species (unspecified), without the easily removed rind, may be boiled in salted water and seasoned in various ways. The dried flowers may be used as a rennet to curdle milk. Additionally, Strike (1994) indicates that the raw or cooked leaves and stems and raw buds were also eaten.

Poverty Weed (*Iva axillaris*)

Description: This is a long-lived perennial with creeping roots. It is a widespread native that is a desirable component of salt marsh and alkali plains. In areas that have been disturbed by overgrazing, it will form large clonal colonies that, once established, are difficult to eradicate.

Interesting Facts: The pollen of this plant is highly allergenic, and plants may cause contact dermatitis in sensitive individuals.

Pussy-toes (*Antennaria*)

Description: In general, pussy-toes are herbaceous, often mat-forming, perennials. The heads are discoid, with small white flowers surrounded by bracts that are typically hairy below with a smooth and membranous portion varying in color from white to pink to dark brown or black. The pappus is composed of numerous hairy bristles. The following eight species in the Sierra Nevada can be found in dry, open habitats or in moist or seasonally wet places from the foothills to alpine areas. It is a rather difficult species to identify in the field.

Quick Key to the Pussy-toes

Flowering heads solitary	Antennaria (*A. dimorpha*)—This species occurs in dry places between 2,400 and 7,200 feet.
Flowering heads several to many	
Plants forming mats and with stolons	
Basal leaves one and a half to three and a half inches long and with petioles	Antennaria (*A. luzuloides*)—This species occurs in dry, open slopes and sagebrush scrub between 3,000 and 5,700 feet in the northern Sierra Nevada.
Basal leaves one-half to one inch long and sessile	Antennaria (*A. geyeri*)—Look for this species in dry, open woods and shrub lands between 2,700 and 7,200 feet.
Plants not forming mats and stolons not present	
Phyllaries dark brown or black-green in color	
Herbage glandular	Antennaria (*A. pulchella*)—This is an uncommon species found in snow basins and high meadows between 8,400 and 11,100 feet.

Herbage usually not glandular	Pussy-toes (*A. media*)—Look for this species in high meadows, snow basins, and ridges between 5,400 and 11,700 feet.
Phyllaries usually white to rose in color	
Phyllaries white tipped and with a dark spot at base	Antennaria (*A. corymbosa*)—This species is found in moist meadows and along streamsides from 5,700 to 9,600 feet.
Phyllaries without the dark spot at the base	
Phyllaries sharp pointed	Pussy-toes (*A. rosea*)—This species occurs in woods, meadow edges, rock barrens, and dry ridges from 3,600 to 11,100 feet.
Phyllaries blunt	Antennaria (*A. umbrinella*)—This uncommon species is found in dry sagebrush scrub and open yellow pine forests in the northern Sierra Nevada between 5,400 and 6,000 feet.

Interesting Facts: The sap from the stem of most species can be chewed like gum and has some nutritive value. Moore (1979) indicates that a tablespoon of the chopped plant steeped in hot water is an excellent remedy for liver inflammation. It has also been used as an astringent to the intestinal tract. Leaves can be poulticed for use on bruises, sprains, and swelling. The blossoms could be boiled and used to bathe sore or ulcerated feet or mashed and applied to sores. A related species, *A. microphylla* (littleleaf pussytoes), was chewed as a cough remedy by the Thompson Indians in British Columbia. The tiny leaves were also dried, stripped, and used as one of the ingredients in Indian tobacco (Teit 1930).

Rabbitbrush (*Chrysothamnus*)

Description: The four species of *Chrysothamnus* in the Sierra Nevada are shrubs with alternate, sessile, entire, and linear leaves. The flowering heads are composed of five to thirty yellow disk flowers and typically bloom in late summer and fall. Rabbitbrush is found in dry, open places at low to middle elevations.

Quick Key to the Rabbitbrushes

Stems of plant glabrous, slightly hairy, or gland-dotted	
Flowers two to three per head	Rabbitbrush (*C. humilis*)—This species is occasionally found in sagebrush scrub above 5,000 feet on the eastern side of the range.
Flowers more than three per head	Sticky-leaved Rabbitbrush (*C. viscidiflorus*)—This shrub is common in dry, open areas from 4,000 to 7,500 feet. It flowers from July to September.
Stems of plant tomentose to felted	
Heads in leafy racemes	Parry's Rabbitbrush (*C. parryi*)—This is a leafy shrub that blooms from July to September.
Heads in rounded, terminal clusters	Common Rabbitbrush (*C. nauseosus*)—This ill-smelling shrub grows in dry areas from 6,000 to 9,500 feet. It flowers from July to September.

Interesting Facts: A tea was reported to be made from the twigs of *C. nauseosus* that provided relief from chest pains, coughs, and toothaches. The leaves and stems were also boiled, and the liquid was used to wash itchy areas.

Great Basin Indians were accustomed to chewing the stems of rabbitbrush to extract the latex. They believed that chewing rabbitbrush relieved both hunger and thirst. The secretion obtained from the top of the roots can also be chewed as gum.

The rubber shortage of World War II stimulated research on rabbitbrush and other rubber-producing plants. Rabbitbrush produces a high-quality rubber called chrysil that vulcanizes easily. Extraction of this rubber for economic reasons at this point is not feasible. Because of its rubber-based compound, rabbitbrush will burn even if it's wet or green. Navajo Indians derived a yellow dye from the flowers, whereas the inner bark yielded a green dye.

Ragweed (*Ambrosia*)

Description: In general, the twelve species of *Ambrosia* in California are annuals or shrubs with leaves that are opposite below and alternate above. The yellow flowers are arranged in spikes or racemes, and the fruit is enclosed in a bur. There is no pappus. Because the windblown pollen is highly allergenic, ragweeds are a notorious cause of hayfever where the plants are common. The genus name is from the Greek and refers to an early name for aromatic plants. It is also the mythic food of the gods.

Interesting Facts: *A. trifida* (great ragweed) was cultivated in prehistoric times for its edible seeds in the midwestern United States. A tea from the leaves of *A. trifida* was formerly used for fevers, diarrhea, dysentery, and nosebleeds and gargled for sore throats. Other species were used in teas for various medicinal purposes. The heated leaves of *A. psilostachia* (cuman ragweed) were used as a poultice to ease aching joints, and a decoction was used to bathe bad sores and burns. Native Americans rubbed the leaves of *A. artemisiifolia* (annual ragweed) on insect bites, infected toes, minor skin eruptions, and hives. A tea was used for fevers, nausea, mucous discharges, and intestinal cramping.

Sagebrush, Wormwood (*Artemisia*)

Description: There are a number of species of *Artemisia* in the Sierra Nevada, including annual, biennial, and perennial herbs and shrubs. They are mostly aromatic with entire or dissected leaves. The flower heads are small, inconspicuous, and composed of disk flowers. The genus name honors Artemisia, wife of Mausolus, who was the king of Caria (a province in Asia Minor). After the king's death in 350 BC, Artemisia built the renowned Mausoleum at Halicarnassus, one of the ancient Seven Wonders of the World.

Quick Key to the Sagebrushes

Plants herbs
    Leaves are entire to shallowly lobed

Leaves are entire and linear, one and a half to three inches long	Tarragon (*A. dracunculus*)—Tarragon occurs in dry, disturbed places below 9,000 feet. It blooms from August to October.
Leaves are ovate to lanceolate, maybe toothed or lobed, three to six inches long	Mugwort (*A. douglasiana*)—Mugwort grows in low places up to 6,000 feet. It flowers from June to October.

Lower leaves are parted into linear or
lanceolate lobes or dissected

    Leaves are green on both surfaces — Mountain Sagebrush (*A. norvegica*)—This species is found in rocky places between 5,000 and 13,000 feet throughout the Sierra Nevada. It blooms from July to September.

    Leaves are dense, matted hair beneath — Western Mugwort (*A. ludoviciana*)—This species is usually found in dry places and dry borders of high-elevation meadows between 4,000 and 11,600 feet throughout the Sierra Nevada.

Plants shrubs

    Leaves are linear and entire — Silver Sagebrush (*A. cana*)—This species occurs in dry, rocky habitats between 6,000 and 10,500 feet.

    Leaves appear wedge shaped and three-toothed

        Plant four to twelve inches tall; leaves less than one-half inch long — Dwarf Sagebrush (*A. arbuscula*)—This sagebrush occurs in dry habitats between 4,000 and 9,500 feet.

    Plants usually more than twelve inches tall; leaves longer than one-half inch

        All leaves usually three-lobed; flower heads less than one-eighth inch in diameter — Great Basin Sagebrush (*A. tridentata*)—This species occurs on open slopes and in woods between 5,000 and 8,000 feet in the central and southern Sierra Nevada. It blooms from July to August.

        Leaves are usually entire on the flowering stems; flowering heads are greater than one-eighth inch in diameter

            Herbage green in color and very aromatic; plants sticky — Timberline Sagebrush (*A. rothrockii*)—Look for this species in rocky places between 6,500 and 11,500 feet from the central Sierra Nevada southward.

            Herbage covered with grayish pubescence; plants not sticky — Snowfield Sagebrush (*A. spiciformis*)—This species appears to be common on open slopes and rocky meadows between 6,500 and 12,000 feet in the north-central Sierra Nevada southward.

201
*Major Plant
Groups*

**WILDERNESS FOOD
STORAGE PIT**
To store foods for extended periods of time, Native Americans used storage pits. After a hole was dug, moisture was removed from the soil by lining the pit with hot rocks and allowing it to steam. With the rocks left in place, the pit was then lined with dried grasses, and food was placed inside. On top of the food, dried bark from junipers or other plants high in tannic acid were placed to repel insects. On top of this, dried, aromatic, nonpoisonous leaves such as sagebrush were placed to disguise the smell of the food. Last, the pit was covered with a thick layer of dirt and heavy rocks to prevent animals from uncovering the food stored.

Interesting Facts: The seeds of many species are edible raw or as flour. The seeds and peeled shoots of *A. douglasiana* and *A. ludoviciana* were eaten raw by Native Americans in California.

Herbage of various *Artemisia* species may be toxic if eaten in large amounts but may be used in small quantities to flavor stews, soups, and other foods. A tea from leaves was a cure for colds and sore eyes and was used as a hair tonic. Some of the "softer" species can be used as toilet paper and foot deodorant. Crushed leaves can be mixed with stored meat to maintain a good odor. Because many species are aromatic, they can be used to store buried food caches by masking the odor of foodstuff and to rub on the body to mask human scent while hunting (see sidebar discussion). The wood of *A. tridentata* is a good material for fire drills. Although cordage can be made from the bark, it is not very strong.

Many species of *Artemisia* have been used as medicine by Native Americans and were used in sweat houses to relieve numerous ailments. A strong tea of *A.*

ludoviciana was used as an astringent for eczema and as a deodorant and anti-perspirant for underarms and feet. A weak tea was used for stomachaches. For sinus ailments, headaches, and nosebleeds, a leaf snuff was used. A leaf or root tea of *A. dracunculus* was used for colds, dysentery, and headaches and to promote an appetite. The leaves were poulticed and used for wounds and bruises. Moore (1979) says that *A. tridentata* is strongly antimicrobial and was used as a disinfectant and cleansing wash. Volatile oils in *A. tridentata* are responsible for its pungent aroma and are so flammable that they can cause even green plants to burn. It should also be noted that the Federal Drug Administration classifies *Artemisia* as an unsafe herb containing "*a volatile oil which is an active narcotic poison*" (Duke 1992a; emphasis added).

Salsify (*Tragopogon porrifolius*)
Description: This is an introduced taprooted biennial herb with milky juice. The leaves are alternate, entire, sessile, clasping at the base, and taper to a long point. The flower heads are solitary and composed of pale yellow or purple ray flowers. The heads open early in the day, close about noon, and remain closed on cloudy, rainy days. They are found in many habitats at lower elevations. The genus name is Greek for goat's beard, probably referring to the thin, tapering, tufted, grasslike leaves.

Interesting Facts: The fleshy roots of salsify can be eaten raw or after cooking. The flavor resembles that of an oyster, an acquired taste! Other species, *T. dubius* (yellow salsify) and *T. pratensis* (meadow salsify), are also edible, but are somewhat smaller, more fibrous, and tough. Salsify root has been cultivated for more than two thousand years in the Mediterranean. The young leaves and stems of all species can be eaten after boiling until tender. The coagulated sap can be used as chewing gum and as a remedy for indigestion.

Sneezeweed (*Helenium bigelovii*))
Description: This stout biennial grows up to thirty-two inches high, and the lower leaves are linear lanceolate to lanceolate. The flowering heads are solitary and showy, and occur on long, leafless peduncles. Both ray and disk flowers are present. The disk flowers are globose and yellowish brown in color. The ray flowers, numbering about thirteen to thirty per head, are with yellow ligules with a three-lobed apex. The ligules are turned downward, giving the disk flowers a globose and prominent appearance. Sneezeweed is common in moist, meadowy places from 5,000 to 8,500 feet. It flowers from June through August. The plant is so named because the appearance of the rays suggests that it has just sneezed.

Interesting Facts: Powdered sneezeweed was used as snuff to induce sweating, which in turn relieved the congestion of head colds.

Sow Thistle (*Sonchus*)
Description: The three species of sow thistle that may be encountered include *S. arvensis* (field sow thistle), *S. asper* (spiny sow thistle), and *S. oleraceus* (common sow thistle). Introduced from Europe, they are weedy perennials and annuals with alternate leaves that are entire to pinnately divided. The leaf bases have ear-shaped lobes, and the margins are prickly. The flower heads are composed of entirely yellow ray flowers. The pappus is bristly. The common name is said to be derived from the observation that pigs eagerly consume the plants. In general, they occur at the lower elevations in gardens and waste places.

Oyster Plant, Salsify (*Tragopogon porrifolius*)

Interesting Facts: The young plants of all three species can be prepared as a potherb. As they get older, they become increasingly bitter. We found that boiling them in at least two changes of water makes them a little more palatable. Because the plants have an abundance of soluble vitamins and minerals, use only a minimum amount of water and boil briefly. The milky gum obtained from *S. oleraceus* was once used in treating opium addiction, and Native Americans used a tea made from the leaves of *S. arvensis* to calm nerves (Foster and Duke 1990). In Europe, a poultice from the leaves was used as an anti-inflammatory.

Sunflower (*Helianthus*)

Description: These are coarse annual and perennial herbs, often with tall stems. Leaves are simple, the lower ones opposite, others sometimes alternate. Flower heads are showy with bright yellow ray flowers. Involucral bracts are green and herbaceous. Other genera, including *Wyethia, Balsamorhiza,* and *Arnica,* are often mistaken for *Helianthus.* The genus name comes from the Greek *helios anthes,* which means sunflower.

Interesting Facts: The largest member of this genus is *H. annuus* (common sunflower), which is a valuable and useful plant. It has been cultivated in the United States since before Columbus. Other species of *Helianthus* may be used similarly.

The seeds may be eaten raw or roasted, then ground into meal and made into bread. The roasted shells can be used as a coffee substitute. To separate large amounts of seeds from shells, first grind them coarsely, then stir vigorously in water. In this way the shells will float, while the seeds sink to the bottom. The tiny unopened flower buds are also edible, with a flavor similar to artichokes. To reduce their bitterness, boil in two to three changes of water. Serve with lemon and melted butter.

Sunflower oil can be extracted from the seeds for cooking and can also be used in making soap, paints, varnishes, and candles. It is extracted by simply boiling the crushed seeds and then skimming the oil from the surface of the water. The pulp remaining after the oil is extracted also provides food for livestock.

Medicinally, the crushed roots can be applied to bruises. Other uses of sunflower include fiber obtained from the stalks for cordage, weaving, and sewing. The Chinese reportedly used the stalk fibers in fabrics and the pulp for paper production; the Russians used the stalks as buoyant material for life preservers. Purple and black dyes can be obtained from the seeds and a yellow dye from the flowers.

Tarweed (*Madia*)

Description: Typically, these are annuals with a tar scent of varying intensity. The leaves are narrow, usually opposite below and alternate above. The flower heads are composed of inconspicuous yellow ray flowers. Tarweed can be found at moderate elevations in open, grassy, or vernally moist areas.

1. Bolander's Madia (*M. bolanderi*)—This species occurs in moist habitats between 3,500 and 8,300 feet. It flowers from July to September.
2. Common Madia (*M. elegans*)—This glandular, sticky, heavy-scented annual is found on dry slopes from 3,000 to 8,000 feet. It flowers from June to August.
3. Threadstem Tarweed (*M. exigua*)—This slender annual is found at low to moderate elevations. It flowers from May to July.
4. Mountain Tarweed (*M. glomerata*)—This strongly ill-scented species occurs in forest openings.

BUTTERFLIES AND EDIBLE PLANTS
From an ecological perspective, all life ultimately depends on other forms of life to survive and reproduce. As John Muir (1838–1914) once said, "When we try to pick out anything by itself, we find it hitched to everything else in the universe" (Vizgirdas 2003b, 1).

There are numerous ways to locate edible and useful plants. One means includes actively searching for the plants. This direct approach requires knowledge of the plants' basic life history, ecology, and distribution. Another way to locate potential useful plants, though indirect but requiring knowledge of the plants, is by watching butterflies (Lepidoptera order).

Most butterflies are closely tied to a specific species or group of plants to complete their life cycle. Specifically, the larvae (caterpillars) feed on those host plants. Adult female butterflies select the proper larval food plant by "smelling" with their with antennae and "tasting" with sensory receptors on their feet. Eggs are then laid on only a specific host plant, so that hatching caterpillars can begin to feed right away.

Here are some butterflies and their host plant(s) and uses:

1. Phoebus Parnassian (*Parnassius phoebus*) feeds on stonecrop (*Sedum*). The leaves, stems, and rhizomes of these plants are edible raw or cooked and are high in vitamins A and C.

2. Western White (*Pontia occidentalis*) are usually associated with mustards (Brassicaceae). Mustards such as *Lepidium, Brassica, Arabis,* and *Sisymbrium* are edible as greens.

5. Slender Tarweed (*M. gracilis*)—This coarsely hairy annual is quite common and grows on wooded hillsides. It blooms from April to August.

6. Madia (*M. madioides*)—Look for this species in forest openings below 4,500 feet from Mariposa County northward. It flowers from May to June.

7. Hemizonella (*M. minima*)—This species is common on gravelly slopes between 3,500 and 8,600 feet. It flowers from May to July.

8. Ramm's Madia (*M. rammii*)—This species is common in open areas below 5,000 feet from Calaveras to Butte Counties. It flowers from May to July.

9. Yosemite Tarweed (*M. yosemitana*)—This is a rather rare tarweed in moist habitats between 4,000 and 7,500 feet in Tuolumne County southward. It flowers from May to July.

Interesting Facts: The seeds of *M. glomerata* (mountain tarweed) may be eaten raw, cooked, or dried and ground into meal. The scalded seeds also yield a nutritious oil. All tarweeds were used medicinally by old Spanish settlers. An oil of excellent quality was made from their seeds in the United States before olives were readily available.

Strike (1994) indicates that tarweed seeds were collected and stored until needed. Seeds were often used in making pinole by many California Natives. The Miwok Indians of California pulverized tarweed seeds and ate them dry (Barrett and Gifford 1933). When tarweed seeds had matured but the plants were still green, Hupa Indians burned the areas where the plants grew. The seeds were then gathered from the scorched plants, and because they needed no further parching, they were crushed into flour. The roots of some species were also eaten.

BUTTERFLIES AND EDIBLE PLANTS (*CONTINUED*)
3. Milbert's Tortoiseshell (*Nymphalis milberti*) is associated with nettle (*Urtica*). These plants provide useful fibers for cordage, and the plants are edible when prepared as potherbs.
4. Fritillary (*Speyeria* spp.) are found on violets (*Viola*), which are edible raw or mixed with other greens in a salad.
5. California Crescent (*Phyciodes oreis*) associates with thistles (*Cirsium*). These plants are edible raw or cooked, the stems can be used in hand drill fire starting, and fibers from the stem can be used as a crude cordage.

Thistle (*Cirsium*)

Description: The many species of thistle that occur in California are characterized as biennial or perennial herbs with alternate leaves that are lobed or cleft with spines. The red, yellow, or white heads are showy, and the involucral bracts are overlapping. The native and introduced species can be found in a wide variety of habitats from the foothills to the higher elevations. *Cirsium* comes from the Greek *kirsos*, meaning swollen vein, for which thistles (*kirsios*) were a reputed remedy.

1. Anderson's Thistle (*C. andersonii*)—This occurs in dry habitats between 5,000 and 11,000 feet.

2. Arizona Thistle (*C. arizonicum*)—This species can be found in dry, stony slopes in Tulare, Inyo, and Alpine Counties.

3. Gray-green Thistle (*C. canovirens*)—Look for this species on dry, open slopes between 5,000 and 12,000 feet and mainly on the east side.

4. Peregrine Thistle (*C. cymosum*)—This can be found in dry habitats below 7,000 feet.

5. Swamp Thistle (*C. douglasii*)—This species occurs below 7,000 feet in the northern Sierra Nevada.

Tarweed (*Grindelia* spp.)

6. Western Thistle (*C. occidentale*)—This slender and spiny biennial is a common plant on dry slopes below 10,000 feet. It blooms from April to July.

7. Elk Thistle (*C. scariosum*)—Look for this species in meadows and moist habitats below 11,500 feet.

8. Bull Thistle (*C. vulgare*)—This common species is usually found in disturbed areas up to 7,000 feet. It is a native of Europe.

Interesting Facts: Thistles were not a major food source in the past, but were used when needed. Here is our favorite story about how useful thistles can be in emergency situations. Truman Everts, a participant in the early explorations of the Yellowstone Park region, became lost for more than a month and subsisted on thistles. He apparently had lost his glasses and was able to identify thistles by touch. Although thistles are difficult to collect, they are well worth the pain.

All species have roots that can be eaten raw, boiled, or roasted. Some have roots that turn sweet when roasted. The immature flower buds (asparagus-like) can be eaten raw or cooked. Young leaves dethorned are edible raw, and a tea can be brewed from all leaves. The peeled young stems may be cooked as greens and resemble celery in taste. The older stalks are also edible but are somewhat more fibrous and bitter. The seeds can be boiled and eaten in the same manner as sunflower seeds, or they can be ground into flour for baking.

Medicinally, thistle stalks were chewed to ease stomach pains. Pounded stalks were used as a salve for facial sores or on infected wounds. A decoction made from thistle roots was used to relieve asthma.

When well dried and dethorned, stems can be used as hand drills for starting fires. The stem fibers of any thistle species can be used as thread or crude cordage. To obtain the fiber, simply soak the stalks in water for a day or more to loosen them from the outer layer. The downy part of seed heads makes good insulating material and a good tinder additive.

Trail Plant (*Adenocaulon bicolor*)

Description and Uses: Trail plant is a slender, single-stemmed perennial herb up to about three feet tall. The leaves are large, arrow-shaped, and arise from near the base of the plant. Disk flowers are white, small, and inconspicuous. There is no pappus. The glandular achenes are sticky. The plant is usually found in moist, shady places at low to midelevations (below 7,000). Although the edibility of this plant is unknown, the Squaxin used the crushed leaves for a poultice (Pojar and MacKinnon 1994).

Thistle (*Cirsium* spp.)

Western Chamomilla (*Chamomilla*)

Description: These are annual or biennial herbs that have a branched habit. Leaves are alternate and pinnately lobed or divided. The small, terminally arranged flower heads are composed of disk or ray flowers.

1. Chamomilla (*C. occidentalis*)—This annual plant is not strongly scented. It is found in wet habitats below 7,500 feet throughout most of the Sierra Nevada.
2. Pineapple Weed (*C. suaveolens*) (= *Matricaria matricarioides*)—This is an erect annual herb that has a branched habit. This is a common species in waste places at the lower elevations.

Interesting Facts: Chamomilla is used as a substitute for chamomile. A delicious tea can be made from the dried flowers of pineapple weed, and the leaves are edible but bitter. The medicinal uses of pineapple weed are identical to that of chamomile (*Anthemis* sp.). It can be used as a tea as a carminative, antispasmodic, and mild sedative.

Western Pearly-everlasting (*Anaphalis margaritacea*)

Description: This perennial grows up to thirty-six inches tall and has white woolly herbage. Leaves are alternate, entire, and lanceolate to linear or oblong in shape. The sessile leaves are also woolly beneath but soon becoming green and glabrous above. The flowering heads form a flat-topped, terminal cluster. The involucral bracts are papery, pearly white, and imbricated in several series.

Pearly-everlasting is found in openings along trails and on talus slopes below 8,500 feet throughout the Sierra Nevada. It blooms from June to August.

Interesting Facts: The herbage of western pearly-everlasting has been used as a tobacco substitute to relieve headaches. As a tea, the plant has been used for colds, bronchial coughs, and throat infections. The whole plant can be used as a wash or poultice for external wounds. It has also been used for rheumatism, burns, sores, bruises, and swellings (Strike 1994).

Western Snakeroot (*Ageratina occidentale*)
Description: This is a genus with about 230 species of annual to perennial herbs and shrubs. The flowers are daisylike and lack distinct ray flowers. This species occurs on rocky slopes and ridges between 6,500 and 11,000 feet throughout the Sierra Nevada. It flowers from July to August.

Interesting Facts: The whole plant of western snakeroot was apparently used externally as a wash for rheumatism and swelling.

White Layia (*Layia glandulosa*)
Description: This annual grows up to twenty-four inches tall and has leaves that are rough, hairy, and linear to lanceolate in shape, with the basal ones being toothed or lobed whereas the upper ones are entire. The flowering heads have both ray and disk flowers present. There are about twenty-five to one hundred disk flowers. *Layia* is found in sandy soil up to 7,800 feet. It blooms from March to June.

Interesting Facts: The seeds of *Layia* were often used in making pinole. The seeds of this species are edible after grinding them into flour for mush.

Wild Lettuce (*Lactuca serriola*)
Description: This is a tall, prickly plant with alternate leaves and milky juice. The yellow, blue, or whitish flowers are all ray. The pappus is white to brownish. This is a rather common weed in fields and waste places in the lower elevations.

Interesting Facts: Collected in the late fall to early spring, the plants should be boiled in a couple of changes of water to reduce the bitterness. The earlier or younger the plant is collected, the better the flavor. Because of the latex sap, raw greens can cause an upset stomach if eaten in quantity. In sensitive people, the latex can cause dermatitis. These wild plants contain more vitamin A than spinach and a good quantity of vitamin C. An extract of the white sap from two species of *Lactuca* in Europe has been used to replace opium in cough remedies. The extract, lactucarium, is reported to be a mild sedative. The plants also contain a mildly narcotic compound in the latex. The active constituents increase during flowering and are relatively low in young plants.

Wire Lettuce, Skeleton Weed (*Stephanomeria*)
Description: These are more or less branched annual or perennial herbs with milky juice. The leaves are small and often scalelike. The flowers are pink and composed of ray flowers. The three species are found in dry, open places at low and midelevations.

1. Large-flowered Stephanomeria (*S. lactucina*)—This species occurs in dry flats and ridges between 4,000 and 8,000 feet in the northern Sierra Nevada. It blooms from July to August.

2. Narrow-leaved Stephanomeria (*S. tenuifolia*)—This species is found on dry slopes between 4,000 and 11,000 feet, especially on the eastern Sierra Nevada. It flowers from July to August.

3. Tall Stephanomeria (*S. virgata*)—This erect, stiff annual is found in dry, disturbed areas below 6,000 feet. It flowers from July to October.

Interesting Facts: Related species have been used medicinally. *Stephanomeria virgata* exudes a milky sap and was used as an eye medication by the Kawaiisu Indians. The sap of *S. pauciflora* was used as a chewing gum.

Woolly Malacothrix (*Malacothrix floccifera*)

Description: This almost leafless annual has branched stems that are four to sixteen inches tall. The stems contain a milky juice. Leaves are mostly basal, oblong, three-quarters to three inches long, pinnatifid, and toothed. The flowering heads contain many flowers that are all ray flowers, with white or pale yellow ligules. These may be tinged with pink. The pappus is of one to eight bristles. Woolly malacothrix grows in sandy and rocky places below 5,000 feet. It flowers from April to June.

Interesting Facts: Several Native American tribes ate the seeds of a related species (*M. californica*).

Woolly Sunflower (*Eriophyllum*)

Description: These are hairy annual to perennial plants that are herbaceous to shrubby. The genus name is from the Greek *erion*, meaning wool, and *phyllon* for leaf.

Quick Key to the Woolly Sunflowers

Plants shrubs or subshrubs; flower heads usually occurring in clusters	
Flower heads many in compact terminal clusters, each less than three-quarters inch in diameter	Golden Yarrow (*E. confertiflorum*)—This perennial plant is common on brushy slopes to 8,000 feet. It flowers from April to August.
Flower heads are solitary or occur in loose clusters, each three-quarters to one and a half inches in diameter	Woolly Yarrow (*E. lanatum*)—This herbaceous perennial is common on brushy slopes up to 10,000 feet. It flowers in June and July.
Plants small annuals; flower heads are usually solitary	
Pappus not present or vestigial	Woolly Sunflower (*E. ambiguum*)—This species occurs below 7,500 feet in Mariposa County. It blooms from April to June.
Pappus composed of scales	
Branches of the plant spread open	Woolly Sunflower (*E. congdonii*)—Look for this species in the vicinity of Iron Mountain.
Branches of the plant are ascending	Yosemite Woolly Sunflower (*E. nubigenum*)—This species occurs in forest openings between 5,000 and 9,000 feet. It flowers from June to July.

Interesting Facts: *Eriophyllum* seeds were parched and ground into a flour by Cahuilla and Luiseno Indians in California. The seeds were also incorporated into pinole.

Wyethia, Mule-ears (*Wyethia*)

Description: All species have leaves on the stems distinguishing them from *Balsamorhiza*, which has leaves only at the base.

Mule-ears (*Wyethia* spp.)

Quick Key to the Wyethias

The basal leaves are usually larger than the stem leaves

The basal leaves are lanceolate in shape; phyllaries are shorter than the flowering head; plants occur mainly below 5,500 feet	Narrow-leaved Mule-ears (*W. augustifolia*)—This species occurs on grassy slopes. It blooms from May to July.

The basal leaves are elliptical in shape

Flowering heads three-quarters to one and a half inches wide; plants found mainly above 5,000 feet	Mountain Mule-ears (*W. mollis*)—Look for this species on dry, wooded slopes and rocky openings between 4,500 and 10,600 feet in the central Sierra Nevada northward.
Flowering heads one and a half to two and a half inches wide; plants found below 5,000 feet	Gray Mule-ears (*W. helenioides*)—Look for this species in open fields and sunny woodlands up to 6,000 feet in the northern Sierra Nevada. It blooms from May to July.

The basal leaves are similar in size to the stem leaves or they may be absent

Leaves are round to deltoid in shape	Southern Wyethia (*W. ovata*)—This species occurs on grassy, open wooded hills below 6,000 feet. It flowers from May to August.

Leaves deltoid to elliptical in shape

Ray flowers are conspicuous, up to two inches long; pappus is present	Hall's Wyethia (*W. elata*)—This species is found on dry, open slopes between 3,000 and 4,600 feet in the central Sierra Nevada. It flowers from May to July.
Ray flowers are short or none; pappus is absent	Coville's Wyethia (*W. invenusta*)—This species is found in open woods between 3,800 and 6,000 feet in the southern Sierra Nevada. It blooms from July to August.

Interesting Facts: The seeds are edible and somewhat resemble sunflower seeds in taste. The roots of *W. helinioides* can be eaten after they have been cooked for a day or two in a steam pit. Regardless of how long the roots are cooked, the smell is almost intolerable to enjoy eating them. The flower stalks of *Wyethia* were also eaten as a vegetable by the Shoshone. A decoction of leaves was used as a bath, producing profuse sweating. The leaves are considered to be *poisonous* and should not be taken internally. The Klamath Indians used the mashed root as a poultice for swellings.

Yarrow (*Achillea millefolium*)

Yarrow (*Achillea millefolium*)

Description: This is a strongly scented perennial herb with alternate leaves that are finely dissected and feathery appearing. The white or sometimes yellow flowers are borne in a flat-topped corymb. Yarrow is widespread and can be found in a variety of habitats from low elevations to above timberline. The generic name honors Achilles. In folklore, his mother supposedly dipped the young Achilles into a yarrow bath to make him invincible. Because she held him by his heels, he was made vulnerable through his "Achilles' heel."

Interesting Facts: Yarrow is often referred to as "poor man's pepper." The leaves can be dried, ground, and used as seasoning. The young leaves can be added to salads. The aromatic leaves were also placed in freshly split fish to expedite drying.

Medicinally, the leaves and stems can be dried, boiled in water, strained, and drunk to remedy a run-down condition or help with an upset stomach. Taken as a hot infusion, yarrow will increase body temperature, open skin pores, and stimulate perspiration, making it a valuable herb for colds and fevers. The juice can be used as an eyewash to reduce redness. Leaves can be used to stop bleeding in small wounds and to heal rashes when applied directly to the skin. Leaves were also chewed to relieve toothaches. A poultice of mashed leaves can be applied to swellings or sores. To date, more than one hundred biologically active compounds have been identified from the species; some are known to be quite toxic (Foster and Duke 1990). Prolonged use of yarrow may cause allergic rashes and make the skin more sensitive to sunlight.

Rubbing the plant on one's clothing and skin was an ancient prescription for repelling biting insects. The stalks burned on coals were said to deter mosquitoes. The leaves were used in herbal snuffs and smoking tobaccos. Yarrow has also been used as a hops substitute for brewing "yarrow beer."

Sycamore Family (Platanaceae)
These are large trees with deciduous bark. There is one genus with about ten species, mostly located in the temperate and subtropical Northern Hemisphere. A few of the species are cultivated as ornamentals.

Western Sycamore (*Platanus racemosa*)
Description: This is a deciduous tree, thirty to seventy-five feet tall with smooth pale bark that easily peels and gives a mottled or camouflaged look. Leaves are large, deeply five lobed, and densely tomentose on both surfaces when young. Its flowers are unisexual and minute, borne in spherical heads about a quarter inch in diameter. Heads occur in chains on slender peduncles on a more or less zigzag axis. The fruit is a dense, spherical head of achenes with tails projecting outward, about one inch in diameter. Western sycamore is common along streams and dry creek beds below 4,000 feet.

Interesting Facts: The presence of sycamore trees is an indication that a stream or underground water source can be found nearby. To ease asthmalike breathing problems, the bark was prepared as a tea and drunk in place of water for about a week. The bark tea was also drunk to aid in childbirth. Sycamore wood is considered to be a good fuel wood.

Teasel Family (Dipsacaceae)
These are mostly herbaceous plants with opposite leaves. There are approximately 10 genera and 270 species found mostly in the Old World. None are native to the United States, although *Dipsacus* is widely naturalized and has become quite weedy.

Teasel (*Dipsacus fullonum*)
Description: This is a stout, prickly, taprooted plant up to six feet tall. The leaves are opposite, lance shaped, and distinctly prickly on the lower surface of the midrib. The small, bluish purple flowers occur in a large, terminal flower head that is egg shaped and armed with numerous sharp-pointed bracts. This non-native plant from Europe is found throughout North America in disturbed soils

with appreciable water-holding capacity. The plant has a tendency to be opportunistic and displaces desirable native plants. The genus name is from the Greek *dipsa*, which means thirst, and refers to the accumulation of water in the cuplike bases of the joined leaves. It is not uncommon to find insects and other invertebrates living in the water found in the leaves.

Interesting Facts: The dried flower spike of teasel looks like a caged "thistle"and is sometimes mistaken for thistle (*Cirsium*) because of its "thorny" character and bluish flowers. However, thistle is a member of the sunflower family (Asteraceae) and has alternate, not opposite, leaves.

Although there are no documented edible or medicinal uses for the plant, they are sought after for use in dry flower arrangements. The dried inflorescence is often sprayed with some interesting colors and sold in craft stores.

Cloth cleaners use the dried flower heads to remove the nap from wool or cloth after beating and cleaning. Apparently, the teasel heads perform the task so well that man-made tools have not replaced them.

Valerian Family (Valerianaceae)
The valerian family has about thirteen genera and four hundred species. Three of the genera are native to the United States. In general, the family is of no economic importance.

Teasel (*Dipsacus fullonum*)

California Valerian (*Valeriana californica*)
Description: These are perennial herbs with aromatic (actually ill-smelling) roots. The stem leaves are opposite and pinnately compound, and the flowers have three stamens. They can be found in open forests and meadows to timberline and above. The genus name comes from the Latin *valere* (to be strong) and refers to medicinal qualities of the plant.

Interesting Facts: The plants commonly contain the alkaloids chatinine and valerine. They are known to act upon the central nervous system as depressants and are prescribed to calm nerves and relieve insomnia. The plants were used by physicians since at least the ninth century. Extracts were used as nerve tonics and may rival the relaxing properties of opium. Valerian was one of seventy-two ingredients Mithridates, king of Pontus, compounded as an antidote to poison, using poisoned slaves as test subjects.

The roots and leaves of a related species, *V. edulis* (edible valerian), can be collected, steamed for a day or two to remove the disagreeable odor, then used in soups as a potato substitute. However, the taste does take a little getting used to and may remain somewhat unpalatable. We found that the steamed roots are better if dried, ground into flour, and then added to other flours. The other species could probably also be used in an emergency, but they do not have the large taproot of *V. edulis*.

Valerian, in general, is best known for its calming qualities. It has been used for more than a hundred years as a remedy for anxiety, muscle tension, and insomnia. The plants contain valepotriates, which are known herbal calmatives, antispasmodics, and nerve tonics and are used for hypochondria, nervous headaches, irritability, and insomnia. Research has confirmed that teas, tinctures, or extracts of this plant are a central nervous system depressant and a sedative for agitation.

Verbena Family (Verbenaceae)

Approximately seventy-five genera and three thousand species occur worldwide, of which fourteen genera are native to the United States. The family is of economic importance because of the highly prized teak wood (*Tectona grandis*) and a number of other ornamentals.

Verbena, Vervain (*Verbena lasiostachys*)

Description: Members of this genus are perennials with opposite, toothed leaves. This is a much-branched perennial with long, hairy stems that grows on dry or moist slopes below 8,000 feet. It flowers from May to September. The tubular flowers with flaring lobes each have a subtending bract. The fruit is a cluster of four nutlets.

Interesting Facts: The seeds of a related species, *V. hastata* (blue vervain), may be gathered, roasted, and ground into a bitter-tasting flour. Leaching the flour may remove the bitter taste. A tea from boiled leaves can be used for a stomachache, and a tea from the roots was used to clear cloudy urine. Moore (1979) says that the plant is used as a sedative, diaphoretic, bitter tonic, and mild coagulant. It promotes sweating, relaxes and soothes, settles the stomach, and gives an overall feeling of relaxed well-being.

Violet Family (Violaceae)

The violet family has approximately 16 genera and 850 species distributed worldwide. Two genera are native to the United States. The family is of little economic importance other than as a source of ornamentals, as many species of *Viola* are cultivated. *Viola* is the classical name for violets.

Violet (*Viola* spp.)

Violet (*Viola*)

Description: There are many species of violets in California. In general, they are low-growing perennial or annual herbs. The leaves are spade shaped and basal. The flowers occur singly on the ends of stems and have five petals. There are two upper and two lateral petals and one lower petal that is prolonged into a nectar-holding pouch at the base of the flower. Most species also have small, self-fertilizing flowers that do not open. Violets can be found in meadows and open forests from the foothills to above timberline.

Quick Key to the Violets

Leaves dissected or compound	
Leaves mostly wider than long	Fan Violet (*V. sheltonii*)—This glabrous perennial grows in the shade of open woods or in brushy areas from 2,500 to 8,000 feet. It flowers from April to July.
Leaves mostly longer than wide	
The upper two petals are yellow on the front and brown to dark on the back	Douglas Violet (*V. douglasii*)—This pubescent perennial grows in open grassy areas in the mountains from 3,500 to 7,500 feet. It flowers from March to May.
The upper two petals are dark red-violet	Great Basin Violet (*V. beckwithii*)—This species occurs in dry, gravelly places, usually among shrubs between 3,000 and 6,000 feet on the eastern side of the range from Inyo County northward.

Leaves mostly entire or deeply lobed
  Flowers are white to blue or purple in
  color, no yellow
    Flowers violet

Western Dog Violet (*V. adunca*)—This slightly pubescent perennial grows on moist banks and at the edge of meadows from 5,000 to 8,000 feet. It flowers from March to July.

    Flowers white

Macloskey's Violet (*V. macloskeyi*)—This white-flowered violet occurs in wet meadows, seeps, and along stream banks from 3,000 to 10,500 feet.

  Flowers are yellow to orange in color, may
  be white with yellow base and spur
    Petals white inside, with yellow base
    and spur, lower three veined purple-
    red

Wedge-leaved Violet (*V. cuneata*)—Look for this violet in moist, open forests below 5,000 feet in Nevada County.

    Petals yellow to orange inside; veined
    dark brown to purple
      Stem leaves restricted to upper
      portion of the stem; basal leaves
      number zero to two
        Upper two petals are yellow
        outside

Smooth Yellow Violet (*V. glabella*)—This species grows in wet, shaded places in woods below 8,000 feet throughout the Sierra Nevada.

        Upper two petals are purple to
        brown-purple outside

Pine Violet (*V. lobata*)—This erect, slightly pubescent perennial grows in elevations from 1,000 to 6,500 feet. It flowers from April to July.

      Stem leaves on lower and upper
      parts of stem; plant prostrate to
      erect; basal leaves number zero to
      six
        Basal leaves two to six times
        longer than wide
          Plant white tomentose

Woolly Violet (*V. tomentosa*)—The woolly violet grows in dry, gravelly places between 5,000 and 6,500 feet from El Dorado to Plumas Counties.

          Plant glabrous

Baker's Violet (*V. bakeri*)—This violet is found growing at elevations between 7,000 and 8,000 feet in Placer and Plumas Counties.

        Basal leaves one to one and a
        half times longer than wide

Mountain Violet (*V. purpurea*)—This pubescent perennial grows on dry slopes below 6,000 feet. It flowers from April to June.

Interesting Facts: The leaves, buds, and flowers of possibly all species are edible raw or cooked, with some being more palatable than others; the leaves make a good tea. Adding the leaves to soups makes them thicker. Violets are high in vitamin C and beta-carotene. Collect the plants by leaving the roots intact; because many species reproduce vegetatively, you will probably not inhibit next year's growth significantly. Many naturalists indicate that all violets are safe for consumption, but there are some experts who insist that some yellow species may be somewhat purgative. All species do have a tendency to be slightly laxative, so proceed slowly (Kirk 1975). The flowers have also been candied or made into jellies and jams.

Violet salve can be made by simmering the entire herb in lard. It was a famous remedy for skin inflammations and abrasions. Violets are also emollients (they soften the skin) and are an excellent ingredient in lotions such as night creams. Flowers and leaves of some species have been used in various herbal remedies as poultices and laxatives and to relieve cough and lung congestion.

WATERLEAF FAMILY (Hydrophyllaceae)
The 20 genera and 270 species within the waterleaf family are distributed worldwide, except for Australia. The western United States appears to be the main center of diversity. Only a few members in the family are cultivated.

Common Fiesta-flower (*Pholistoma auritum*)
Description: This is an annual plant with straggling, loosely branched stems. The lower leaves are oblong in outline, and the flowers are lavender to purple in color with darker markings. It occurs on shaded slopes and in deep canyons from Calaveras County south.

Interesting Facts: The leaves and stem of a related species (*P. membranaceum*), were eaten. Essentially, the plant was rolled in the palm of the hand with salt grass and eaten.

Nemophila (*Nemophila*)
Description: These are mostly annuals with stems that are diffuse, weak, and sometimes prostrate. Leaves are mainly opposite, and the flowers occur in the upper axils of the leaves. The style is deeply bifid. The genus name is from the Greek for grove loving, referring to the woodland habitat of many species in this genus.

1. Variable-leaved Nemophila (*N. heterophylla*)—This species prefers light shade, slopes, and canyons below 5,000 feet. It flowers from April to July.
2. Fivespot (*N. maculata*)—Look for this species on moist slopes and flats below 9,000 feet on the western side of the range.
3. Small-flowered Nemophila (*N. parviflora*)—There are two subspecies, occurring between 2,500 and 7,000 feet.
4. Meadow Nemophila (*N. pedunculata*)—This species occurs in moist habitats below 5,000 feet from Calaveras County northward. It flowers from May to August.
5. Pretty Nemophila (*N. pulchella*)—This species is found in partial shade and moist places below 5,500 feet.
6. Sierra Nemophila (*N. spatulata*)—Look for this species in shaded, damp places between 4,000 and 10,500 feet from Plumas County southward.

Interesting Facts: The roots were used to prepare a decoction to cure asthma by some Native tribes (Strike 1994).

Phacelia (*Phacelia*)
Description: There are about twenty-one species of *Phacelia* in the Sierra Nevada. They include herbaceous annuals, biennials, and perennials with various degrees of hairiness. Their flowers are five-parted, spirally coiled, with stamens extending beyond the corolla.

Interesting Facts: At least one species, *P. ramosissima* (branching phacelia), can be cooked and used as greens. However, Strike (1994) suggests that the stems and leaves of *Phacelia* may have been eaten raw but were most likely cooked. The boiled roots of *P. ramosissima* were used to cure coughs and colds and to alleviate lethargy. A decoction was also used as an emetic and to relieve stomachaches.

Purple Mat (*Nama*)

Description: These are annual or perennial plants with leaves well distributed along the stems. The flowers occur in cymes and are axillary, and the calyx is deeply divided.

1. Nama (*N. aretioides*)—This species is found below 6,000 feet.
2. Matted Purple Mat (*N. densum*)—This species occurs in dry, loose, sandy soil between 3,000 and 11,000 feet.
3. Lobb's Purple Mat (*N. lobbii*)—Look for this species on dry, rocky, or sandy slopes and ridges between 4,000 and 7,000 feet. It flowers from June to August.
4. Rothrock's Purple Mat (*N. rothrockii*)—This species grows on dry, sandy flats and benches between 7,000 and 10,000 feet. It flowers from July to August.

Interesting Facts: The seeds of a related species (*N. demissum*) were dried, pulverized, and boiled with water to make mush or porridge.

Turricula (*Turricula parryi*)

Description: This is a tall, stout perennial plant, with coarse, glandular, hairy stems that grows three to seven feet high. The leaves are alternate, lanceolate, toothed or entire, and without petioles. The flowers are numerous in a coiled inflorescence. The purple corolla is funnel shaped and shallowly five lobed. Turricula grows in dry or disturbed places up to 8,000 feet and is common along roadsides. It flowers from June to August.

Interesting Facts: The Kawaiisu used a leaf infusion externally to relieve rheumatic pains and to reduce swellings (Zigmond 1981).

Caution: Beware of picking or touching turricula, as it can cause a dermatitis similar to poison oak on some people.

Waterleaf (*Hydrophyllum*)

Description: This is a somewhat fleshy perennial herb with leaves that are pinnately divided. Its flowers are white to bluish. It is found in moist soils from the foothills to an alpine environment.

1. Woolen-breeches (*H. capitatum*)—This species is found in moist habitats between 3,000 and 7,000 feet from Placer County north. It flowers from May to June.
2. Western Waterleaf (*H. occidentale*)—This species occurs on dry or moist, more or less shaded slopes between 2,500 and 9,000 feet throughout the Sierra Nevada. It flowers from May to July.

Interesting Facts: The young shoots, leaves, and flowers of *Hydrophyllum* can be eaten raw, or these and the roots may be cooked and eaten. We find them exceptionally good in salads or when eaten as a trail nibble. They do have a texture that takes some getting used to.

The leaves can be used as a protective dressing for minor wounds and are slightly astringent. As a poultice, it can be used for insect bites and other minor skin irritations.

Yerba Santa (*Eriodictyon californicum*)

Description: The genus *Eriodictyon* is made up of nine species of aromatic shrubs that grow in the southwestern United States and Mexico. The common name of yerba santa is often used when referring to any one of the several different species that were used medicinally. It is found on slopes, in fields, along roadsides, and in woodlands and chaparral below 6,000 feet.

Interesting Facts: This is one of the most important medicinal plants used by California Natives. The leaves were brewed into a tea to cure stomachaches, colds, coughs, inflammation of the throat, rheumatic pains, paralysis, and fevers. This tea also purified the blood. Used externally, a leaf decoction relieved sores, reduced fevers, and cured paralysis. A poultice of mashed leaves healed sores, wounds, cuts, abrasions, insect bites, sprains, and rashes caused by poison oak (Sweet 1976; Callegari and Durand 1977).

In some places, the leaves served as tobacco for smoking or chewing or both. The leaves were chewed as a thirst quencher. A weak solution of boiled *Eriodictyon* leaves can be drunk as a refreshing tea.

Note: It is reported that yerba santa proved to be so useful in treating bronchitis that it was listed by the Pharmacopoeia of the United States from 1894 to 1905 and 1916 to 1947 and was listed by the National Formulary from 1926 to 1960 (Callegari and Durand 1977).

WATER-LILY FAMILY (Nymphaeaceae)
There are six genera and sixty-eight species found throughout the world in aquatic habitats. Four of the genera are native to the United States and are a source of food for birds and aquatic animals. The family name (Nymphaeceae) translates as water nymph or water virgin. It was once believed that the plants in this family had anti-aphrodisiacal properties and were then used in art to represent virginity. A few species are used in cultivation.

Cow-lily, Yellow Pond-lily (*Nuphar luteum*)
Description: This plant has mostly floating leaves. The flowers are of obvious bright-yellow sepals, and the petals are inconspicuous and yellowish green to purplish tinged. It is common in shallow portions of many ponds and lakes of low to middle elevations.

Interesting Facts: *Nuphar luteum* is easier to identify than to harvest. The rhizomes are prime during the early spring and fall. These starchy rhizomes can be boiled and then peeled and eaten, placed in soup or stew, or dried and ground into meal and used as flour. The plant reproduces by seeds and rhizomes and is very easy to culture.

The seeds can be collected and, when dry, will keep indefinitely. They can also be treated like popcorn. Simply pop them and eat or grind them into meal. The seeds can also be steamed as a dinner vegetable or cooked like oatmeal—one part seeds to two parts water. In Turkey, the flowers of another species are distilled into a beverage called *pufer cicegi* (Saunders 1976).

The leaves and stalks have been used as poultices for boils, ulcerous skin conditions, and swelling. An infusion of the root is useful as a gargle for mouth and throat sores. Some aboriginal people still use a root medicine for numerous illnesses, including colds, internal pains, rheumatism, chest pains, and heart conditions. According to Spellenberg, pond-lilies produce alcohol: "[W]hen the mud in which the stems grow loses oxygen, a small amount of alcohol instead of carbon dioxide is produced" (1979, 623).

WATER-MILFOIL FAMILY (Haloragaceae)
This family occurs throughout the world but mostly in the Southern Hemisphere. In general, members of this family are aquatic herbs with simple or pinnatifid leaves that are opposite, alternate, or whorled.

Water-milfoil (*Myriophyllum spicatum*)

Description: The genus name is from the Greek *myrios* (many) and *phyllon* (leaf), referring to the finely divided leaves. There are about forty species of these submerged aquatic and terrestrial herbs with pinnate leaves and spikes of small wind-pollinated flowers. This species occurs in quiet waters below 8,000 feet throughout the Sierra Nevada. It flowers from June to September.

Interesting Facts: The rhizomes were frozen for future use, eaten raw, fried in grease, or roasted. The rhizomes are sweet, crunchy, and a much-relished food and were an important food during periods of low food supplies.

WATERSHIELD FAMILY (Cabombaceae)

Two genera and eight species in the watershield family are distributed in temperate and tropical America, Africa, east Asia, and Australia. They are aquatic perennials typically found in freshwater. The genus is sometimes placed in the water-lily family (Nymphaeaceae) but differs in having simple pistils.

Watershield (*Brasenia schreberi*)

Description: Watershield is anchored to muddy substrates by slender rootstocks. All exposed portions of the plant are covered with a gelatinous sheath. The leaves are nearly round and arise near the tops of the stems. The flowers have purplish petals and sepals.

Interesting Facts: The starchy rhizome of watershield can be peeled, boiled, and eaten; dried and stored; or ground into flour. The unexpanded young leaves and leaf stems can also be eaten in a salad, slime and all (Harrington 1967). The rhizomes were used to cure dysentery and stomachaches.

WATER STARWORT FAMILY (Callitrichaceae)

Members of this family are small annual or perennial herbs with slender, usually lax, stems. The leaves are simple, entire, and opposite or whorled. The minute unisexual flowers are borne in the axils of the leaves. There are no sepals or petals. The small four-lobed fruit splits into four sections upon maturity. The plants are inconspicuous in standing water or drying mud. There is only one genus (*Callitriche*) and approximately forty species in this family.

Water Starwort (*Callitriche verna*)

Description and Interesting Facts: This is the only genus of flowering plants in which aerial, floating, and subsurface pollination systems have all been reported. This species occurs in shallow water or on mud up to 11,000 feet in the Sierra Nevada. It flowers from May to August. The genus name is from the Greek *kallos*, which means beautiful, and *trichos*, which means hair, referring to the slender stems. Strike (1994) indicates that the Maidu Indians used *Callitriche* to relieve urinary problems. However, the method is not reported.

WAX-MYRTLE FAMILY (Myricaceae)

This is a family with three genera and fifty species of aromatic shrubs and trees. The leaves are simple or pinnately cut leaves and unisexual flowers that are borne in catkinlike spikes. Both sexes are on the same plant. The genus name is derived from *myrike*, the classical Greek name for tamarisk.

FIRE-BY-FRICTION
To early humans, the ability to start a fire to cook food and use as a tool was an important skill. In fact, a look at human history reveals fire as the greatest invention of all time. There are several methods of starting a fire by friction. The drill and hearth method was used by a great many Native Americans. The hearth was a flat piece of wood, usually a softer wood than the drill. It had a small hole reamed in it and a notch leading to the edge of the hearth. The drill was a piece of wood about two feet long and a half inch or less in diameter. Holding the drill vertically and rapidly rotating it between the palms of the hands, embers were formed in the hole. Embers spilled through the notch onto the tinder placed beside the hearth. By cradling the tinder in the hands and gently blowing on the embers, a flame was produced. Depending on craftsmanship, humidity, and other factors, a fire could be created in just a few seconds—or never. To help increase the friction, sand grains were sometimes placed in the hole. Also, "points" of ember-producing wood were inserted into straight canelike sticks when appropriate wood could not be found in long straight pieces. The wood these points were made of was more efficient in starting a glowing ember. This method of making fire could be accomplished by a single person or two people taking turns rotating the drill. Another variation used a "bow" to rotate the drill. Some woods are better than others in starting a fire. A few choice woods in the Sierra Nevada include sagebrush (*Artemisia tridentata*), buckeye (*Aesculus*), certain species of willow (*Salix* spp.), juniper (*Juniperus* spp.), aspen (*Populus tremuloides*), and cottonwoods (*Populus* spp.).

Sierra Sweet-bay (*Myrica hartwegii*)

Description: This shrub has aromatic leaves that appear after the flowers. It grows along stream banks between 1,000 and 5,000 feet from Fresno to Yuba Counties. It flowers from June to July.

Interesting Facts: The fruit of related species are edible, and other species have been used medicinally.

WILLOW FAMILY (Salicaceae)

This family has two to three genera and more than five hundred species distributed worldwide. *Salix* and *Populus* are native to the United States. The family is of little economic importance, except as a source of ornamentals.

Cottonwoods and Quaking Aspen (*Populus*)

Description: These are trees with sticky, resinous leaf buds and deciduous leaves. Older trees of some species have gray, rough bark; young bark is smooth and whitish. The flowers are borne in catkins that appear before the leaves. They prefer moist soils, and *P. tremuloides* is usually found along streams. They grow rapidly and are planted for quick shade or wind protection. The soft wood of some species is used for veneers, boxes, matches, excelsior, and paper.

Quick Key to the Cottonwoods and Aspen	
Leaves ovate to deltoid; petioles nearly round in cross section	Black Cottonwood (*P. balsamifera*)—This deciduous tree is common along streams below 9,000 feet. At the higher elevations the leaves are more lanceolate than ovate or deltoid. It flowers from February to April.
Leaves round-ovate; petioles flattened laterally	Quaking Aspen (*P. tremuloides*)—This deciduous tree is recognized by its fluttering leaves. It occurs in moist places along stream banks from 6,000 to 10,000 feet. It flowers from April to June.

Interesting Facts: The catkins may be eaten raw or boiled in stews and are a source of vitamin C. The inner bark can also be eaten as a spring tonic or dried and ground into a flour substitute or extender. The fresh or dried plant can be used in poultices for muscle aches, sprains, or swollen joints. The primary action of *Populus* is that of an analgesic, used topically and internally. It contains varying amounts of populin and salicin, compounds related to early forms of aspirin. The leaves and bark are the most effective parts for tea and aid in diarrhea problems. The wood makes for an excellent bow and drill fire set. Cottonwoods are considered to be botanical indicators of water, and trappers often used aspen as bait in beaver sets.

Willow (*Salix*)

Description: Many species of willow are found in California. They are mostly shrubs with numerous stems. Their flowers are in catkins that appear before, with, or after the leaves. The various species of willows in the Sierra Nevada generally grow along streams or other moist habitats. Willow roots easily and occasionally form dense thickets. They are often planted to reduce stream-bank erosion.

Interesting Facts: The young shoots and leaves can be eaten raw. The bitter inner bark can also be eaten raw, although it is better dried and ground into a flour

OXALIC ACID

The tart, lemony taste of wood-sorrel (*Oxalis* spp.), cacti (*Opuntia* spp.), lambs's-quarters (*Chenopodium* spp.), amaranth (*Amaranthus* spp.), knotweed (*Polygonum* spp.), dock (*Rumex* spp.), and other species is due to the presence of soluble oxalic acid. Without proper preparation, these plants when eaten in substantial amounts should be considered toxic. However, when properly prepared, these plants are an excellent food source.

The soluble oxalic acid, also known as salt of lemon, is what makes the plants tasty as well as dangerous. The oxalic acid is dangerous because of its solubility and its affinity for calcium. The solubility allows the acid to enter the blood-stream, where it promptly combines with calcium to form nonsoluble calcium oxalate. This precipitates in the kidneys where it both plugs the tubules and "burns" all cells in contact with it, potentially leading to renal failure and death.

Oxalic acid is readily dissolved in heated water and will combine with calcium as readily in that water as in the bloodstream. Adding bone fragments, eggshells, or some other sources of calcium to cooking water will transform the oxalic acid to nonsoluble calcium oxalate in the pot, retaining the full flavor but rendering harmless the acid. If you have no bone fragments or eggshells, just pour out the first water after boiling for a time and replace with fresh water.

substitute or extender. The plant contains salicin, which is similar to aspirin and useful as a substitute. Any part of the willow can be used to produce a tea for use as an aspirin replacement for headache and body pain. The highest concentrations of salicin, however, are found in the inner bark. Because it is not nearly as strong as aspirin, you may have to drink quite a bit of it. The leaves have astringent properties that are effective when placed on wounds and cuts. Bark was chewed as a toothache remedy. Bark, leaves, twigs, and roots produced medicinal teas, powders, washes, and poultices to relieve pain, swelling, infection, bleeding, and many other ailments. Willows, like the cottonwoods, are botanical indicators of water. The branches of many willow species are very flexible and make them very useful for traps, arrow shafts, and other needs, such as basketry. The bark can also be used as crude cordage.

Willows were also an important basketry plant and were often used as a foundation material and twining material for twined baskets. Other uses of the wood include framework for dwellings, fish dams and weirs, racks for drying and cooking food, and light hunting bows. Fiber from the bark was used for cordage, nets, and clothing. Also, willows root easily due to the large amounts of indole acetic acid (IAA), a plant hormone in their stems. IAA can be extracted in cold water from one-inch sections of the stem and used to induce rooting of other species for transplanting.

WOOD-SORREL FAMILY (Oxalidaceae)

There are seven genera and more than one thousand species distributed worldwide. Only *Oxalis* is native in the United States. In general, they are small plants with leaf blades divided into three heart-shaped segments. The flowers are five-merous and yellow or purple. The seedpods split explosively, scattering seeds some distance from the plant. The family is of little economic importance.

Wood-sorrel (*Oxalis*)

Description: The genus description is similar to what has been described above. Two species (*O. corniculata* and *O. pes-caprae*) that commonly occur in the lower elevations as weeds in lawns and gardens are beginning to become established in some developed areas in the Sierra Nevada, as well as elsewhere. The genus name is derived from the Greek word *oxys*, meaning sour.

Wood-sorrel (*Oxalis* spp.)

Interesting Facts: The leaves and stems of *Oxalis* may be eaten raw. To make a tasty dessert, collect and allow a mass of the plants to ferment for a while. The plants also contain a high percent of oxalic acid; therefore, it is recommended that one eat the plants sparingly until accustomed to them. One symptom of too much oxalate is painful or swollen taste buds. The plants are also high in vitamin C and were used to remedy scurvy. A drink can be made by steeping the leaves in hot water, followed by chilling and sweetening it.

Flowering Plants
Monocots

Most of the luxuries, and many of the so-called
comforts of life are not only indispensable,
but positive hindrances to the elevation of mankind.
—Henry David Thoreau, *Walden*

Monocots are distinguished by a number of features. In monocots, the embryos of the seeds have only one cotyledon (seed leaf). The plant leaves usually
have parallel veins. The vascular bundles of the stems are irregularly arranged,
and the cambium is lacking. The flower parts are arranged in threes or sixes,
never in fives (or fours). Monocots are usually herbs, rarely shrubby.

ARROWGRASS FAMILY (Juncaginaceae)

There are four genera and twenty-six species in the arrowgrass family. Two genera, including *Triglochin*, occur in the United States. The family is of no economic
importance.

Arrowgrass (*Triglochin maritima*)

Description: This is a slender, grasslike plant with fleshy basal leaves that arises
from a rhizome. The flowering stems are long and smooth, and the flowers
occur in terminal bractless racemes. Arrowgrass occurs in mountain swamps
and around lakes. It flowers from April to August. Another species, *T. palustris*
(marsh arrow-grass), also occurs in the Sierra Nevada and is found growing in
mudflats and springy habitats between 7,500 and 11,500 feet in Tulare and Inyo
Counties.

Interesting Facts: The seeds of arrowgrass can be parched and ground into
flour. Roasted, they can be used as a coffee substitute. Seeds need to be parched
because they contain cyanogenetic toxins that have caused death in livestock
(Muenscher 1962). Parching or roasting the seeds renders them safe because the
poison is volatile (Harrington 1967; Kirk 1975).

The young white leaf bases were collected around April or May from the inner
leaves of the basal cluster. These leaf bases, when eaten raw at the right stage, have
a mild, sweet, cucumber-like taste. They are generally better if cooked. In springtime, the leaf bases contain few toxic compounds, whereas the mature leaves and
flower stalks should never be eaten. The leaves contain hydrocyanic acid, a toxin
that interferes with the uptake of oxygen. Symptoms include headache, heart
palpations, dizziness, and convulsions.

Caution: These plants are *toxic* when fresh. Several references list these plants
as livestock poisoners until they dry and then cyanogenic properties evaporate,
break down, or dissipate.

ARROWHEAD FAMILY (Alismataceae)

Thirteen genera and ninety species in this family are found worldwide, of which
five genera are native to the United States. Members of this family are aquatic and
marsh perennial herbs. Their long-stalked leaves and even longer flowering stems

are well adapted to wetland habitats. The flowers comprise three green sepals, three white petals, numerous stamens, and many pistils. This family is of little economic importance.

Caution: Aquatic plants such as arrowhead and water plantain are sometimes found growing in polluted or contaminated water.

Quick Key to the Arrowhead Family

Emergent or floating leaves with arrowhead-shaped blades (two lobes at the base)	Wapato, Arrowhead, Tule-potato (*Sagittaria*)
Emergent or floating leaves with elliptical or lance-shaped blades	Water Plantain (*Alisma plantago-aquatica*)

Wapato, Arrowhead, Tule-potato (*Sagittaria*)

Description: These are aquatic perennials with glabrous stems that grow eight to forty-eight inches high and contain a milky juice. The leaves are large and arrow shaped, with the basal lobes spreading outward. The stem is mostly leafless and is often branched. The flowers occur in whorls of three and are located near the ends of the stem. The sepals are ovate, whereas the petals are white. Tule-potato grows in marshy places along the edge of ponds or streams or in wet meadows below 7,000 feet. It flowers in July and August. *Sagitta* is Latin for arrow, referring to the shape of the leaves.

Quick Key to the Tule-potatoes

Beak of mature achene less than one-sixteenth inch long and erect	Tule-potato (*S. cuneata*)—This perennial occurs in ponds, slow streams, and ditches below 7,500 feet.
Beak of mature achene one-eighth inch long or more and horizontal	Wapato (*S. latifolia*)—This perennial occurs in ponds, slow streams, and ditches below 5,500 feet.

Interesting Facts: All species produce starchy, white tubers that can be roasted or boiled and then eaten. An important source of carbohydrates, the tubers contain a milky juice with a bitter flavor that is destroyed by heat. In the Pacific Northwest and northern Rockies, the Lewis and Clark expedition is said to have utilized these tubers extensively while exploring the Columbia River region (Hart 1996). To many Native American tribes, arrowhead was a primary vegetable. The small tubers are located at the ends of the long underwater rhizomes, perhaps a meter or more from the plant. They can be carefully removed without pulling up the whole plant. To collect the tubers, you can use your hands, a forked stick, or, if the water is deep, your feet. Wade into the pond where the plants are growing and feel around for the tubers in the mud with your toes. They should feel like round lumps varying in size from a peanut to a potato. After dislodging them, the tubers usually float to the surface for easy harvesting. They are best developed in late summer or autumn. Boil or bake them like a potato to remove the poisonous properties, then peel and eat them (Anderson 1939). They can also be dried for future use. The tubers were used in a tea for indigestion and in a poultice for wounds and sores. The leaves and stem contain alkaloids and may be *poisonous*.

Water Plantain (*Alisma plantago-aquatica*)

Description: This is a perennial plant from fleshy, bulblike stems. The basal leaves are long stalked and egg shaped, and the flowers are white. Water plantain is usually found in marshes and ponds at lower elevations.

Interesting Facts: The starchy, bulbous bases of water plantain are edible as a starchy vegetable (potato) after drying. Drying is said to remove the strong flavor. *Alisma* has a long history of use in Chinese medicine and is mentioned in texts dating back to about AD 200. It was also used by early herbalists as a diuretic and by the Cherokee Indians for application to sores, wounds, and bruises (Hamel and Chiltoskey 1975). It is described as a sweet, cooling herb that lowers blood pressure, cholesterol, and blood sugar levels. The root was also used as a diuretic in the treatment of dysuria, edema, distention, diarrhea, and other ailments (Foster and Duke 1990). Water plantain also furnishes food for waterbirds and muskrats.

BUR-REED FAMILY (Sparganiaceae)

One genus with about twenty species occurs in this family. Bur-reeds are found primarily in the cooler regions of the north temperate zone, Australia, and New Zealand. Several species are native to the United States. The plants arise from creeping rhizomes, and the roots are fibrous. The species have no direct economic importance to humans. According to Jepson, this family is now incorporated into the cattail family (Typhaceae).

Water Plantain
(*Alisma plantago-aquatica*)

Bur-reed (*Sparganium*)

Description: These are aquatic perennials with unbranched, erect or floating stems. The leaves are linear and sheath the stem. The flowers are borne in dense, round clusters. Bur-reeds can be found in shallow waters of marshes, ponds, and slow-moving streams.

Interesting Facts: The bulbous bases of the stems and tubers of *S. eurycarpum* (broadfruit bur-reed), *S. simplex* (simplestem bur-reed), and *S. angustifolium* (narrowleaf bur-reed) can be used as food in much the same way as cattails (*Typha* spp.) and bulrushes (*Scirpus* spp.)—dried and pounded into flour.

CATTAIL FAMILY (Typhaceae)

There is one genus (*Typha*) with about fifteen species worldwide in this family. The members are marsh or aquatic perennials with creeping rhizomes. Two kinds of flowers are borne in crowded spikes, with the staminate (male) flowers above and the pistillate (female) flowers below. The family is of little economic importance. This family now incorporates the previous family, Sparganiaceae.

Cattail (*Typha*)

Description: The two species found in California, *T. angustifolia* (narrowleaf cattail) and *T. latifolia* (broadleaf cattail), are also found over much of North America. *Typha* is Greek for cattail. It is often referred to as "Cossack asparagus." Cattails reproduce rapidly in marshy areas that are often unsuited for agriculture.

Interesting Facts: In his book *Mountain Man*, Vardis Fisher illustrates how Indians utilized cattails:

> "And what be that?" asked Sam, staring at the mold. He knew that Indians
> ate just about everything in the plant world, except such poisons as

toadstools, larkspurs, and water parsnips. It was a marvel what they did with the common cattail—from the spikes to the root, they ate most of it. The spikes they boiled in salt water, if they had salt; of the pollen they made flour; of the stalk's core they made kind of a pudding; and the bulb sprouts on the ends of the roots they peeled and simmered. (1965, 76)

Virtually every part of the plant has a use, from food to fiber. Euell Gibbons considered the cattail the "supermarket of the swamps." Although both cattail species have edible rhizomes, the rhizomes should never be eaten raw because they may cause vomiting (Jencks 1919). The rhizomes should be boiled, roasted, or dried and then ground into meal or flour. Another way to obtain the starch from the rhizomes is to follow a technique described by Gibbons:

> [A]fter scrubbing the root [that is, the rhizome] and peeling off the spongy outer layer surrounding the white stiff core, cut the core into small sections, and place the pieces in a bowl of cold water. Work the core with your hands, separating the fibers and scraping out the starch. Slosh the fiber around in the water until you have removed all the starch. Pour off the water through a course sieve to extract the fibers. Allow the water to settle for a little while the starch settles to the bottom of the container. Then carefully pour off the water, leaving the starch in the bowl. For a cleaner starch, pour in some more water and let it settle again. Then pour off the water. After this, you can use the starch almost immediately to make pancakes, breads, and biscuits. (1962, 58)

Harrington (1967) reports that one acre of cattails can yield more than three tons of nutritious flour, although extraction techniques need to be refined for commercial exploitation.

When pulling up the rhizome, you may notice newly emerging buds. These can be scrubbed, peeled, and eaten raw or boiled. The swollen joint between bud and rhizome is also starchy. Peel it, then roast or boil for a potato-like vegetable. Like the rhizomes, this part should not be eaten raw. The young green shoots can be peeled of their green outer layer and eaten raw or cooked. It is always good to boil them in a couple of changes of water if there is any bitterness. The peeled core can also be sliced and added to salads.

While the flower spikes are still green, remove the papery sheath and boil the cluster for a few minutes. The flower spikes can then be eaten like corn on the cob, although the core of the cluster is inedible. Cattail pollen is high in protein and can be used in flour for breads or eaten raw. However, if you are allergic to pollen, it should be avoided. The seeds from the female portion of the flower spike can be pulverized to make a nutritious, protein-rich flour. Seeds can be extracted from the fluff by parching them.

Useful fibers can be derived from cattails. Fibers in stems can be loosened by soaking plant material in water for several days. The silky fluff on the seeds is buoyant and water repellent and makes a good insulator, especially in boots. The silk can be used for stuffing items from pillows to down vests. It can also be used for tinder. The fuzz will explode into flame with a spark from a flint-and-steel set. Leaves can be woven to make mats, sandals, baskets, and the like. The stems provide a good coil foundation for baskets. Additionally, the stalks have been used as arrows and hand drills. A toothbrush can be fashioned from the fuzzy stem with the flowers removed.

Medicinally, the chopped or pounded rhizome was applied to the skin for

Cattail (*Typha* spp.)

CATTAIL JELLY
Jelly from cattails? The following is a recipe from Jan Phillips, author of *Wild Edibles of Missouri*, who says she makes jelly after the first flour has been rubbed out. The jelly is made by boiling the roots (rhizomes) for ten minutes in enough water to cover them. For every cup of liquid, add an equal amount of sugar and a package of pectin for every four cups of juice. The jelly somewhat resembles honey in both taste and color.

minor wounds and burns. Cattail down was used as dressings for wounds. Brown (1985) indicates that a sticky juice derived from between the young leaves can be used as a styptic, antiseptic, and anesthetic. The jelly from between the young leaves can be applied to wounds, sores, external inflammations, and boils to soothe pain. Brown also indicates that the jelly was rubbed on the gums as a novocaine substitute for dental extraction.

DUCKWEED FAMILY (Lemnaceae)

Plants in this family are small and float on slow or stagnant waters. They have small threadlike root hairs that obtain nutrients from the water. All duckweeds are used as food by wildlife and have been recorded from the stomachs of ducks. Two genera can be encountered: *Lemna* and *Spirodela*.

Quick Key to the Duckweed Genera	
Roots solitary on each plant	Duckweed (*Lemna*)
Roots usually two or more on each plant	Duckmeat, Great Duckweed (*Spirodela polyrhiza*)

Duckmeat, Great Duckweed (*Spirodela polyrhiza*)

Description and Interesting Facts: This species is frequently associated with *Lemna*. Duckmeat is a coarse species with a purplish-tinged lower side. The thalli are egg shaped or round and are solitary or in colonies. As with *Lemna*, duckmeat can also provide copious and palatable material for salads. The Chinese use this species to treat hypothermia, flatulence, and acute kidney infections (Culley and Epps 1973).

Duckweed (*Lemna*)

Description: These are small plants, often not much larger than a pinhead. The plant body (thallus) is flattened with a single root. They float on the surface or submerged in the water.

Interesting Facts: Under survival conditions, duckweeds can provide copious and palatable material for salads. Additionally, the Maidu and Miwok Indians in California used duckweed as a diuretic and a general tonic (Barrett and Gifford 1933; Strike 1994).

GRASS FAMILY (Poaceae; formerly Gramineae)

With 600 genera and 10,000 species of grasses worldwide, the grass family is the most common family of flowering plants found in practically all habitats and on all continents. There are more than 180 genera and nearly 1,000 species in the United States. Grasses have round, hollow stems with linear, sheathing leaves. Because grasses are wind pollinated, they have no showy flowers to attract insects. The flowers have been reduced to scaly bracts that enclose the male and female parts. The grains form within the papery bracts after pollination. The grass family contains many of the most economically important plants in the world, including all the important cereal crops and forage grasses essential to raising domesticated livestock. Grasses can be found from alpine meadows to sea level.

All grasses in the Sierra Nevada have edible grains that are generally small and tedious to collect. The small seeds are tightly enclosed in scales that are hard to remove. The larger grains of some grass species were a staple among aboriginal

AQUATIC PLANTS AND WATER
The days of safely drinking water straight out of the mountain stream with a Sierra cup are long gone. That also goes for eating aquatic plants such as cattail (*Typha*), duckweed (*Lemna* and *Spirodela*), watercress (*Rorippa*), pond-lily (*Nuphar*), and bulrush (*Scirpus*). Even in the most remote areas of the country, there are a number of disease-causing bacteria, protozoans, and viruses. A fairly common intestinal disorder among backcountry hikers is caused by the protozoan *Giardia lamblia*, which is carried in the intestines and feces of muskrats, beavers, moose, voles, and other water-loving mammals. Humans too are responsible for spreading *Giardia* to other areas by careless or improper disposal of human waste.

Other intestinal diseases can also be found in what appear to be clean water sources. Cases of *Campylobactor, E. coli*, and type A hepatitis have been traced back to drinking untreated water in the back-country. Additionally, cities, farms, and suburbs can experience clean water problems resulting from oil spills, pesticides, and toxic pollutants from mining operations.

In areas where edible plants are found, it is best to treat the water and the plants that grow in it. One method would be to soak the fresh greens in a disinfectant if pollution is suspected. You can use any of the water purification tablets on the market or a teaspoon to tablespoon of bleach in a quart of water. Then rinse the plants well and prepare. Obviously, if the waters are polluted with oils, pesticides, or other toxic pollutants, this method will not work. In any case, extreme care should be exercised.

peoples. The grains were harvested with a beater and ground for mush and flour. The grains are rich in protein and can be eaten raw but are better if roasted, ground into flour, or boiled into mush. They may also be boiled in the same way as rice and added to soups or stews. The reported toxicity of some grass species may be that of a fungus (that is, *Claviceps purpurea*) associated with the grasses. Any inflorescences containing black or purple-black grains should be discarded because they may have a harmful fungus infection.

All bladed grasses are edible and rich in vitamins and minerals. Animals often consume grasses to get nutrients they cannot get elsewhere. The young shoots are edible raw and are not as fibrous and therefore easier to digest than mature leaves. The green or dried leaves can be steeped to make a tea.

Note: Grasses that are infected with *Claviceps purpurea* develop purple sclerotia (ergots) in place of the healthy grain. The sclerotia contain a number of toxic alkaloids and if eaten can cause severe illness and sometimes death. One effect of the toxins is constriction of the blood vessels, whereby the impaired circulation may cause gangrene or loss of limbs. Another effect is on the nervous system, resulting in convulsions and hallucinations (Webster 1980). The sclerotia of *C. pupurea* are used medicinally to hasten uterine contractions during childbirth. The ergot of commerce is produced by cultivating the fungus on rye (*Secale cereale*) and other plants. Attempts are being made to extract the medically important alkaloids from pure cultures of the fungus (Webster 1980). The following key is very simplified and has been included as a means to identify some of the genera that are known to occur in the Sierra Nevada. However, the discussion of uses for the various species has been limited due to the difficulty of working with these plants in the field. Suffice it to say, a grass is a grass, and many have similar uses.

WILD FLOURS (PINOLE)
Pinole was made using small grains that were parched by tossing in a basket with glowing coals or hot pebbles, keeping the grains in constant motion. Grains were then pulverized and eaten. Sometimes the pulverized grains were pressed into cakes held together by the grains' natural oil, with no other liquids added. Grains from several different species of plants were often mixed together to enhance the flavor of the pinole.

Quick Key to Some Grasses

Panicle open
    Plants of wet habitats
        Leaf tips boat-shaped   *Poa*
        Leaf tips not boat-shaped
            Spikelets one-flowered   *Agrostis*
            Spikelets two or more flowered   *Glyceria*
    Plants of dry habitats
        Leaf tips boat-shaped   *Poa*
        Leaf tips not boat-shaped
            Spikelets one-flowered
                Leaves greater than 12 inches long; ligule pointed   *Oryzopsis*
                Leaves shorter; ligule not pointed   *Agrostis*
            Spikelets two or more flowered
                Spikelets many-flowered   *Bromus*
                Spikelets two to eight flowered   *Deschampsia*
                    Ligule reduced to a tuft of hairs   *Danthonia*
                    Ligule not as above   *Festuca*
Panicle compact
    Plants of wet or moist areas
        Leaf tips boat-shaped; spikelets two-flowered   *Poa*
        Leaf tips not boat-shaped; spikelets one-flowered
            Spikelets sessile, forming a dense spike less than ½ inch thick   *Alopecurus*
            Spikelets stalked; spike more than ½ inch thick   *Calamagrostis*
    Plants of dry habitats
        Spikelets one-flowered
            Awns more than two inches long
                Awns straight   *Hordeum*
                Awns bent and twisted   *Stipa*

Awns less than two inches long	
Panicle cylindrical; awn less than ⅛ inch long	*Phleum*
Panicle not cylindrical; awn more than ⅛ inch long	*Oryzopsis*
Spikelets two or more flowered	
Leaf tip boat-shaped	*Poa*
Leaf tip not boat-shaped	
Awns bent	*Trisetum*
Awns straight	
Panicle with short branches	
Spikelets two-flowered; fanicle four to twelve inches long	*Deschampsia*
Spikelets with two or more flowers; panicle less than eight inches long	*Festuca*
Panicle not branched; spikelets sessile	
Spikelets solitary at each node	*Agropyron*
Spikelets more than one at each node	*Elymus*

225

Major Plant Groups

CEREALS
Cereals are an excellent source of proteins. However, the proteins they contain are incomplete, as they do not have all the essential amino acids. When cereals (lacking lysine but rich in methionine) and legumes (lacking methionine but rich in lysine) are eaten together, the amino acids found in cereals complement those in legumes, providing a person with all the necessary proteins.

Common Reed (*Phragmites communis*)

Description: Common reed is native to every continent except Antarctica. It often forms dense thickets along streams, ditches, and marshes at low elevations. The record of human uses for this species is extensive. The reeds have been used for roof insulation since before the time of Christ. Because it is light, it is best known for its use in arrow shafts. The roots can be eaten raw or cooked.

Interesting Facts: Meal can be obtained from pulverized stems. In the fall, the leaves and stems may become encrusted with grayish exudate. This exudate, actually honeydew (excreta of whitefly and aphids), was obtained from stalks. Stalks were cut and flayed to remove honeydew crystals, which were winnowed and cooked into stiff dough. Dough was formed into cakes, sun dried, and stored. Split culms provided fiber. Common reed was used to make flutes and other musical instruments in addition to carrying nets and cordage. The honeydew was given to pneumonia patients to loosen phlegm and soothe pain in lungs.

Ricegrass (*Oryzopsis hymenoides*)

Description: A perennial bunchgrass forms large tufts with the remains of old leaves at the base. The stems (culms) are from eight to twenty inches tall. The spikelets are borne on the ends of long, greatly divided, spreading or reflexed branches in the open inflorescence. Ricegrass is found in dry rocky or sandy ground, grasslands, and valleys from low to midelevations.

Interesting Facts: The relatively large grains of this species have been used by Native Americans as food for centuries (Doebley 1984). The grains were collected with the stems and held over a fire to singe off the fine white hairs. They can also be collected in a pan or basket with hot coals or rocks and shaken to burn off the hairs. The grains were then ground into flour and used as mush or to thicken soup or made into cakes.

IRIS FAMILY (Iridaceae)

There are about seventy genera and fifteen hundred species in this family worldwide, with the chief center of distribution in South America and tropical America. Five genera are native to the United States. In the Sierra Nevada, members of this family are perennials with narrow, grasslike leaves and thick rhizomes or fibrous roots. The flowers have three petals and three petal-like sepals joined on top of the ovary. The family is of economic importance as a source of ornamentals and saffron dye.

Quick Key to the Iris Family	
Petal-like sepals larger than the petals and reflexed; flowers irislike	Iris (*Iris*)
Sepals and petals similar in size and shape; flowers star shaped	Blue-eyed Grass (*Sisyrinchium*)

Blue-eyed Grass (*Sisyrinchium*)

Description: The three species are perennial herbs with generally tufted, narrow leaves. The flowers have three petals and three sepals that are alike, pinkish purple to blue in color. They are found in meadows and are often inconspicuous because of their tufted, grasslike leaves. The flowers open only in bright sunshine.

1. Blue-eyed Grass (*S. bellum*)—This species is found growing in wetlands and in the foothills, as well as the mixed conifer zone.
2. Yellow-eyed Grass (*S. elmeri*)—This slender perennial grows in boggy and wet places from 4,000 to 8,500 feet. It flowers from July to August.
3. Idaho Blue-eyed Grass (*S. idahoensis*)—This perennial has a mostly leafless stem and grows in wet meadows and in moist areas from 4,500 to 10,800 feet. It flowers in July and August.

Interesting Facts: Although the uses of most species are unknown, *S. bellum* was known among the Spanish Californians as "azulea" and "villela." It was made into a tea considered to be a valuable remedy in treating fevers. It was thought that a patient could subsist for many days upon it alone (Sweet 1976).

Iris (*Iris*)

Description: These perennials arise from a creeping rhizome. The stems are erect, and the leaves are sword shaped or linear. The flowers are large, occurring singly or in panicles. The perianth is composed of six clawed segments that are united below into a tube. The outer three segments are spreading or reflexed, whereas the inner are smaller and erect. *Iris* is the Greek word for rainbow.

1. Wild Iris (*I. hartwegii*)—This perennial with slender stems grows in dry woods from 5,000 to 7,500 feet. It flowers in May and June.
2. Western Blue Flag (*I. missouriensis*)—This slender perennial grows in moist areas as in meadows from 3,000 to 11,000 feet. It flowers in May and June.
3. Slender Iris (*I. tenuissima*)—Look for this iris in dry, sunny woods from Sierra County north.

Interesting Facts: Members of the genus *Iris* contain irisin, an acrid resin concentrated mainly in the rhizomes and present in the foliage and flowers. People who raise irises sometimes develop a skin rash from handling the rhizomes. Cattle have died as a result of eating relatively large quantities of the plants. The rootstock produces a burning sensation when chewed. If eaten in quantity, irises will cause diarrhea and vomiting. The poisonous rhizomes were used by Native Americans in a mixture of bile to poison arrowpoints (Kingsbury 1964).

Wild Iris (*Iris* spp.)

Iris was an important plant for Native Americans. The leaves were woven into mats and lined with cattail "down" for use as a diaper for babies.

LILY FAMILY (Liliaceae)

This family contains many beautiful wildflowers. It is a large and varied family with approximately 250 genera and 4,000–6,000 species worldwide. About 75 genera are native to the United States. The family is characterized by a perianth of six parts, with a superior ovary and a three-lobed stigma. The fruit is a capsule that splits open when ripe. The family is a source of many ornamentals, several important fibers, fermented and distilled beverages, and steroidal compounds. Although some members of this family are edible, there are many poisonous species. For example, you can boil and eat the bulbs of the true lilies (*Lilium*), but some genera contain highly toxic alkaloids. Nearly two hundred alkaloids and numerous glycosides occur in the family.

Bear-grass, Indian Basket-grass (*Xerophyllum tenax*)

Description: This is stout perennial growing up to five feet tall. The stems arise from large clumps of bluish green, wiry, saw-edged, grasslike leaves that are up to two feet long. The cream-colored flowers are borne in a hemispheric terminal inflorescence. Common bear-grass is found in mid- to lower alpine open slopes and forests. The thick, shallow rhizome can remain vegetative for many years, then flowers and dies.

Interesting Facts: The fibrous roots of bear-grass can be eaten after roasting or boiling. Although the sharp leaves are not very pleasant to handle, when dried and bleached they can be used for weaving baskets and clothing. The baskets are particularly pliable and durable. Lather from the roots was used to bathe sores.

Brodiaea (*Brodiaea*)

Description: These are erect herbs from corms. The leaves are linear, and the flowers are in umbels. They can be found in dry to moist soil, in rocky areas, and in meadows from low to midelevations.

1. Brodiaea (*B. coronaria*)—Look for this species in open, dry habitat below 6,000 feet from Tuolumne County north.
2. Harvest Brodiaea (*B. elegans*)—This is a common species in dry or occasionally moist flats and on slopes up to 7,000 feet on the western side of the range from Tehama County to northern Tulare County.

Brodiaea (*Brodiaea* spp.)

Interesting Facts: The corms of most *Brodiaea* species are edible raw but are somewhat mucilaginous. It is better if they are boiled for a few minutes or roasted. They can also be mashed and dried for future use in stews. Because the corms grow deep, it is usually easiest to harvest them with digging sticks (Elliott 1976).

The crushed corms were used as a paste that was smoothed over the sinew backing on bows. The paste was also used to bind paint pigments to hunting bows. *Brodiaea* corms and flowers were used as soap and shampoo.

Caution: When collecting bulbs, be aware that the poisonous *Zigadenus* (death camas) may be in the area too.

California Greenbriar (*Smilax californica*)

Description: This plant with woody stems has long, ovate leaves that are two to four inches long. The flowers are greenish or yellowish and occur in umbels. The berries are black. Greenbriar is found in thickets and streambanks below 5,000 feet from Butte County northward. It flowers from May to June.

DIGGING STICKS
Roots and bulbs of many plants are best collected with the aid of a digging stick. A digging stick is about two to three feet long, one to two inches thick, beveled at one end, and fire hardened. To use, thrust the stick into the ground beside the plant and pry upward while pulling on the plant from above.

To fire harden your digging stick, simply hold the point a few inches above a bed of hot coals and slowly turn it as you would a skewer. Take care not to char the wood, but let it turn to a light brown color.

Interesting Facts: The roots may be used in soups and stews or dried and ground into flour. The flour mixed with water and sugar makes a good drink. The young shoots in early season may be eaten raw or cooked. The berries are edible raw or cooked.

Common Camas (*Camassia quamash*)

Description: Arising from a bulb, it has bright-blue to violet, six-parted flowers in a showy, spikelike raceme. The leaves are basal and grasslike. The plant can be found in meadows, marshes, grassy slopes, and fields from low to midelevations. The plant's common name is derived from the Nootka Indian word *chamas*, which means sweet.

Interesting Facts: Camas was perhaps the most important food of Native Americans in western North America. As Gunther writes: "[E]xcept for choice varieties of dried salmon there was no article of food that was more widely traded in western Washington then camas" (1973, 24).

Camas was an important staple food to many aboriginal peoples, and the bulbs were eaten wherever available. In fact, some tribes managed or "owned" areas where the plant grew. They would intentionally set fires to ensure a bountiful harvest. As testimony to the species' former abundance, Meriwether Lewis in his journal entry of June 12, 1806, stated: "The quawmash is now in blume and from the colour of its bloom at a short distance it resembles lakes of fine clear water, so complete in this deseption that on first sight I could have sworn it was water."

Camas bulbs were dug out of the ground from late July through September with digging sticks. The black outer covering of the bulb is removed, and the white bulbs are then steamed or cooked in pits for twenty-four hours or more. When cooked this way, the bulbs turn dark brown and become quite moist, soft, and sweet. Cooking is required because the plant contains a carbohydrate called inulin. Inulin is not very digestible or very palatable in its "raw" form. Cooking is necessary to chemically break down the inulin into its component fructose sugar. Common in fruits and honey, fructose is both easily digested and sweet tasting. After cooking, the camas bulbs can be mashed and dried into cakes for storage. They are considered to be more nutritious than potatoes. The bulbs can also be boiled down into a syrup. Too much camas, however, is both an emetic and purgative (Willard 1992).

Caution: When collecting bulbs, be aware that the *poisonous* death camas (*Zigadenus*) may be in the area too.

Corn Lily (*Veratrum californicum*)

Description: This is a tall, stout perennial that has leafy stems and grows three to six feet tall. The leaves are broad and clasping, strongly veined, and ovate to elliptic in shape. The cream to white flowers occur in a dense, terminal, branched cluster. The six perianth segments are all alike, with a green gland near the base. Corn lily is common in wet meadows and along streambanks, particularly at higher elevations, but usually found below 11,000 feet. It flowers in July and August. These plants are often shredded by hail during storms.

Interesting Facts: *These plants are very poisonous if ingested and have an inconsistent mixture of several powerful alkaloids.* Some of the symptoms include depressed heart action, salivation, headache, burning sensation in the mouth, slowing of respiration, and death from asphyxia. These violent symptoms of poisoning may occur within ten minutes. Avoid any use of the plant that involves ingestion. In some cases, just handling *Veratrum* can cause severe itchiness and irritation. The water in which the roots of *Veratrum* were boiled is considered effective in killing head lice. The powdered roots have also been used as an insecticide (Sweet 1976). Even nectar in the flowers is poisonous to insects and can cause serious losses among honeybees.

Death Camas, Zygadene (*Zigadenus*)

Description: The three species of *Zigadenus* are glabrous perennials with bulbs and grasslike leaves. The cream-colored to greenish white flowers are stalked and subtended by narrow bracts in an elongated inflorescence. Zygadene and death camas occur in grasslands, meadows, and forest openings into the alpine zone.

Death Camas
(*Zigadenus* spp.)

Quick Key to the Death Camases	
Flowers occurring in racemes	Death Camas (*Z. venenosus*)—This species grows in moist habitats below 9,200 feet throughout the Sierra Nevada.
Flowers occurring in panicles	
Stamens equal to or longer than the perianth	Sand-corn (*Z. paniculatus*)—This plant occurs mainly on the eastern side of the range between 4,000 and 7,000 feet.
Stamens shorter than the perianth	Giant Zygadene (*Z. exaltatus*)—This species is found below 4,000 feet.

Interesting Facts: Death camas is a very *poisonous* plant if ingested. The alkaloids, primarily concentrated in the bulbs, can cause muscular weakness, slow heartbeat, subnormal temperature, stomach upset with pain, vomiting, diarrhea, and excessive watering of the mouth. Death camas should not be confused with the edible camas (*Camassia*), which formed a staple food for aboriginal peoples in the Northwest. It is also difficult to distinguish death camas from other edible plants, including wild onion (*Allium*), sego lilies (*Calochortus*), fritillaries (*Fritillaria*), and brodiaeas (*Brodiaea*) prior to flowering.

Crushed death camas bulbs were used by some Native Americans as poultices for boils, bruises, strains, rheumatism, and in some cases rattlesnake bites.

Fairy Bells (*Disporum hookeri*)

Description: These are rhizomatous perennials with branched, leafy stems. The sessile leaves are prominently veined and slightly rotated where they meet the stem. The flowers hang from the tips of the branches and are tubular to bell shaped. They can be found in moist to somewhat dry forest areas from the foothills to the subalpine zone.

Interesting Facts: The velvet-skinned yellow or orange berries of a related species, *D. trachycarpum*, can be eaten raw. The fruit of *D. hookeri* was used to relieve kidney ailments.

False Solomon's-seal (*Smilacina;* also known as *Maianthemum*)

Description: These are perennial herbs with extensive, horizontal rhizomes. Leaves are alternate and sessile or on short petioles. There are several to numerous small, white flowers borne in a branched or simple inflorescence. The petals and sepals are similar (tepals). There are six stamens, and the fruit is a globose berry with few seeds.

Quick Key to the False Solomon's-seals

Perianth segments less than one-eighth inch long; fruit a red berry	Feathery False Solomon's-seal (*S. racemosa* = *M. racemosum*)—This is an erect perennial that grows from 5,000 to 8,000 feet. It flowers in June and July.
Perianth segments one-quarter inch long; fruit a purple to black berry	Starry False Solomon's-seal (*S. stellata* = *M. stellatum*)—This erect perennial grows in wet or brushy places from 4,000 to 8,000 feet. It flowers from April through June.

Interesting Facts: Both species have edible berries that are not especially palatable. If eaten in quantity they can act as a laxative. Cooking the berries removes much of the purgative elements, making them a bit more palatable. They are also high in vitamin C. The young shoots and leaves can be used like asparagus or eaten as a potherb.

False solomon's-seals have starchy rootstocks that may be eaten. However, the rootstocks must be soaked overnight in lye. The Ojibwa Indians of Ontario, Canada, used the white ashes from their fire pits instead of lye, which supposedly removed the bitterness. The roots are then boiled and rinsed several times to remove the lye (Turner and Szczawinski 1991). A tea made from the roots was used for headaches.

A tea from starry false solomon's-seal was used by some Native Americans as a contraceptive (Willard 1992). The powdered roots were used on wounds to stop bleeding, and a root decoction was used internally as a tonic or externally as an antiseptic wash for infected sores or wounds. The mashed root of starry false solomon's-seal was thrown into a stream as a fish stupefier, making the fish easier to catch. A decoction of feathery false solomon's-seal was used to regulate menstrual disorders, to relieve kidney problems, to heal wounds, and as a heart tonic.

Fritillary (*Fritillaria*)

Description: These are glabrous, perennial herbs from bulbs with numerous white bulblets around the base. The flowers are usually nodding with similar petals and sepals (tepals). The genus name comes from the Latin *fritillus*, meaning dice box, in reference to the short, broad capsule characteristic of the genus. Fritillary can be found in open areas and forests, meadows, and grassland habitats at low and middle elevations.

Quick Key to the Fritillaries

Style not parted	
Flowers yellow to orange	Yellow Mission Bells (*F. pudica*)—Look for this yellow flowered species in grassy or brushy habitats or slopes below 5,000 feet.
Flowers pink or purple	Brandegee's Fritillary (*F. brandegei*)—This species has nodding pink- to purplish-colored flowers and grows on granitic soils and in open forests between 5,000 and 7,000 feet.
The style is three-parted at the apex	
Flowers are not mottled	Brown Bells (*F. micrantha*)—This plant has purplish or greenish white flowers and is found on dry benches and slopes below 6,000 feet from Plumas County southward.

Flowers are mottled with yellow		
Flowers scarlet colored; petals generally about one or more inches long	Scarlet Fritillary (*F. recurva*)—This species occurs on dry hillsides between 2,000 and 6,000 feet.	231 *Major Plant* *Groups*
Flowers purple; petals less than one inch long		
Flowers nodding; stems solid	Purple Fritillary (*F. atropurpurea*)—This plant with nodding purplish brown flowers is found under trees between 6,000 and 10,500 feet throughout the Sierra Nevada.	
Flowers erect; stems hollow	Davidson's Fritillary (*F. pinetorum*)—This purplish-flowered species is found on shaded granitic slopes between 6,000 and 10,500 feet from Alpine County southward.	

Interesting Facts: The bulbs of this genus have been a staple for Native peoples since prehistoric times. Bulbs of all species are edible raw or cooked but are relatively rare and should be considered only in an emergency. The fruit pods of yellow mission bells can be eaten as a potherb.

Lily (*Lilium*)

Description: The true lilies have large, showy flowers that are usually spotted. The flowers may be funnel shaped, or the tepals may be curved backward. The color varies from yellow to orange, with one to several flowers appearing at the top of a single stalk up to three feet tall. Leaves are alternate or whorled. The roots are covered with scales, an important identifying characteristic.

Quick Key to the Lilies

Flowers are white	Washington Lily (*L. washingtonianum*)—Look for this lily among bushes and in dry, granitic, and loamy soils between 4,000 and 7,400 feet from Fresno County northward.
Flowers are orange-reddish	
Plants growing in dry, shaded forests	Humboldt Lily (*L. humboldtii*)—This lily with orange-yellow flowers is found in dry, open habitats and forests below 4,500 feet from Fresno to Butte Counties.
Plants of wet, shaded areas	
Flowers ascending of horizontal	Alpine Lily (*L. parvum*)—This orange- to dark red–flowered lily is found in boggy habitats, usually among alders or willows, between 4,000 and 8,000 feet throughout the Sierra Nevada.
Flowers nodding	
Flowers are not really fragrant	Leopard Lily (*L. paradalinum*)—This stout perennial grows in large colonies along stream banks and in springy places below 6,000 feet. It flowers from May through July.
Flowers are fragrant	Kelley's Lily (*L. kelleyanum*)—This lily with fragrant orange and yellow flowers occurs between 4,000 and 10,500 feet throughout the Sierra Nevada.

Interesting Facts: The bulbs of all members of this genus are edible. They can be eaten raw or steamed and have a bitter or peppery taste. Some suggest that the plants are better used as a flavoring agent and that the bulbs taste better after flowering. One way to locate bulbs in the fall is to flag the plants during the summer. However, because of their relative rarity and beauty, they should not be harvested except in an emergency.

Found a lovely lily (*Calochortus albus*) in a shady adenostoma thicket near Coulterville, in company with *Adiantum chilense*. It is white with a faint purplish tinge inside at the base of the petals, a most impressive plant, pure as snow crystal, one of the plant saints that all must love and be made so much the purer by every time it is seen. It puts the roughest mountaineer on his good behavior.
—John Muir, *My First Summer in the Sierra*

Mariposa Lily, Sego Lily (*Calochortus*)

Description: These are characterized as perennials from bulbs, with tuliplike flowers that are few and showy. Mariposa is the Spanish name for butterfly. These species can be found in dry, open places from low to midelevations.

Quick Key to the Mariposa Lilies

Flowers nodding; the corolla is narrow or barely widened at apex	
Flowers deep rose and bell shaped	Rosy Fairy Lantern (*C. amoenus*)—Look for this species in the leafy loam of grassy slopes below 5,000 feet from Madera County southward.
Flowers are white	Fairy Lantern (*C. albus*)—This species is found growing in shaded, often rocky places below 5,000 feet from Madera to Butte Counties.
Flowers erect; the petals are spreading wide at the apex	
Petals without a central dark spot	
Petals blue in color and bearded above gland	Beavertail-grass (*C. coeruleus*)—This plant grows in open, gravelly places between 3,500 and 7,500 feet.
Petals white	Lesser Star-tulip (*C. minimus*)—Look for this plant in moist, grassy places between 4,000 and 9,500 feet from eastern El Dorado County southward.
Petals with a central dark spot	
Sepals usually greater than one inch long	
Petals with dark red spot, with sometimes a second lighter spot above the first	Butterfly Mariposa Lily (*C. venustus*)—This species is found in light, sandy soils, often decomposed granite, below 8,000 feet from El Dorado County southward.
Petals' spot surrounded by yellow	Superb Mariposa Lily (*C. superbus*)—This species is found on open or wooded slopes below 5,000 feet throughout the Sierra Nevada.
Sepals one-half to one inch long	
Petals white or tinged lilac, sometimes purple spotted below the nectary	Plain Mariposa Lily (*C. invenustus*)—This perennial plant with slender, simple, and erect stems grows on dry soils from 4,500 to 9,000 feet. It flowers from May to August.
Petals white to smoky blue, often tinged with pink, red to black spotted above the nectary	Leichtlin's Mariposa Lily (*C. leichtlinii*)—This species is found in open, gravelly places between 4,000 and 11,000 feet throughout the Sierra Nevada.

Mariposa or Sego Lily
(*Calochortus* spp.)

Interesting Facts: Although the entire *Calochortus* plant is edible and can be used as a potherb, the highly nutritious bulbs are usually sought. They are smaller than walnuts and may be eaten raw, boiled, or roasted in hot ashes in pits or steamed before eating. The bulbs are dug in the early spring, usually before flowering. The bulbs can also be threaded on a string and dried with or without cooking first. They can also be dried and ground into flour or cooked and mashed into cakes for preservation. *Calochortus* bulbs were eaten by many Native Americans and were considered an important food source (Olsen 1990). When numerous bulbs were collected, they were usually pit cooked. The flower buds can be eaten and have a sweet taste. The seeds are also edible.

However, care should be taken not to overharvest these plants. This is particularly true in Utah, where *C. nuttallii* (sego lily) is the state flower.

Note: These plants are becoming increasingly rare in some areas, primarily due to habitat destruction and overgrazing. Harvesting the corms destroys the plants. Because of the plants' rarity and beauty, their use today is not recommended.

ONION VALLEY
The name Onion Valley has been used throughout the Sierra Nevada for meadows where early settlers found edible onions.

Onion (*Allium*)

Description: The many species of *Allium* in the Sierra Nevada arise from bulbs, and all have the characteristically distinct onion odor. The odor is apparently caused by the presence of volatile sulfur compounds in all parts of the plant (causing their strong flavor and irritation to eyes). In all, there are about three hundred species of onions in the world. Some other common names of *Allium* include leeks, garlic, and chives. The small flowers are clustered together in umbels. Onions are found in a variety of habitats from low elevations to the alpine zone. *Allium* is the ancient Latin name for garlic. The derivation of this name may be from the Celtic *all*, which means pungent.

Onion (*Allium* spp.)

1. Fringed Onion (*A. abramsii*)—This onion with yellow anthers grows in gravelly habitats between 4,500 and 10,000 feet from Madera County south.
2. Narrow-leaved Onion (*A. amplectens*)—This white- to pinkish-flowered onion is found on dry slopes below 6,000 feet throughout the Sierra Nevada.
3. Sierra Onion (*A. campanulatum*)—This pale rose–colored onion with reddish anthers is found in dry woods between 2,000 and 8,900 feet throughout the Sierra Nevada.
4. Paper-flowered Onion (*A. hyalinum*)—This onion with white or pinkish flowers occurs in rather moist places, on grassy and rocky slopes below 5,000 feet.
5. Lemmon's Onion (*A. lemmonii*)—This white- to pale rose–flowered onion is found in heavy soils.
6. Membranous Onion (*A. membranaceum*)—This onion with whitish- to pink-colored flowers is found in wooded areas on the west side of the range from Mariposa to Plumas Counties.
7. Red Sierra Onion (*A. obtusum*)—This onion is found growing on sandy or gravelly slopes and benches between 7,000 and 12,000 feet from Plumas County south.
8. Broad-stemmed Onion (*A. platycaule*)—This onion with deep-rose flowers occurs on dry gravelly slopes and knolls between 4,000 and 9,000 feet.
9. Three-bracted Onion (*A. tribracteatum*)—This onion species is found growing in volcanic soils within Yosemite National Park.
10. Swamp Onion (*A. validum*)—This purple-violet– to almost white–flowered onion grows in wet meadows between 4,000 and 11,000 feet throughout the Sierra Nevada.
11. Yosemite Onion (*A. yosemitense*)—This onion with pale-rose flowers grows in open forests within Yosemite National Park.

Interesting Facts: All *Allium* species are known to be edible. The bulbs may be eaten raw, boiled, steamed, creamed, and in soup, and are especially good when used as a seasoning. Ingestion of large amounts of onions, including the culti-vated ones, can cause poisoning or cause goiter but are otherwise not known to be harmful. Regardless, eating them in moderation is the key. The plants are valuable in all seasons and can be used as greens and as flavoring. The seeds and leaves can also be eaten. Onions will keep a long time, because the skin dries and preserve the flesh inside. Wild onions do contain large amounts of some impor-tant micronutrients, more vitamin C than an equal weight of oranges, and more than twice as much vitamin A as an equal weight of spinach (Kindscher 1987). Additionally, onions contain a significant amount of a starch called inulin, which is not easily digested by humans.

Medicinally, onions have a number of uses. Soldiers during World War I took advantage of their natural antiseptic properties by applying *Allium* juice to wounds to prevent infection. The juice of wild onions can be boiled down until it is thick and used as a treatment for colds and throat irritations. The juice was also used as an insect repellent when rubbed over the body. The onion smell appar-ently has some beneficial effects on the circulatory, digestive, and respiratory systems (Moore 1979; Tull 1987).

Warning: Wild onions should not be confused with the so-called poison onions or death camas (*Zigadenus venenosus*). Death camas is most likely to be confused with *Allium*. These are bulb-bearing plants with grasslike leaves, also in the lily family. They have upright, more elongated (not umbrella-like) clusters of white or cream flowers. They contain highly toxic alkaloids, and all plant parts, including the bulbs, can be fatal if ingested in any quantity. They also lack the characteristic strong odor of onions.

Soap Plant, Amole (*Chlorogalium pomeridianum*)

Description: The stems of this plant grow two to five feet and are nearly leafless. There are many basal leaves that are keeled and wavy edged. The white flowers occur in a long, spreading cluster and are purple veined. The bulb, which can grow to about two inches in diameter, has a dense coat of coarse fibers. The species is found in dry, open, and stony ground up to 5,500 feet. It flowers from May to August.

Interesting Facts: The delicate flowers open only in the afternoon. The bulbs can be eaten after being cooked in an earthen oven. The young shoots can be eaten in the spring.

The baked bulbs were used as a poultice on skin sores. The scales of the bulb form a lather with water. Fish can be caught by throwing crushed amole bulbs into a stream that has been dammed. This stupefies the fish, causing them to float to the surface to be collected (Sweet 1976).

Triteleia (*Triteleia*)

Description: These are perennials that arise from corms. The leaves are grasslike, few in number, and withering by the time the plants flower. The flowers occur in umbels. The genus name comes from the Greek *tri* (three) and *teleios* (perfect), referring to the floral parts that occur in threes.

1. Dudley's Triteleia (*T. dudleyi*)—This is a rare species found in subalpine forests between 9,500 and 11,500 feet in Tulare County.
2. White Brodiaea (*T. hyacinthina*)—This species is common in moist to dryish habitats below 7,200 feet on the western side of the range.

3. Golden Brodiaea (*T. ixioides*)—This yellow-flowered species is common in dryish habitats throughout the Sierra Nevada.
4. Grass Nut (*T. laxa*)—This white- to purple-flowered species occurs in heavy soils below 4,600 feet from Tehama County south.

Interesting Facts: The corms are edible after cooking. *Triteleia* is now a member of the Themidaceae family, which also includes *Brodiaea* and *Dichelostemma*. Until recently, this group of bulbous plants was included in the closely allied onion family or Alliaceae. All three genera were sometimes referred to as the brodiaea complex and sometimes still are. There are fourteen species of *Triteleia* found in western North America and one species in Mexico. Most are native to California and are found in a variety of habitats including rocky cliffs, open conifer forests, grasslands, and vernally wet meadows. The time of bloom occurs from March to August depending on the taxa and habitat.

Twisted-stalk (*Streptopus amplexifolius*)

Description: Twisted-stalk is a perennial herb with creeping rootstocks. The sessile or clasping leaves are alternate, elliptical to ovate in shape, and the flowers are yellowish green. The one to two pendant flowers hang from the axils of the upper leaves on stalks that are bent in the middle. Common in moist soil and along streams and thickets in the montane and lower subalpine zone, they are often associated with genera such as *Smilacina* (false solomon's-seal) and *Actaea* (baneberry).

Interesting Facts: In terms of edibility, these plants have escaped mention in many guides but are indeed safe. The new spring shoots and clasping young leaves can be eaten raw or added to salads and taste somewhat like cucumbers. The berries, often referred to as watermelon berries, are somewhat laxative if eaten in excess but may be eaten raw or cooked in soups and stews. They are sometimes referred to as "scooter berries," because if you eat too much you can find yourself "scooting" to the bathroom. The species are easy to grow in wild gardens. The stems were used in poultices for cuts (Schofield 1989).

Warning: Anyone wishing to use the young shoots of twisted-stalk should be very careful to identify it correctly. At the shoot stage, these plants resemble the highly toxic *Veratrum* spp. (corn lily).

Wake-robin, Trillium (*Trillium*)

Description: These are low, glabrous herbs from short, fleshy rootstocks. The stems bear a whorl of three sessile, broadly egg-shaped leaves, and the single stalked flower is borne above them. The sepals are green and narrower and shorter than the snow-white petals. *Trillium* is common in moist, rich soils of forests from the foothills to the subalpine zone.

1. Trillium (*T. album*)—This species with white to pink flowers is common below 6,500 feet.
2. Wake-robin (*T. angustipetalum*)—This is a purple-flowered plant that occurs in dry to moist areas below 6,500 feet throughout the Sierra Nevada.

Interesting Facts: The root of wake-robin was used to make a tonic to help stop nosebleeds and the bleeding of menstruation and childbirth.

The stems and leaves of related species may be boiled and eaten as greens. The leaves should be collected before they fully unfold because when the flowers appear, the leaves become bitter. The berries and roots are inedible, possibly possessing emetic properties (Fernald and Kinsey 1958; Peterson 1978). Because of their relative rarity and beauty, refrain from using them except in an emergency situation.

The juice from another related species, *T. ovatum* (Pacific trillium), and a poultice made from the leaves and roots were used for boils (Strike 1994). Decoctions of the plant were used to treat nearly every illness and to prevent deep sleep.

White-flowered Schoenolirion (*Hastingsia alba*)

Description: This plant arises from a coated bulb and has flat leaves. The flowers are white and are tinged with green or pink. Look for white-flowered schoenolirion in meadows and swampy places between 1,500 and 8,000 feet from Nevada County northward.

Interesting Facts: The Karok put the leaves of this plant over their teeth to make a snapping sound for amusement (Baker 1981).

Wild Hyacinth (*Dichelostemma multiflorum*)

Description: This is a perennial herb with a slender scape up to two feet tall. The blue-violet flowers occur in a dense, headlike cluster of four to ten flowers. The cluster is subtended by four ovate, purplish bracts. This is a common plant on dry hills and plains. It flowers from March to May.

Interesting Facts: The flowers and bulbs are edible raw or cooked.

Yucca, Spanish Dagger (*Yucca whipplei*)

Description: This tree occurs on dry, rocky slopes in the Mojave Desert and the southern desert areas, mainly below 5,000 feet. It flowers from April to May. Yuccas were once included in the agave family (Agavaceae).

Interesting Facts: There is a unique relationship between yuccas and the yucca moth (*Tegeticula maculata*). This is the only pollinator for the plant. The mouthparts of the female moth are such that they gather the pollen and form it into a ball that is then scraped across the next stigma. In addition, the moth lays its eggs only in the seedpod of the yucca. Without each other, both of these organisms could not exist.

All species of *Yucca* provided a plentiful and dependable food source for many Native Californians. The large pulpy fruits were eaten raw or roasted or cooked and formed into cakes or dried for future use. The flowers buds were also roasted and eaten, and they have a high sugar content that served as a sweet treat for the children. A tea can be made from the seeds, or the seeds could be pounded and cooked as mush.

The leaves can be pounded to release the fibers for use in basketry and cordage making. One method used by the Diegueno was to bury the leaves until the fleshy part rotted away, yielding a fine white fiber. The tough fibers were also used in making nets, hats, sandals, and mattresses (Hinton 1975).

The chopped or pounded roots yield a soaplike substance. It can be used for general washing. Paintbrushes can be made by fringing the smaller leaves.

Orchid Family (Orchidaceae)

This is one of the largest families in the world, with the greatest concentration of species found in the Tropics. There are more than fifteen thousand species in several hundred genera. The flowers are irregular in shape, with three sepals and three petals. One of the petals forms a lip, sac, or pouch on the lower side of the flower. The flower structure is highly specialized for insect pollination. The family is an outstanding source of ornamentals, and although a few orchid species have utilitarian uses, the majority are rare and should be considered for use only in emergency situations.

Bog Orchid (*Platanthera;* known also as *Habenaria*)

Description: These are perennials, often with fleshy or tuberous roots. The small white to yellowish green flowers are in spikelike racemes. At the base of the lip is a spur. *Habenaria* is Latin for reins or narrow strap, which refers to the lip of some species. Rein orchids can be found in forest understories or in wet areas and meadows. Three species can be found in the Sierra Nevada.

Quick Key to the Bog Orchids	
Flowers white to cream in color	Bog Orchid (*P. leucostachys*)—This species grows in wet, springy, or boggy places below 11,000 feet. It blooms from May through August.
Flowers green to yellow-green in color Spur tip blunt, less than one-quarter inch long	Green-flowered Bog Orchid (*P. hyperborea*)—This species grows in bogs up to elevations of 10,000 feet.
Spur tip acute, greater than one-quarter inch long	Few-flowered Bog Orchid (*P. sparsiflora*)—This bog orchid grows along streams and in boggy places from 4,000 to 11,000 feet. It flowers from June through August.

Interesting Facts: Research on species of *Habenaria* in the eastern United States indicates that mosquitos are the pollinating agent for this genus. An interesting activity to verify if this is true for western species would be to examine the mosquitoes visiting the plants. What you would be looking for are the pollinia (pollen) sacs attached to the head or body of the mosquitoes.

The tuberlike roots of many species may be eaten raw or cooked. However, it is recommended that anyone experimenting with these plants take a cautious approach until the poisonous nature of the plants is clarified. Some Native Americans used extracts from these plants as poison to sprinkle on baits for coyotes and grizzly bears (Willard 1992).

Coralroot (*Corallorrhiza*)

Description: Members of this genus are saprophytic and lack the green leaves. They obtain all their nutrients from soil organic matter. The branched roots are bumpy and resemble coral. The yellow to reddish purple stems are covered with membranous sheathing bracts. Short-stalked flowers are borne in narrow inflorescences. The sepals are longer than the petals, and the upper petals and upper sepal are curved over the top of the column. The lower petal (lip) is egg shaped or elliptical and three lobed or undivided at the tip. *Corallorrhiza* means coral root in Greek.

237
Major Plant Groups

AN ENDANGERED FAMILY
The entire orchid family is listed in appendix 2 to the Convention on Trade in International Species of Flora and Fauna (CITES), meaning that trade in these species is restricted. Nine species of orchids are listed in appendix 1 to CITES, meaning trade is severely restricted for these species because they are in danger of extinction. The World Conservation Union lists 325 species of orchids as endangered in the 1997 *Red List of Threatened Plants*. Orchids have been affected mainly by habitat destruction but also by collection.

Sepals and petals with three to five distinct red to purple stripes	Striped Coralroot (*C. striata*)—This pinkish yellow- or whitish-flowered plant occurs in woods below 7,500 feet from Sierra County northward.
Sepals and petals sometimes with reddish veins but without distinct stripes	
Lip mostly less than one-quarter inch long; lateral sepals with only one nerve	Northern Coralroot (*C. trifida*)—Look for this species in wet habitats within coniferous forests between 5,000 and 6,000 feet.
Lip mostly greater than one-quarter inch long; lateral sepals usually with more than one nerve	Spotted Coralroot (*C. maculata*)—This species has crimson-purple to greenish flowers and grows in woods below 9,000 feet throughout the Sierra Nevada.

Interesting Facts: Consumption of the toxic rhizome can cause hyperthermia and profuse perspiration. The rhizome has been used as a diaphoretic, febrifuge, and sedative, and the dried stems were used by the Paiute and Shoshone of Nevada to make a tea to build up blood in pneumonia patients (Coffey 1993). Strike (1994) indicates that the plant was used to reduce fevers or as a sedative.

Giant Helleborine (*Epipactis gigantea*)

Description: Giant helleborine is a perennial with one to several erect stems arising from rhizomes. The stems are covered with numerous, sheathing leaves. The showy flowers form a raceme with all flowers on one side of the stem. The sepals are coppery green with light-brown venation. The name *Epipactis* derives from a classical name used by Theophrastus (ca. 350 BC) for a plant used to curdle milk. There are about twenty-five species found worldwide, mostly in Europe (fifteen species) and Asia. Two species are found in the United States: *E. gigantea* and *E. helleborine*. The latter is thought to have arrived in North America from Europe in the nineteenth century and grows in the northeastern United States. This species occurs along moist streambanks below 7,500 feet. It flowers from May to June.

Interesting Facts: Native Americans made a decoction of the fleshy roots for internal use when they felt "sick all over." The flowers have a sweet aroma that attracts flies in the Syrphidae family. This aroma mimics that of honeydew, the sugary secretion given off by aphids. Syrphid flies normally lay their eggs among aphids, which become food for their larvae. In this case, as the fly lays its eggs, it picks up pollen that is then deposited on the next flower.

Hooded Ladies'-tresses (*Spiranthes romanzoffiana*)

Description: This is a small herb with fleshy roots. The white flowers are in a dense, spirally twisted spike, with the sepals and lateral petals appearing to be fused and forming a hood around the column. Hooded ladies'-tresses can be found in dry to moist areas, meadows, lakeshores, bogs, and marshes, up to timberline. The species was named after Nikolei Rumliantzev, Count Romanzoff, a Russian patron of science. This species is found in springy places below 8,000 feet. It flowers from July to August.

Interesting Facts: This species has strong diuretic properties, making it undesirable for eating. Additionally, Foster and Duke (1990) indicate that Native Americans used a plant tea of *S. cernua* (nodding ladies'-tresses) of the eastern United States as a diuretic for urinary disorders and venereal disease and as a wash to strengthen infants. They also state that other North American, European, and South American species have also been used as diuretics and aphrodisiacs.

Lady's Slipper (*Cypripedium*)

Description: Easily recognized by the large, inflated lip, they are among the rarest plants. The flower attracts insects into the pouch from which there is only one exit. The insect must pass over the stigma where pollen from previously visited flowers is brushed off, then go under one of the anthers where new pollen is picked up. This highly advanced pollination strategy reduces the chance of self-fertilization.

1. California Lady's Slipper (*C. californicum*)—This lady's slipper is found in moist habitats in the northern Sierra Nevada.
2. Mountain Lady's Slipper (*C. montanum*)—Look for this lady's slipper in moist woods below 5,000 feet from Mariposa County north.

Lady's Slipper
(*Cypripedium* spp.)

Interesting Facts: Lady's slippers are showy orchids and relatively scarce. They are rarely successful in gardens and should be left in their natural environment. Theodore Niehaus, author of *Sierra Wildflowers,* indicates that he studied a species of lady's slipper (*C. fasciculatum*) that was at least ninety-five years old. This adds additional support for not picking these or other orchid species (Niehaus 1974). Lady's slippers emit a peculiar odor that evokes a strong allergic reaction in some people. Many species appear to cause dermatitis.

The rhizome of *C. montanum* is said to be toxic but has been used medicinally as a stimulant and antispasmodic. It has been used with good results in reflex functional disorders or chorea, hysteria, nervous headache, insomnia, low fevers, nervous unrest, hypochondria, and nervous depression accompanying stomach disorders (Willard 1992). The plants were also used by Native Americans as tranquilizers. *C. calceolus* var. *pubescens* (= *C. pubescens*) was also widely used as a sedative for nervous headaches, hysteria, insomnia, and nervous irritability (Foster and Duke 1990).

Rattlesnake-plantain (*Goodyera oblongifolia*)

Description: This is a perennial herb with glandular-hairy stems. The persistent basal leaves are dark green and have winged petioles and broadly lance-shaped blades that are mottled with white along the midvein. The flowers are greenish white. It is found in open or deep forests up to the subalpine elevations.

Interesting Facts: The herbage of rattlesnake-plantain is slightly toxic if eaten (Weiner 1972). The plant was also known by some aboriginal peoples as a medicine for childbirth and as a poultice for cuts and sores whereby the leaves were split open and the moist inner part of the plant was placed over the wound (Pojar and MacKinnon 1994).

Rein Orchid (*Piperia*)

Description: These are perennials with erect, unbranched stems. The two to five leaves are basal, but the cauline leaves are much reduced in size. In general, these orchids are found in wet habitats. The genus name honors Charles Vancouver Piper (1867–1926), an agronomist with the U.S. Department of Agriculture and an expert on Pacific Northwest flora.

Quick Key to the Rein Orchids	
Spur one-quarter to three-quarters inch long	Royal Rein Orchid (*P. transversa*)—This species grows mainly on dry sites below 8,500 feet throughout the Sierra Nevada.
Spur less than one-quarter inch Lip lanceolate	Dense-flower Rein Orchid (*P. leptopetala*)— This is an uncommon species found in dry places below 7,000 feet throughout the Sierra Nevada.
Lip oblong to triangular in shape	Slender-spire Orchid (*P. unalascensis*)—Look for this species along streams or in boggy places between 4,000 and 11,000 feet.

Interesting Facts: The bulbs of *P. unalascensis* were baked and eaten like baked potatoes by the Pomo and Kashaya (Balls 1970; Barrett 1952).

PONDWEED FAMILY (Potamogetonaceae)

Of the two genera and one hundred species worldwide, *Potamogeton* is native to the United States. The family is of no direct economic importance to humans, but provides a valuable source of food for ducks and other wildlife.

Pondweed (*Potamogeton*)

Description: Pondweeds are perennial aquatic plants with extensive, slender rhizomes and simple, branched stems that often root at the nodes. The leaves have stipules that clasp the stem. The lower leaves are alternate and submerged, and the upper often floating leaves are wider and opposite. The small, greenish flowers are clustered in a spike that arises from the upper leaf axils. Most pondweed species can be identified without flowers and fruits, but the floating or submersed leaves may be necessary. Hybridization among species is fairly common. There are many species of *Potamogeton* in the Sierra Nevada.

Interesting Facts: Probably all pondweeds have starchy, edible rhizomes, but species with larger rootstocks are preferred for gathering. *Potamogeton diversifolius* (waterthread pondweed) was an important source of strong fibers that were rolled into cordage to make carrying nets, rabbit-trap nets, and other items (Strike 1994).

RUSH FAMILY (Juncaceae)

Nine genera and four hundred species of rushes are found in damp and wet sites of the cool temperate and subarctic regions. Two genera are native to the United States. These are grasslike annual and perennial plants with solid rounded or flattened stems at wet and damp sites. The leaves are basal or alternate and may be flat, folded, or round and taper to a point. The flowers are small and have six undifferentiated sepals and petals (often termed tepals), three to six stamens, and a three-parted ovary with many seeds. The family is of no direct economic importance to humans, although a few are ornamentals.

Quick Key to the Rush Family	
Seeds many per capsule; leaves usually with open sheaths and hairless blades	Rush (*Juncus*)
Seeds three per capsule; leaves with closed sheaths and often with marginal hairs	Wood Rush (*Luzula*)

Rush, Wire-grass (*Juncus*)

Description and Interesting Facts: Many rush species occur in the Sierra Nevada. In general, they are annual or perennial herbs often found in water or wet places. The flowers are in heads or panicles, and the tough, fibrous stems are inedible. However, they are useful in weaving baskets and mats.

Wood Rush (*Luzula*)

Description: The six species in the Sierra Nevada are tufted perennials with slender unbranched stems. In all, there are about eighty species of annual or perennial herbs with grasslike leaves throughout the world. The brown-green or white flowers occur in panicles.

Interesting Facts: Although there are no known uses for the Sierra Nevada species, related species have been used medicinally as emetics and decoctions. Additionally, *L. luciola* was used as a wick in candles. The seeds of *L. campestris* (field wood rush) are dispersed by ants because of their juicy outgrowths.

SEDGE FAMILY (Cyperaceae)

There are ninety genera and four thousand species worldwide, particularly in the cool temperate and subarctic regions. Twenty-four genera are native to the United States. In the Sierra Nevada, sedges are a large family of grasslike plants often found growing in wet places. Sedges make up a large proportion of the plants found at higher elevations. Most of the species have triangular stalks with the flowers arranged in spikelets. Species of this family are often confused with grasses (Poaceae) and rushes (Juncaceae). Grasses generally have round, hollow (except at the nodes) stems and a characteristic flower consisting of two glumes subtending one to many flowers, each with a lemma or palea. Members of the rush family have round stems and small, lilylike flowers with three petals, three sepals, and six stamens.

The family is of little economic importance, although several members are edible and medicinal. Sedge family members, however, should not be used medicinally without medical supervision because their properties and effectiveness are variable.

Bulrush (*Scirpus*)

Description: The many species of bulrush in California are perennials with grasslike or scalelike leaves. The various species can be found in marshy areas around lakes and ponds and in other moist or wet areas from low to high elevations.

Interesting Facts: The edible rhizomes of all species are quite starchy. They may be eaten raw or baked, dried, or ground into a flour. As Olsen (1990) describes:

> [T]he young shoots just protruding from the mud are a delicacy raw or cooked. Furthermore, one can harvest them by wading into the water and feeling down along the plant until you come across the last shoot in a string of shoots that protrudes above the water. You then push your hands into the mud until the lateral rootstalk is encountered. By feeling along the rootstalk in a direction leading away from the last shoot, one can find a protruding bulb from which the new shoot is starting. This is easily snapped off and is edible on the spot.

The young roots when crushed and boiled also yield a sweet syrup. The pollen may be gathered and pressed into cakes and baked. The seeds may be used whole,

BASKETRY
Baskets were an integral part of many Native American lives. They touched every aspect of their lives and were used in cooking and serving food, storage of goods and water, harvesting and winnowing seeds, and trapping fish. Weaving baskets demanded skill and patience. In many instances, the baskets were made according to tradition rather than by written guides.

There are two principal techniques used in basket weaving: coiling and twining. In coiling, a foundation material composed of a single stem or bundle of grass stalks was coiled around itself. Each coil was then bound and fastened to the one below it by stitching with a second type of pliable element, the wrapping. An awl made from bone or a spine from a cactus was used to make an opening in the lower coil through which the wrapping material was passed. Coiled baskets were often water-proofed by smearing hot pine pitch on the inside and outside for storing water.

In the twining technique, a fairly rigid material was the foundation around which two or three pliable fibers were interlaced. The start of a twined basket looked like a starburst, with the foundation sticks radiating out from a common center. As the twining progressed, the sticks would be bent upward and pulled together to form the final shape of the basket. Twining was used to make coarse, undecorated work baskets and to make open work baskets such as winnowers and seed beaters.

parched, ground, or in mush. The stem bases may be eaten raw and are good for quenching thirst.

Scirpus stems, roots, and leaves can be used as the foundation and twining material in twined baskets. They were used extensively by Native Americans for making cordage, sandals, baskets, and mats. Waterproof "water bottles" can be made from the baskets by coating the inside with asphaltum. The stems were used for sleeping mats, padding, thatching dwellings, skirts, and sandals. Duck-shaped decoys were also made when hunting.

Leaves of *S. acutus* (also known as *Schoenoplectus acutus*) were used to poultice wounds and burns (Strike 1994).

Dulichium (*Dulichium arundinaceum*)

Description: This is a perennial herb found with rounded stems arising from deep-seated rhizomes. It can be found growing in wet meadows, and margins of ponds and streams at low to middle elevations.

Interesting Facts. The rhizome is edible, but it is usually found among other plants of higher food value (Weedon 1996).

Flat-sedge, Nut-grass (*Cyperus*)

Description: Members of this genus are annual or perennial and grasslike. Leaves are mainly basal and three-ranked, and the stems are triangular in cross section. The spikelets are clustered in ball-shaped heads, and the inflorescence consists of numerous heads borne on stalks radiating from the top of the stem and subtended by long, leaflike bracts. The various species occur in wet, open soils along riverbanks and the margins of lakes and ponds.

Interesting Facts: The rhizome of *C. esculentus* (chufa flatsedge) bears small, nutlike, underground tubers that are edible. These may be eaten raw, boiled, dried and ground into flour, or roasted to a dark brown and ground into coffee. The species was so valued in ancient times that its tubers were placed in Egyptian tombs dating back to more than 2000 BC (Saunders 1976). Other species may also have tubers or tuberlike structures. Volatile oils and astringent substances are found in a number of species that are used in perfumery and as remedies for digestive problems.

Lipocarpha (*Lipocarpha*)

Description and Interesting Facts: The genus was formally known as *Hemicarpha*. The two species in the Sierra Nevada include *L. occidentalis* and *L. micrantha*. Both are small annuals with three-angled leafy stems found in damp locations. The pollen and seeds are edible (Weedon 1996).

Sedge (*Carex*)

Description: Numerous species within this genus occur in the Sierra Nevada. They are normally difficult to identify in the field. A handlens and mature fruit are often necessary for positive identification of most species. In general, they are perennial grasslike herbs with creeping rhizomes, short rhizomes or fibrous roots, three-sided stems, and three-ranked leaves. Sedges occur in a great variety of habitats from the foothills to the alpine zone.

Interesting Facts: The young shoots and tender leaf bases of almost all species are sweet and furnish a tasty nibble. The fruits are also edible. *Carex* is a widespread genus with more than one thousand species, making it available as an important emergency food in many parts of the country.

NUT-GRASS DRINK
This is a Spanish recipe for a refreshing drink. Soak about a half pound of tubers in water for forty-eight hours. Then mash the tubers and add one quart of water and one-third of a pound of sugar. Then strain the liquid through a sieve and serve as a drink (Fernald and Kinsey 1958).

Sedge roots were used extensively as basketry material, particularly in coiled baskets. Some Native Americans would spend considerable time untangling roots in order to remove them in one piece. In fact, some areas were cleared of other plants to allow sedge to grow long and untangled.

Slender Cotton-grass (*Eriophorum gracile*)

Description: These are perennials with solid triangular or round stems from rhizomes. The lower leaves are grasslike. The flowers are spirally arranged in spikes that resemble tufts of cotton at maturity. *Eriophorum* translates as wool (*erion*) and bearing (*phoros*). The species can be found in cold swamps and bogs from mid- to high elevations.

Interesting Facts: The pinkish bases of these plants can be collected in the early summer and eaten raw or added to soups. Dug in the early spring or fall, the corms were prepared by pouring boiling water over them to remove the thick, black outer covering. The corms were then eaten raw or boiled (Arnason, Hebda, and Johns 1981). A decoction from the rootstock was used in treating colds and coughs, but its use was not widespread (Schofield 1989). The woolly down from the flowers was spun as fiber, although it was inferior to cotton. However, the fiber can be used as candlewicks.

Sierra Nevada Habitats and Their Associated Plants and Uses

Common Name Scientific Name	Food, Beverage	Medicine Personal Care	Dye, Soap	Gum, Rubber, Smoking	Fire, Fiber	Tools Abrasives Crafts	Containers Baskets	Poisons
Foothill Plant Communities Valley Grassland—Grasses and Wildflowers								
Wild Oats								
Genus *Avena*	X	X	–	–	X	–	X	–
Bromes								
Genus *Bromus*	X	X	–	–	X	–	X	–
Purple Needlegrass								
Stipa pulchra	X	X	–	–	X	–	X	–
Foxtail Barley								
Hordeum jubatum	–	X	–	–	–	–	X	–
Foothill Scabrella								
Poa scabrella	X	X	–	–	–	–	–	–
Filaree								
Genus *Erodium*	X	X	–	–	–	–	–	–
Turkey Mullein								
Eremocarpus setigerus	–	X	–	–	X	X	–	X
Burclover								
Medicago lupulina	X	X	–	–	–	–	–	–
Common Vetch								
Vicia sativa	–	X	–	–	–	–	–	X
Larkspurs								
Genus *Delphinium*	–	X	–	–	–	–	–	X
Carolina Geranium								
Geranium carolinianum	X	–	–	–	–	–	–	–
California Buttercup								
Ranunculus californicus	X	–	–	–	–	–	–	X
Lacepod								
Thysanocarpus curvipes	X	–	–	–	–	–	–	–
Showy Milkweed								
Asclepias speciosa	X	X	X	X	X	X	X	X
Baby Blue-eyes								
Nemophila menziesii	X	–	–	–	–	–	–	–
Chick Lupine								
Lupinus microcarpus	–	–	–	–	–	–	–	X
Spider Lupine								
Lupinus benthamii	–	–	–	–	–	–	–	X
Buckwheats								
Genus *Eriogonum*	X	X	–	–	–	X	X	–
Common Madia								
Madia elegans	–	X	–	X	–	–	–	X

Common Name Scientific Name	Food, Beverage	Medicine Personal Care	Dye, Soap	Gum, Rubber, Smoking	Fire, Fiber	Tools Abrasives Crafts	Containers Baskets	Poisons
Menzie's Fiddleneck								
Amsinckia menziesii	–	–	–	–	–	–	–	X
American Wintercress								
Barbarea orthoceras	X	X	–	–	–	–	X	X
Meadowfoam								
Limnanthes alba	–	–	–	X	–	X	–	–
Pink Paintbrush								
Castilleja exserta	X	X	–	–	–	–	–	X
Red Maids								
Calandrinia ciliata	X	–	–	–	–	–	–	–
Fremont's Layia								
Layia fremontii	X	–	–	–	–	–	–	–
California Poppy								
Eschscholzia californica	–	X	–	–	–	–	–	X
Harvest Brodiaea								
Brodiaea elegans	X	X	–	–	–	–	–	–
Blue Dicks								
Dichelostemma capitatum	X	X	–	–	–	–	–	–
Ithuriel's Spear								
Triteleia laxa	–	X	–	–	–	–	–	–
Charming Mariposa Lily								
Calochortus invenustus	X	X	X	–	–	X	X	–
Williamson's Clarkia								
Clarkia williamsonii	X	–	–	–	–	–	–	–

Foothill Woodland—Trees and Shrubs

Common Name Scientific Name	Food, Beverage	Medicine Personal Care	Dye, Soap	Gum, Rubber, Smoking	Fire, Fiber	Tools Abrasives Crafts	Containers Baskets	Poisons
Macnab's Cypress								
Cupressus macnabiana	–	X	–	–	–	X	X	X
Valley Oak								
Quercus lobata	X	X	X	–	–	X	X	X
Blue Oak								
Quercus douglasii	X	X	X	–	–	X	X	X
Canyon Live Oak								
Quercus chrysolepis	X	X	X	–	–	X	X	X
Interior Live Oak								
Quercus wislizenii	X	X	X	–	–	X	X	X
Gray Pine								
Pinus sabiniana	X	X	–	X	X	X	X	–
Buckeye								
Aesculus californica	–	X	–	–	X	–	–	X
Tanbark Oak								
Lithocarpus densiflora	X	X	X	–	–	X	X	–
Bigleaf Maple								
Acer macrophyllum	X	X	–	–	X	X	X	–
California Coffeeberry								
Rhamnus californica	–	–	–	–	–	–	–	–
Bitter Gooseberry								
Ribes amarum	X	X	–	–	–	X	X	–
Parry Manzanita								
Arctostaphylos manzanita	X	X	–	–	X	X	X	X
California Bladdernut								
Staphylea bolanderi	–	X	–	–	–	–	–	–
California Blackberry								
Rubus ursinus	X	X	X	–	–	X	–	–

Common Name Scientific Name	Food, Beverage	Medicine Personal Care	Dye, Soap	Gum, Rubber, Smoking	Fire, Fiber	Tools Abrasives Crafts	Containers Baskets	Poisons
Skunkbrush								
Rhus trilibata	X	X	–	–	–	X	X	X
Blue Elderberry								
Sambucus caerulea	X	X	–	X	X	X	–	X
Chaparral Currant								
Ribes malvaceum	X	X	–	–	–	X	–	–
Buck Brush								
Ceanothus cuneatus	X	X	–	–	–	X	–	X
Pacific Ninebark								
Physocarpus capitatus	X	X	–	–	X	X	–	X
Redbud								
Cercis occidentalis	X	X	–	–	–	–	X	–
Poison Oak								
Toxicodendron diversiloba	–	–	–	–	–	–	X	X

Foothill Woodland—Wildflowers

Common Name Scientific Name	Food, Beverage	Medicine Personal Care	Dye, Soap	Gum, Rubber, Smoking	Fire, Fiber	Tools Abrasives Crafts	Containers Baskets	Poisons
Wild Oats								
Avena fatua	X	–	–	–	X	–	–	X
Slender Wild Oats								
Avena barbata	X	–	–	–	X	–	–	X
Foxtail Barley								
Hordeum jubatum	X	–	–	–	X	–	–	X
Larkspurs								
Genus *Delphinium*	–	X	–	–	–	–	–	X
Carolina Geranium								
Geranium caroliniana	X	X	–	–	–	–	–	–
Redstem Filaree								
Erodium cicutarium	X	X	–	–	–	–	–	–
Lupines								
Genus *Lupinus*	–	–	–	–	–	X	–	X
Miner's Lettuce								
Claytonia perfoliata	X	X	–	–	–	–	–	–
Chinese Caps								
Euphorbia crenulata	–	X	–	–	–	–	–	X
Shooting Star								
Dodecatheon hendersonii	X	X	–	–	–	–	–	–
Self-heal								
Prunella vulgaris	X	X	–	–	–	–	–	–
Yarrow								
Achillea millefolium	–	X	–	X	–	–	–	X
Iris								
Genus *Iris*	–	–	–	–	–	X	X	X
Chinese Houses								
Collinsia heterophylla	–	X	–	–	–	–	–	–
Pink Paintbrush								
Castilleja exserta	X	–	–	–	–	–	–	X
White Globe Lily								
Calochortus albus	X	–	–	–	–	–	–	–
California Silene								
Silene californica	?	–	–	–	–	X	–	–
Menzies Fiddleneck								
Amsinckia menziesii	?	–	–	–	–	–	–	?
Baby Blue-eyes								
Nemophila menziesii	–	?	–	–	–	–	–	–

Common Name Scientific Name	Food, Beverage	Medicine Personal Care	Dye, Soap	Gum, Rubber, Smoking	Fire, Fiber	Tools Abrasives Crafts	Containers Baskets	Poisons
Charming Mariposa Lily *Calochortus invenustus*	X	–	–	–	–	–	–	–
Popcorn Flower *Plagiobothrys nothofulvus*	?	X	X	–	–	X	–	–

Foothill Riparian and Freshwater Areas—Trees, Shrubs, and Vines

White Alder *Alnus rhombifolia*	X	X	X	–	X	X	–	–
Black Cottonwood *Populus balsamifera*	X	X	–	–	X	X	X	–
California Black Walnut *Juglans californica*	X	X	X	–	–	X	–	X
Valley Oak *Quercus lobata*	X	X	–	–	–	X	X	X
Willows Genus *Salix*	X	X	–	–	X	X	X	–
Western Sycamore *Platanus racemosa*	X	X	–	X	–	–	X	–
Oregon Ash *Fraxinus latifolia*	?	–	–	–	–	X	–	–
California Blackberry *Rubus ursinus*	X	X	–	–	–	–	X	–
Maples Genus *Acer*	X	X	–	–	X	X	X	–
Western Azalea *Rhododendron occidentale*	–	–	–	–	–	–	–	X
Honeysuckle Genus *Lonicera*	X	–	X	–	X	X	–	–
Mountain Dogwood *Cornus nuttallii*	X	X	–	–	X	X	X	–
Spice Bush *Calycanthus occidentalis*	?	–	–	–	–	X	X	–
Snowberry *Symphoricarpus albus*	X	X	X	–	–	X	X	X
Wild Grape *Vitis californica*	X	X	–	–	X	–	–	–

Foothill Riparian and Freshwater Areas—Herbaceous Plants

Larkspurs Genus *Delphinium*	–	–	–	–	–	–	–	X
Stinging Nettle *Urtica dioica*	X	X	–	–	X	X	X	X
Miner's Lettuce *Claytonia perfoliata*	X	X	–	–	–	–	–	–
Canyon Dudleya *Dudleya cymosa*	X	X	–	–	–	X	–	–
Western Dog Violet *Viola adunca*	X	X	–	–	–	–	–	–
Monkeyflowers Genus *Mimulus*	X	X	X	–	–	–	–	–
White Hedge Nettle *Stachys albens*	X	X	–	–	–	–	–	–
Stream Orchid *Epipactis gigantea*	–	X	–	–	–	–	–	–

Common Name / Scientific Name	Food, Beverage	Medicine Personal Care	Dye, Soap	Gum, Rubber, Smoking	Fire, Fiber	Tools Abrasives Crafts	Containers Baskets	Poisons
Elk Clover / *Aralia californica*	–	X	–	–	–	–	–	–
Watercress / Genus *Rorippa*	X	X	–	–	–	–	–	–
Cattail / *Typha latifolia*	X	X	–	X	X	X	X	–
Common Tule / *Scirpus acutus*	X	X	–	–	–	–	–	–
Water Buttercup / *Ranunculus aquatilus*	X	X	–	–	–	–	–	X
Water Plantain / *Alisma plantago-aquatica*	X	X	–	–	–	–	–	X
Horsetails / Genus *Equisetum*	X	X	–	–	–	–	–	X

Chaparral—Trees and Shrubs

Common Name / Scientific Name	Food, Beverage	Medicine Personal Care	Dye, Soap	Gum, Rubber, Smoking	Fire, Fiber	Tools Abrasives Crafts	Containers Baskets	Poisons
California Buckeye / *Aesculus californica*	X	–	–	–	–	–	–	X
Gray Pine / *Pinus sabiniana*	X	X	–	–	–	X	X	–
Canyon Live Oak / *Quercus chrysolepis*	X	X	X	–	X	X	X	X
Interior Live Oak / *Quercus wislizenii*	X	X	X	–	X	X	X	X
Chamise / *Adenostoma fasciculatum*	X	X	–	–	–	–	–	–
Poison Oak / *Toxicodendron diversiloba*	–	–	–	–	–	–	X	X
Fremontia / *Fremontodendron californicum*	–	X	–	–	X	X	X	–
Interior Live Oak / *Quercus wislizenii*	X	X	X	–	–	X	X	X
Chaparral Virgin's Bower / *Clematis lasiantha*	–	X	X	–	X	–	–	X
Chaparral Currant / *Ribes malvaceum*	X	X	–	–	–	–	–	–
Buck Brush / *Ceanothus cuneatus*	X	X	X	–	–	–	–	X
Deer Brush / *Ceanothus integerrimus*	X	X	X	–	–	–	–	X
Toyon / *Heteromeles arbutifolia*	X	X	–	–	–	X	X	–
Manzanita / Genus *Arctostaphylos*	X	X	–	X	–	X	X	–
Redberry / *Rhamnus crocea*	X	X	–	–	–	–	–	–
Skunkbrush / *Rhus trilobata*	X	X	–	–	X	–	X	–
Birch-leaf Mountain-mahogany / *Cercocarpus betuloides*	–	–	–	–	–	X	X	–
Chaparral Yucca / *Yucca whipplei*	X	X	X	–	X	X	X	–
Redbud / *Cercis occidentalis*	X	X	–	–	X	–	X	–

Common Name Scientific Name	Food, Beverage	Medicine Personal Care	Dye, Soap	Gum, Rubber, Smoking	Fire, Fiber	Tools Abrasives Crafts	Containers Baskets	Poisons
California Coffeeberry								
Rhamnus californica	–	X	–	–	–	–	–	–
Bush Poppy								
Dendromecon rigida	–	?	–	X	–	–	–	–

Chaparral—Wildflowers

Snake Lily								
Dichelostemma volubile	X	–	–	–	–	–	–	–
Charming Mariposa Lily								
Calochortus invenustus	X	–	–	–	–	–	–	–
Indian Warrior								
Pedicularis densiflora	–	X	–	–	–	–	–	X

Forest Plant Communities
Mixed Coniferous Forest Trees—Westside

Ponderosa Pine								
Pinus ponderosa	X	X	–	–	–	X	X	–
Jeffrey Pine								
Pinus jeffreyi	X	X	–	–	–	X	X	–
Sugar Pine								
Pinus lambertiana	X	X	–	–	–	X	X	–
Douglas-fir								
Pseudotsuga menziesii	X	X	–	–	–	X	X	–
White Fir								
Abies concolor	–	X	–	–	X	X	X	–
Giant Sequoia								
Sequoiadendron giganteum	–	X	X	–	X	–	X	–
Incense Cedar								
Calocedrus decurrens	–	X	X	–	X	X	X	–
Black Oak								
Quercus kelloggii	X	X	X	–	–	X	X	X
Canyon Live Oak								
Quercus chrysolepis	X	X	X	–	–	X	X	X
Interior Live Oak								
Quercus wislinzenii	X	X	–	–	–	X	X	X
Tanbark Oak								
Lithocarpus densiflora	X	X	–	–	–	X	X	X
Mountain Dogwood								
Cornus nuttallii	X	X	–	X	–	X	X	–
Madrone								
Arbutus menziesii	X	–	–	–	–	X	–	–
Cascara								
Rhamnus purshiana	–	X	–	–	–	–	–	X
Western Yew								
Taxus brevifolia	?	X	–	?	–	X	–	X
California Nutmeg								
Torreya californica	X	–	–	–	–	X	–	–
Quaking Aspen								
Populus tremuloides	X	X	–	–	–	X	X	–
Black Cottonwood								
Populus balsamifera	X	X	–	–	–	X	X	–
Willows								
Genus *Salix*	X	X	–	–	X	X	X	–

Common Name Scientific Name	Food, Beverage	Medicine Personal Care	Dye, Soap	Gum, Rubber, Smoking	Fire, Fiber	Tools Abrasives Crafts	Containers Baskets	Poisons
White Alder								
Alnus rhombifolia	X	X	–	–	X	X	X	–
Mountain Alder								
Alnus teniufolia	X	X	–	–	X	X	X	–
Bigleaf Maple								
Acer macrophyllum	X	X	–	X	X	X	X	–
California Laurel								
Umbellularia californica	X	X	–	X	–	X	X	X
Oregon Ash								
Fraxinus latifolia	X	X	–	–	–	X	–	–

Mixed Coniferous Forest Trees—Eastside

Jeffrey Pine								
Pinus jeffreyi	X	X	X	X	X	X	X	–
White Fir								
Abies concolor	–	X	–	–	–	X	X	–
Quaking Aspen								
Populus tremuloides	X	X	–	–	X	X	X	–
Black Cottonwood								
Populus balsamifera	X	X	–	–	X	X	X	–
Water Birch								
Betula occidentalis	X	X	–	–	–	X	X	–
Western Juniper								
Juniperus occidentalis	X	X	X	X	X	X	X	X

Red Fir Forest Trees

Red Fir								
Abies magnifica	–	X	–	–	–	X	X	X
Lodgepole Pine								
Pinus contorta	X	X	–	X	X	X	X	–
Jeffrey Pine								
Pinus jeffreyi	X	X	–	X	X	X	X	–
Western White Pine								
Pinus monticola	X	X	–	X	X	X	X	–
Mountain Hemlock								
Tsuga mertensiana	X	X	–	–	X	X	–	–
Western Juniper								
Juniperus occidentalis	X	X	X	X	X	X	X	X
Quaking Aspen								
Populus tremuloides	X	X	–	–	X	X	X	–

Lodgepole Pine Forest Trees

Lodgepole Pine								
Pinus contorta	–	–	–	–	–	–	–	–
Western White Pine								
Pinus monticola	–	–	–	–	–	–	–	–
Whitebark Pine								
Pinus albicaulis	X	X	–	–	–	X	X	–
Red Fir								
Abies magnifica	–	X	–	–	–	X	X	–
Quaking Aspen								
Populus tremuloides	X	X	–	–	X	X	X	–

Common Name Scientific Name	Food, Beverage	Medicine Personal Care	Dye, Soap	Gum, Rubber, Smoking	Fire, Fiber	Tools Abrasives Crafts	Containers Baskets	Poisons
Subalpine Forest Trees								
Whitebark Pine								
Pinus albicaulis	X	X	–	–	–	X	X	–
Foxtail Pine								
Pinus balfouriana	X	X	–	–	–	X	X	–
Lodgepole Pine								
Pinus contorta	X	X	–	–	–	X	X	–
Limber Pine								
Pinus flexilis	X	X	–	–	–	X	X	–
Western White Pine								
Pinus monticola	X	X	–	–	–	X	X	–
Mountain Hemlock								
Tsuga mertensiana	X	X	–	–	X	X	X	–
Western Juniper								
Juniperus occidentalis	X	X	–	–	X	X	X	–
Forest Shrubs and Wildflowers *Shrubs of dry shady forests*								
Sierra Coffeeberry								
Rhamnus rubra	–	X	–	–	–	–	–	–
Mountain Misery								
Chamaebatia foliosa	–	X	–	–	–	–	–	–
Currants and Gooseberries								
Genus *Ribes*	X	X	–	–	–	X	–	X
Deer Brush								
Ceanothus integerrimus	X	X	X	–	–	X	–	X
Tobacco Brush								
Ceanothus velutinus	X	X	X	–	–	X	–	X
Pinemat Manzanita								
Arctostaphylos nevadensis	X	X	–	X	–	X	X	X
Greenleaf Manzanita								
Arctostaphylos patula	X	X	–	X	–	X	X	X
Snowberries								
Genus *Symphoricarpus*	X	X	X	–	X	X	X	X
Shrubs of open forests								
Bush Chinquapin								
Chrysolepis sempervirens	X	–	–	–	–	–	–	–
Currants and Gooseberries								
Genus *Ribes*	X	X	–	–	–	X	X	–
Mountain Misery								
Chamaebatia foliosa	–	X	–	–	–	–	–	–
Huckleberry Oak								
Quercus vaccinifolia	X	X	–	–	–	X	–	X
Bitter Cherry								
Prunus emarginata	X	X	–	–	X	X	X	X
Snow Brush								
Ceanothus cordulatus	X	X	X	–	–	–	–	X
Deer Brush								
Ceanothus integerrimus	X	X	X	–	–	–	–	X
Squaw Carpet								
Ceanothus prostratus	X	X	X	–	–	–	–	X

Common Name Scientific Name	Food, Beverage	Medicine Personal Care	Dye, Soap	Gum, Rubber, Smoking	Fire, Fiber	Tools Abrasives Crafts	Containers Baskets	Poisons
Small-leaved Ceanothus								
Ceanothus parvifolius	X	X	X	–	–	–	–	X
Sierra Coffeeberry								
Rhamnus rubra	–	X	–	–	–	–	X	X
Mountain Snowberry								
Symphoricarpus vaccinioides	X	X	X	–	X	–	X	X
Greenleaf Manzanita								
Arctostaphylos patula	X	X	–	X	–	X	X	–
Pinemat Manzanita								
Arctostaphylos nevadense	X	X	–	X	–	X	X	–

Shrubs of streamsides or damp forests

Common Name Scientific Name	Food, Beverage	Medicine Personal Care	Dye, Soap	Gum, Rubber, Smoking	Fire, Fiber	Tools Abrasives Crafts	Containers Baskets	Poisons
Willows								
Genus *Salix*	X	X	–	–	X	X	X	–
California Hazelnut								
Corylus californica	X	–	?	–	X	X	X	–
Spice Bush								
Calycanthus occidentalis	?	–	–	–	–	X	X	–
Sierra Wax Myrtle								
Myrica hartwegii	?	?	–	–	–	–	–	–
Western Virgin's Bower								
Clematis lingustifolia	–	X	X	–	X	X	X	X
Currants and Gooseberries								
Genus *Ribes*	X	X	–	–	–	–	X	–
Serviceberry								
Amelanchier alnifolia	X	X	–	–	–	–	–	–
Bitter Cherry								
Prunus emarginata	X	X	–	–	–	X	–	X
Sierra Plum								
Prunus subcordata	X	X	–	–	–	X	–	X
Western Chokecherry								
Prunus demissa	X	X	–	–	–	X	–	X
Wild Roses								
Genus *Rosa*	X	X	–	–	–	X	–	–
Thimbleberry								
Rubus parviflorus	X	X	–	–	–	–	–	–
Mountain Maple								
Acer glabrum	X	X	–	X	X	X	X	–
Mountain Ash								
Sorbus scopulina	X	X	–	–	–	X	–	–
Dogwood								
Cornus sericia	X	X	–	?	–	X	X	?
Labrador Tea								
Ledum glandulosum	–	?	–	–	–	X	–	X
Western Azalea								
Rhododendron occidentale	–	–	–	–	–	–	–	X
Red Mountain Heather								
Phyllodoce breweri	–	?	–	–	–	–	–	–
Double Honeysuckle								
Lonicera conjugalis	X	X	–	–	X	X	X	–
Twinberry								
Lonicera involucrata	X	X	–	–	X	X	X	–
Red Elderberry								
Sambucus microbotrys	?	X	X	–	–	X	X	X

Common Name Scientific Name	Food, Beverage	Medicine Personal Care	Dye, Soap	Gum, Rubber, Smoking	Fire, Fiber	Tools Abrasives Crafts	Containers Baskets	Poisons
Shrubs of rocky habitats								
Bush Chinquapin								
Castanopsis sempervirens	X	X	–	–	X	–	–	–
Alpine Gooseberry								
Ribes lasianthum	X	X	–	–	–	X	X	X
Alpine Prickly Currant								
Ribes montigenum	X	X	–	–	–	X	X	X
Whitesquaw Currant								
Ribes cereum	X	X	–	–	–	X	X	X
Sierra Plum								
Prunus subcordata	X	X	–	–	–	X	–	X
Pinemat Manzanita								
Arctostaphylos nevadensis	X	X	–	–	–	X	–	X
Mountain Spray								
Holodiscus microphyllus	X	X	–	–	–	X	–	–
Red Elderberry								
Sambucus microbotrys	X	–	X	–	–	X	–	X
Granite Gilia								
Leptodactylon pungens	–	X	–	–	–	–	–	–
Huckleberry Oak								
Quercus vaccinifolia	X	X	X	–	X	X	X	X
Wildflowers of dry shady forests								
Heartleaf Arnica								
Arnica cordifolia	–	X	–	–	–	–	–	X
California Indian Pink								
Silene californica	?	?	–	–	–	–	–	–
California Waterleaf								
Hydrophyllum occidentale	X	X	–	–	–	–	–	–
One-sided Wintergreen								
Orthilia secunda	–	X	–	–	–	–	–	–
Little Prince's Pine								
Chimaphila umbellata	X	X	–	X	–	–	–	X
Western Prince's Pine								
Chimaphila occidentalis	X	X	–	X	–	–	–	X
Snow Plant								
Sarcodes sanguinea	X	X	–	–	–	–	–	–
Pinedrops								
Pterospora andromedea	–	X	–	–	–	–	–	–
Pine Woods Lousewort								
Pedicularis semibarbata	?	X	–	–	–	–	–	X
California Skullcap								
Scutellaria tuberosa	–	X	–	–	–	–	–	X
Sierra Nevada Pea								
Lathyrus nevadensis	?	–	–	–	–	–	–	X
Hartweg's Ginger								
Asarum hartwegii	X	X	–	–	–	–	–	–
California Harebell								
Campanula prenanthoides	?	–	–	–	–	–	–	–
False Solomon's-seal								
Smilacina racemosa	X	X	X	–	–	–	–	X
Washington Lily								
Lilium washingtonianum	X	X	–	–	–	–	–	–

Common Name Scientific Name	Food, Beverage	Medicine Personal Care	Dye, Soap	Gum, Rubber, Smoking	Fire, Fiber	Tools Abrasives Crafts	Containers Baskets	Poisons
Humboldt Lily								
Lilium humboldtii	X	X	–	–	–	–	–	–
Rattlesnake Orchid								
Goodyera oblongifolia	–	X	–	–	–	–	–	X
Spotted Coralroot								
Corallorrhiza maculata	–	X	–	–	–	–	–	X
Striped Coralroot								
Corallorrhiza striata	–	X	–	–	–	–	–	X

Wildflowers of open forests

	Food, Beverage	Medicine Personal Care	Dye, Soap	Gum, Rubber, Smoking	Fire, Fiber	Tools Abrasives Crafts	Containers Baskets	Poisons
Larkspurs								
Genus *Delphinium*	–	–	–	–	–	–	–	X
Wild Peony								
Paeonia brownii	X	X	–	–	–	–	–	–
Sidalceas								
Genus *Sidalcea*	?	?	–	–	–	–	–	–
Pine Violet								
Viola lobata	X	X	–	–	–	–	–	–
Campions								
Genus *Silene*	?	?	–	–	–	–	–	–
Naked-stemmed Eriogonum								
Eriogonum nudum	X	X	–	–	–	–	–	–
Sulphur Buckwheat								
Eriogonum umbellatum	X	X	–	–	–	–	–	–
Wright's Eriogonum								
Eriogonum wrightii	X	X	–	–	–	–	–	–
Common Knotweed								
Polygonum aviculare	X	X	–	–	–	–	–	X
Green Gentian								
Swertia radiata	X	?	–	–	–	–	–	–
Waterleaf Phacelia								
Phacelia hydrophylloides	?	?	–	–	–	–	–	–
Pretty Nemophila								
Nemophila pulchella	–	X	–	–	–	–	–	–
Showy Penstemon								
Penstemon speciosus	–	X	–	–	–	–	–	–
Indian Paintbrush								
Castilleja applegatei	X	X	–	–	–	–	–	–
Mountain Pennyroyal								
Monardella odoratissima	X	X	–	–	–	–	–	–
Brewer's Lupine								
Lupinus breweri	–	–	–	–	–	–	–	X
Anderson's Lupine								
Lupinus andersonii	–	–	–	–	–	–	–	X
Fireweed								
Epilobium angustifolium	X	X	–	X	X	–	–	X
Bedstraw								
Galium aparine	X	X	–	–	–	X	X	–
Mountain Mule's Ears								
Wyethia mollis	X	X	X	–	–	X	–	X
Single-stem Butterweed								
Senecio integerrimus	?	?	–	–	–	–	?	X
Short-beaked Agoseris								
Agoseris glauca	X	–	–	X	–	–	X	–

Common Name Scientific Name	Food, Beverage	Medicine Personal Care	Dye, Soap	Gum, Rubber, Smoking	Fire, Fiber	Tools Abrasives Crafts	Containers Baskets	Poisons
White-flowered Hawkweed								
Hieracium albiflorum	–	X	–	X	–	–	–	–
Brewer's Golden Aster								
Chrysopis breweri	X	–	–	–	–	–	–	–
Bear Grass								
Xerophyllum tenax	X	X	X	–	X	–	X	–
Sierra Onion								
Allium companulatum	X	X	–	–	–	–	–	X

Wildflowers of streamsides and damp forests

Common Name Scientific Name	Food, Beverage	Medicine Personal Care	Dye, Soap	Gum, Rubber, Smoking	Fire, Fiber	Tools Abrasives Crafts	Containers Baskets	Poisons
Larkspurs								
Genus *Delphinium*	–	–	–	–	–	–	–	X
Columbia Monkshood								
Aconitum columbianum	–	–	–	–	–	–	–	X
Crimson Columbine								
Aquilegia formosa	X	X	–	–	–	–	–	X
Plantain-leaf Buttercup								
Ranunculus alismaefolius	X	X	–	–	–	–	–	X
Waterfall Buttercup								
Ranunculus hystriculus	X	X	–	–	–	–	–	X
Fendler's Meadow-rue								
Thalictrum fendleri	–	X	–	–	–	–	–	X
Geraniums								
Genus *Geranium*	X	X	–	–	–	–	–	–
MacCloskey's Violet								
Viola mackloskeyi	X	X	–	–	–	–	–	–
Western Dog Violet								
Viola adunca	X	X	–	–	–	–	–	–
Bleeding Heart								
Dicentra formosa	–	X	–	–	–	–	–	X
Watercresses								
Genus *Rorippa*	X	X	–	–	–	–	–	–
Alaska Whitlowgrass								
Draba stenoloba	–	X	–	–	–	–	–	–
Chamisso's Starwort								
Stellaria crispa	?	?	–	–	–	–	–	–
Miner's Lettuce								
Claytonia perfoliata	X	X	–	–	–	–	–	–
Shooting Stars								
Genus *Dodecatheon*	X	–	–	–	–	–	–	–
Sierra Nemophila								
Nemophila spatulata	–	X	–	–	–	–	–	–
Common Monkeyflower								
Mimulus guttatus	X	X	–	–	–	–	–	–
Primrose Monkeyflower								
Mimulus primuloides	X	X	–	–	–	–	–	–
Scarlet Monkeyflower								
Mimulus cardinalis	X	X	–	–	–	–	–	–
Speedwell								
Veronica americana	X	X	–	–	–	–	–	–
Great Red Paintbrush								
Castilleja miniata	X	X	–	–	–	–	–	X
Leopard Lily								
Lilium pardalinum	X	X	–	–	–	–	–	–

Common Name / Scientific Name	Food, Beverage	Medicine Personal Care	Dye, Soap	Gum, Rubber, Smoking	Fire, Fiber	Tools Abrasives Crafts	Containers Baskets	Poisons
Hooker's Fairy Bell / *Disporum hookeri*	X	X	–	–	–	–	–	–
Arrowhead Butterweed / *Senecio triangularis*	?	?	–	–	–	–	–	X
California Coneflower / *Rudbeckia californica*	X	?	–	–	–	–	–	?
Elk Clover / *Aralia californica*	?	?	–	–	–	–	–	?
Evening-primrose / *Oenothera elata*	X	X	–	X	X	–	–	–
Narrowleaf Lotus / *Lotus oblongifolius*	–	–	–	–	–	–	–	X
Broadleaf Lupine / *Lupines latifolius*	–	–	–	–	–	–	X	X
Strawberry / *Fragaria virginiana*	X	X	–	–	–	–	–	X
Bud Saxifrage / *Saxifraga bryophora*	X	X	–	–	–	–	–	–
White Hedge Nettle / *Stachys albens*	X	X	–	–	–	–	–	–

Wildflowers of rocky habitats

Common Name / Scientific Name	Food, Beverage	Medicine Personal Care	Dye, Soap	Gum, Rubber, Smoking	Fire, Fiber	Tools Abrasives Crafts	Containers Baskets	Poisons
Pasque Flower / *Anemone occidentalis*	–	X	–	–	–	–	–	X
Mountain Violet / *Viola purpurea*	X	X	–	–	–	–	–	–
Steer's Head / *Dicentra formosa*	–	?	–	–	–	–	–	X
Mountain Jewel Flower / *Streptanthus tortuosus*	?	–	–	–	–	–	–	–
Rock Cresses / Genus *Arabis*	?	X	–	–	–	–	–	–
Western Wallflower / *Erysimum capitatum*	X	X	–	–	–	–	–	–
Sierra Wallflower / *Erysimum perenne*	X	X	–	–	–	–	–	–
Drabas / Genus *Draba*	–	X	–	–	–	–	–	–
Sandworts / Genus *Arenaria*	–	X	–	–	–	–	–	–
Campions / Genus *Silene*	?	?	–	–	–	–	–	–
Pussypaws / *Calyptridium umbellatum*	?	–	–	–	–	–	–	–
Bitterroot / *Lewisia rediviva*	X	X	–	X	–	–	–	–
Eriogonums / Genus *Eriogonum*	X	X	–	–	–	–	–	–
Mountain Sorrel / *Oxyria digyna*	X	X	–	–	–	–	–	–
Sierra Primrose / *Primula suffrutescens*	?	–	–	–	–	–	–	–
Green Gentian / *Frasera speciosa*	X	X	–	–	–	–	–	–

Common Name Scientific Name	Food, Beverage	Medicine Personal Care	Dye, Soap	Gum, Rubber, Smoking	Fire, Fiber	Tools Abrasives Crafts	Containers Baskets	Poisons
Showy Polemonium								
Polemonium pulcherrimum	–	X	–	–	–	–	–	–
Collomia								
Genus *Collomia*	–	X	–	–	–	–	–	–
Spreading Phlox								
Phlox diffusa	?	?	–	–	–	–	–	–
Sierra Cryptantha								
Cryptantha nubigena	X	–	–	–	–	–	–	–
Davidson's Penstemon								
Penstemon davidsonii	–	X	–	–	–	–	–	–
Sierra Penstemon								
Penstemon heterodoxus	–	X	–	–	–	–	–	–
Mountain Pride								
Penstemon newberryi	–	X	–	–	–	–	–	–
Gray Paintbrush								
Castilleja pruinosa	X	X	–	–	–	–	–	–
Narrowleaf Stonecrop								
Sedum lanceolatum	X	X	–	–	–	–	–	–
Rosy Stonecrop								
Sedum rosea	X	X	–	–	–	–	–	–
Brewer's Lupine								
Lupinus breweri	–	–	–	–	–	–	–	X
Lotuses								
Genus *Lotus*	–	–	–	–	–	–	–	X
Locoweed								
Genus *Astragalus*	–	–	–	–	–	–	–	X
Fireweed								
Epilobium angustifolium	X	X	–	–	X	–	X	–
Rock Fringe								
Epilobium obcordatum	X	X	–	–	X	–	X	–
Terebinth Pteryxia								
Pteryxia terebintha	X	–	–	–	–	–	–	X
Mountain Mule's Ears								
Wyethia mollis	X	X	–	–	–	–	–	–
Loose Daisy								
Erigeron vagus	–	X	–	–	–	–	–	–
Cutleaf Daisy								
Erigeron compositus	–	X	–	–	–	–	–	–
Butterweeds								
Genus *Senecio*	–	–	–	–	–	–	–	X
Shaggy Hawkweed								
Hieracium horridum	?	?	–	–	–	–	–	–
Rosy Everlasting								
Antennaria rosea	X	X	–	X	–	–	–	–
Low Everlasting								
Antennaria dimorpha	X	X	–	X	–	–	–	–
Chaenactises								
Genus *Chaenactis*	–	?	–	–	–	–	–	–
Leichtlin's Mariposa Lily								
Calochortus leichtlinii	X	–	–	–	–	–	–	–

Montane Meadows
Meadow Sedges and Grasses

Sedges								
Genus *Carex*	X	X	–	–	X	–	–	–

Common Name / Scientific Name	Food, Beverage	Medicine Personal Care	Dye, Soap	Gum, Rubber, Smoking	Fire, Fiber	Tools Abrasives Crafts	Containers Baskets	Poisons
Brewer's Shorthair Grass								
Calamagrostis breweri	X	X	–	–	–	–	–	–
Hairgrass								
Deschampsia caespitosa	X	X	–	–	–	–	X	–
Spiked Trisetum								
Trisetum spicatum	X	–	–	–	–	–	–	–
Squirreltail								
Sitanion hystrix	X	–	–	–	–	–	–	–
Hansen's Bluegrass								
Poa hanseni	X	–	–	–	–	–	–	–
Columbia Needlegrass								
Stipa columbiana	X	–	–	–	–	–	–	–
Western Needlegrass								
Stipa occidentalis	X	–	–	–	–	–	–	–
Alpine Fescue								
Festuca brachyphylla	X	–	–	–	–	–	–	–
Gray Wild Rye								
Elymus glaucus	X	–	–	–	–	–	–	–

Meadow Wildflowers

Common Name / Scientific Name	Food, Beverage	Medicine Personal Care	Dye, Soap	Gum, Rubber, Smoking	Fire, Fiber	Tools Abrasives Crafts	Containers Baskets	Poisons
Marsh Marigold								
Caltha leptosepala	X	X	–	–	–	–	–	X
Buttercups								
Genus *Ranunculus*	X	X	–	–	–	–	–	X
Creeping Sidalcea								
Sidalcea reptans	?	?	–	–	–	–	–	–
Geraniums								
Genus *Geranium*	X	X	–	–	–	–	–	–
Starworts								
Genus *Stellaria*	X	X	–	–	–	–	–	–
Knotweeds								
Genus *Polygonum*	X	X	–	–	–	–	–	X
Shooting Stars								
Genus *Dodecatheon*	X	–	–	–	–	–	–	–
Gentians								
Genus *Gentiana*	–	X	–	–	–	–	–	–
Hiker's Gentian								
Gentianopsis simplex	–	X	–	–	–	–	–	–
Monkeyflowers								
Genus *Mimulus*	X	X	–	–	–	–	–	–
Blue-eyed Mary								
Collinsia torreyi	–	?	–	–	–	–	–	–
Sierra Penstemon								
Penstemon heterodoxus	–	X	–	–	–	–	–	–
Meadow Penstemon								
Penstemon rydbergii	–	X	–	–	–	–	–	–
Little Elephant's Head								
Pedicularis attolens	–	X	–	–	–	–	–	X
Greater Elephant's Head								
Pedicularis groenlandica	–	X	–	–	–	–	–	X
Lemmon's Paintbrush								
Castilleja lemmonii	X	X	–	–	–	–	–	–
Brewer's Paintbrush								
Castilleja breweri	X	X	–	–	–	–	–	–

Common Name Scientific Name	Food, Beverage	Medicine Personal Care	Dye, Soap	Gum, Rubber, Smoking	Fire, Fiber	Tools Abrasives Crafts	Containers Baskets	Poisons
Bud Saxifrage								
Saxifraga bryophora	X	–	–	–	–	–	–	–
Grass of Parnassus								
Parnassia fimbriata	–	X	–	–	–	–	–	–
Slender Cinquefoil								
Potentilla gracilis	–	X	–	–	–	–	–	–
Drummond's Cinquefoil								
Potentilla drummondii	–	X	–	–	–	–	–	–
Sierra Lupine								
Lupinus confertus	–	–	–	–	–	–	–	X
Coville's Lupine								
Lupinus covillei	–	–	–	–	–	–	–	X
Carpet Clover								
Trifolium monanthum	X	X	–	–	–	–	–	–
Nuttall's Gayophytum								
Gayophytum nuttallii	–	?	–	–	–	–	–	–
California Valerian								
Valeriana californica	?	?	–	–	–	–	–	X
Anderson's Alpine Aster								
Aster alpigenus	?	?	–	–	–	–	–	X
Wandering Daisy								
Erigeron peregrinus	–	X	–	–	–	–	–	–
Bigelow's Sneezeweed								
Helenium bigelovii	–	–	–	–	–	–	–	–
Sierra Butterweed								
Senecio scorzonella	–	–	–	–	–	–	–	X
Camas								
Camassia quamash	X	–	–	–	–	–	–	–
Leichtlin's Camas								
Camassia leichtlinii	X	–	–	–	–	–	–	–
Death Camas								
Zigadenus venenosus	–	–	–	–	–	–	–	X
Corn Lily								
Veratrum californicum	–	–	–	–	–	–	–	X
Swamp Onion								
Allium validum	X	X	–	–	–	–	–	–
Western Blue Flag								
Iris missouriensis	–	?	–	–	–	–	–	X

Meadow Shrubs

Willows								
Genus *Salix*	X	X	–	–	X	X	X	–
Bush Cinquefoil								
Potentilla fruticosa	–	X	–	–	–	–	–	–
Alpine Laurel								
Kalmia polifolia	–	–	–	–	–	X	–	X
Sierra Bilberry								
Vaccinium nivictum	X	X	–	–	–	–	–	–
Western Blueberry								
Vaccinium occidentale	X	X	–	–	–	–	–	–

Common Name / Scientific Name	Food, Beverage	Medicine Personal Care	Dye, Soap	Gum, Rubber, Smoking	Fire, Fiber	Tools Abrasives Crafts	Containers Baskets	Poisons
Alpine Environment								
Alpine Meadow Wildflowers								
Alpine Buttercup								
Ranunculus eschscholtzii	X	–	–	–	–	–	–	X
Dwarf Knotweed								
Polygonum minimum	X	X	–	–	–	–	–	–
Alpine Shooting Star								
Dodecatheon alpinum	X	–	–	–	–	–	–	–
Subalpine Shooting Star								
Dodecatheon subalpinum	X	–	–	–	–	–	–	–
Gentians								
Genus *Gentiana*	–	X	–	–	–	–	–	–
Little Elephant's Head								
Pedicularis attolens	–	?	–	–	–	–	–	X
Greater Elephant's Head								
Pedicularis groenlandica	–	?	–	–	–	–	–	X
Alpine Paintbrush								
Castilleja nana	X	–	–	–	–	–	–	–
Rosy Stonecrop								
Sedum rosea	X	X	–	–	–	–	–	–
Sierra Saxifrage								
Saxifraga aprica	X	–	–	–	–	–	–	–
Drummond's Cinquefoil								
Potentilla drummondii	–	X	–	–	–	–	–	–
Anderson's Alpine Aster								
Aster alpigenus	–	X	–	–	–	–	–	X
Alpine Goldenrod								
Solidago multiradiata	X	–	X	X	–	–	–	X
Alpine Meadow Shrubs								
Alpine Willow								
Salix arctica	X	X	–	–	X	X	X	–
Snow Willow								
Salix nivalis	X	X	–	–	X	X	X	–
Alpine Laurel								
Kalmia polifolia	–	–	–	–	–	X	–	X
Sierra Bilberry								
Vaccinium caespitosum	X	X	–	–	–	–	–	–
Western Blueberry								
Vaccinium uliginosum	X	X	–	–	–	–	–	–
Alpine Rock Communities								
Gravel flats and scree slopes								
Rockcresses								
Genus *Arabis*	?	X	–	–	–	–	–	–
Drabas								
Genus *Draba*	?	?	–	–	–	–	–	–
Sandworts								
Genus *Arenaria*	–	X	–	–	–	–	–	–
Pussypaws								
Calyptridium umbellatum	–	?	–	–	–	–	–	–
Dwarf Lewisia								
Lewisia pygmaea	X	–	–	–	–	–	–	–

Common Name Scientific Name	Food, Beverage	Medicine Personal Care	Dye, Soap	Gum, Rubber, Smoking	Fire, Fiber	Tools Abrasives Crafts	Containers Baskets	Poisons
Eriogonums								
Genus Eriogonum	X	X	–	–	–	–	–	–
Mountain Sorrel								
Oxyria digyna	X	X	–	–	–	–	–	–
Kellogg's Knotweed								
Polygonum kelloggii	X	X	–	–	–	–	–	–
Dwarf Knotweed								
Polygonum minimum	X	X	–	–	–	–	–	–
Showy Polemonium								
Polemonium pulchrrimum	–	–	–	–	–	–	X	–
Spreading Phlox								
Phlox diffusa	–	X	–	–	–	–	–	–
Coville's Phlox								
Phlox covillei	–	X	–	–	–	–	–	–
Little Elephant's Head								
Pedicularis attolens	–	?	–	–	–	–	–	X
Alpine Saxifrage								
Saxifraga tolmei	X	–	–	–	–	–	–	–
Ivesias								
Genus Ivesia	–	?	–	–	–	–	–	–
Alpine Lupine								
Lupinus lyalli	–	–	–	–	–	–	–	X
Locoweeds								
Genus Astragalus	–	–	–	–	–	–	–	X
Gray's Cymopteris								
Cymopteris cinerarius	X	X	–	–	–	–	–	–
Dwarf Daisy								
Erigeron pygmaeus	–	X	–	–	–	–	–	–
Alpine Everlasting								
Antennaria alpina	X	X	–	–	–	–	–	–

Rock crannies—shrubs

Alpine Prickly Currant								
Ribes montigenum	X	X	–	–	–	X	X	–
Whitesquaw Currant								
Ribes cereum	X	X	–	–	–	X	X	–
Creambush								
Holodiscus microphyllus	X	X	–	–	–	X	–	–
Bush Cinquefoil								
Potentilla fruticosa	–	X	–	–	–	–	–	–
Red Mountain Heather								
Phyllodoce breweri	–	X	–	–	–	–	–	–
Granite Gilia								
Leptodactylon pungens	–	X	–	–	–	–	–	–
Snowberry								
Symphoricarpos rotundifolius	X	X	–	–	X	X	X	X
Goldenbushes								
Genus Haplopappus	X	X	–	–	–	–	–	–

Rock crannies—wildflowers

Alpine Columbine								
Aquilegia pubescens	X	X	–	–	–	–	–	X
Broad-seeded Rockcress								
Arabis platysperma	X	X	–	–	–	–	–	–

Common Name / Scientific Name	Food, Beverage	Medicine Personal Care	Dye, Soap	Gum, Rubber, Smoking	Fire, Fiber	Tools Abrasives Crafts	Containers Baskets	Poisons
Lemmon's Draba								
Draba lemmonii	?	?	–	–	–	–	–	–
Sargent's Campion								
Silene sargentii	–	?	–	–	–	–	–	–
Pussypaws								
Calyptridium umbellatum	–	?	–	–	–	–	–	–
Eriogonums								
Genus *Eriogonum*	X	X	–	–	–	–	–	–
Mountain Sorrel								
Oxyria digyna	X	X	–	–	–	–	–	–
Sierra Primrose								
Primula suffrutescens	X	–	–	–	–	–	–	–
Sky Pilot								
Polemonium eximium	–	?	–	–	–	–	–	–
Showy Polemonium								
Polemonium pulcherrimum	–	?	–	–	–	–	–	–
Spreading Phlox								
Phlox diffusa	–	X	–	–	–	–	–	–
Timberline Phacelia								
Phacelia frigida	?	?	–	–	–	–	–	–
Sierra Cryptantha								
Cryptantha nubigena	–	X	–	–	–	–	–	–
Davidson's Penstemon								
Penstemon davidsonii	–	X	–	–	–	–	–	–
Sierra Penstemon								
Penstemon heterodoxus	–	X	–	–	–	–	–	–
Stonecrops								
Genus *Sedum*	X	X	–	–	–	–	–	–
Alpine Saxifrage								
Saxifraga tolmei	X	–	–	–	–	–	–	–
Drummond's Cinquefoil								
Potentilla drummondii	–	X	–	–	–	–	–	–
Club-moss Ivesia								
Ivesia lycopodioides	–	X	–	–	–	–	–	–
Rock Fringe								
Epilobium obcordatum	X	X	–	–	X	–	–	–
Cutleaf Daisy								
Erigeron compositus	–	X	–	–	–	–	–	–
Alpine Everlasting								
Antennaria alpina	–	X	–	–	–	–	–	–
East Side of the Range								
Pinyon-Sagebrush Trees								
Dry slopes or flats								
Singleleaf Pinyon Pine								
Pinus monophylla	X	X	–	–	–	X	X	–
Western Juniper								
Juniperus occidentalis	X	X	–	–	X	X	X	–
Streamsides or other moist areas								
Jeffrey Pine								
Pinus jeffreyi	X	X	–	–	–	X	X	–
Lodgepole Pine								
Pinus contorta	X	X	–	–	–	X	X	–

Common Name Scientific Name	Food, Beverage	Medicine Personal Care	Dye, Soap	Gum, Rubber, Smoking	Fire, Fiber	Tools Abrasives Crafts	Containers Baskets	Poisons
White Fir								
Abies concolor	X	X	–	–	–	X	X	–
Arroyo Willow								
Salix lasiolepis	X	X	–	–	X	X	X	–
Yellow Willow								
Salix lasiandra	X	X	–	–	X	X	X	–
Quaking Aspen								
Populus tremuloides	X	X	–	–	X	X	X	–
Black Cottonwood								
Populus balsamifera	X	X	–	–	X	X	X	–
Canyon Live Oak								
Quercus chrysolepis	X	X	X	–	–	X	X	X
Black Oak								
Quercus kelloggii	X	X	X	–	–	X	X	X
Water Birch								
Betula occidentalis	X	X	–	–	–	X	X	X
Arizona Ash								
Fraxinus velutina	X	X	–	–	–	–	–	–

Pinyon-Sagebrush Shrubs
Dry slopes or flats

Common Name Scientific Name	Food, Beverage	Medicine Personal Care	Dye, Soap	Gum, Rubber, Smoking	Fire, Fiber	Tools Abrasives Crafts	Containers Baskets	Poisons
Curl-leaf Mountain-mahogany								
Cercocarpus ledifolius	–	X	–	–	–	X	–	X
Antelope Bitterbrush								
Purshia tridentata	–	X	X	–	X	X	X	–
Hoary Sagebrush								
Artemisia cana	X	X	–	X	X	X	–	X
Great Basin Sagebrush								
Artemisia tridentata	X	X	–	X	X	X	–	X
Low Sagebrush								
Artemisia arbuscula	X	X	–	X	X	X	–	X
Common Rabbitbrush								
Chrysothamnus viscidiflorus	–	X	X	X	–	–	–	–
Rubber Rabbitbrush								
Chrysothamnus nauseosus	–	X	X	X	–	–	–	–
Mormon Tea								
Ephedra viridis	X	X	–	–	–	–	–	–
Winter Fat								
Eurotia lanata	–	X	–	–	–	–	X	–
Plateau Gooseberry								
Ribes velutinum	X	X	–	–	–	–	–	–
Desert Peach								
Prunus andersonii	X	X	–	–	–	–	–	X
Mohave Ceanothus								
Ceanothus vestitus	X	X	X	–	–	–	X	–
Beavertail Cactus								
Opuntia basilaris	X	X	X	–	–	X	–	–
Alpine Sagebrush								
Artemisia rothrockii	X	X	X	–	X	–	X	X
Littleleaf Brickellbush								
Brickellia microphylla	–	X	–	–	–	–	–	–

Common Name / Scientific Name	Food, Beverage	Medicine Personal Care	Dye, Soap	Gum, Rubber, Smoking	Fire, Fiber	Tools Abrasives Crafts	Containers Baskets	Poisons
Streamsides or other moist areas								
Arroyo Willow *Salix lasiolepis*	X	X	–	–	X	X	X	–
Yellow Willow *Salix lasiandra*	X	X	–	–	X	X	X	–
Narrowleaf Willow *Salix exigua*	X	X	–	–	X	X	X	–
Golden Currant *Ribes aureum*	X	X	–	–	–	X	–	–
California Wild Rose *Rosa californica*	X	X	–	–	–	X	–	–
Buffalo Berry *Shepherdia argentea*	X	X	X	–	–	–	–	X
Water Birch *Betula occidentalis*	X	X	–	–	–	X	–	–
Pinyon-Sagebrush Wildflowers *Dry slopes or flats*								
Sagebrush Buttercup *Ranunculus glaberrimus*	X	X	–	–	–	–	–	X
Blazing Star *Mentzelia laevicaulis*	X	–	–	–	–	–	–	–
Violet *Viola sororia*	X	X	–	–	–	–	–	–
Prickly Poppy *Argemone munita*	–	X	–	–	–	–	–	X
Dagger Pod *Phoenicaulis cheiranthoides*	?	?	–	–	–	–	–	–
Rock Cresses Genus *Arabis*	?	X	–	–	–	–	–	–
Sierra Tansy *Descurainia pinnata*	X	–	–	–	–	–	–	–
Sandworts Genus *Arenaria*	–	X	–	–	–	–	–	–
Mountain Campion *Silene montana*	?	–	–	–	–	–	–	–
Pussypaws *Calyptridium umbellatum*	–	X	–	–	–	–	–	–
Bitterroot *Lewisia rediviva*	X	X	X	–	–	–	–	–
Eriogonums Genus *Eriogonum*	X	X	–	–	–	–	–	–
Douglas' Knotweed *Polygonum douglasii*	X	X	–	–	–	–	–	–
Showy Milkweed *Asclepias speciosa*	X	X	–	–	X	X	X	X
Stansbury's Phlox Phlox *stansburyi*	–	X	–	–	–	–	–	–
Nuttall's Linanthus *Linanthus nuttallii*	–	?	–	–	–	–	–	–
Bridges' Gilia *Gilia leptalea*	X	–	–	–	–	–	–	–
Ballhead *Ipomopsis congesta*	–	X	X	–	–	–	–	–

Common Name Scientific Name	Food, Beverage	Medicine Personal Care	Dye, Soap	Gum, Rubber, Smoking	Fire, Fiber	Tools Abrasives Crafts	Containers Baskets	Poisons
Scarlet Trumpet Flower								
Ipomopsis aggregata	–	X	X	–	–	–	–	–
Two-leaved Phacelia								
Phacelia bicolor	?	X	–	–	–	–	–	–
Lowly Penstemon								
Penstemon humilis	–	X	–	–	–	–	–	–
Bridges' Penstemon								
Penstemon bridgesii	–	X	–	–	–	–	–	–
Desert Paintbrush								
Castilleja chromosa	X	X	–	–	–	–	–	–
Prairie Smoke								
Geum ciliatum	–	X	–	–	–	–	–	–
Common Silverweed								
Potentilla anserina	–	X	–	–	–	–	–	–
Dusky Horkelia								
Horkelia fusca	–	?	–	–	–	–	–	–
Nevada Pea								
Lathyrus lanzwertii	?	?	–	–	–	–	–	X
Tahoe Lupine								
Lupinus meionanthus	–	–	–	–	–	–	–	X
Locoweeds								
Genus *Astragalus*	–	–	–	–	–	–	–	X
Low Evening-primrose								
Oenothera caespitosa	X	X	–	–	X	–	–	–
Mountain Mule's Ears								
Wyethia mollis	X	X	–	–	–	–	–	–
Nevada Daisy								
Erigeron nevadincola	–	X	–	–	–	–	–	–
Eaton's Daisy								
Erigeron eatonii	–	X	–	–	–	–	–	–
Sego Lily								
Calochortus nuttallii	X	–	–	–	–	–	–	–

Streamsides or other moist areas

Common Name Scientific Name	Food, Beverage	Medicine Personal Care	Dye, Soap	Gum, Rubber, Smoking	Fire, Fiber	Tools Abrasives Crafts	Containers Baskets	Poisons
Glaucous Larkspur								
Delphinium glaucum	–	–	–	–	–	–	–	X
Western Plantain Buttercup								
Ranunculus alismaefolius	X	–	–	–	–	–	–	X
Western Shooting Star								
Dodecatheon pulchellum	X	–	–	–	–	–	–	–
Brewer's Navarretia								
Navarretia breweri	X	X	–	–	–	–	–	–
Suksdorf's Monkeyflower								
Mimulus suksdorfi	X	X	–	–	–	–	–	–
Inyo Meadow Lupine								
Lupinus pratensis	–	–	–	–	–	–	–	X
Evening-primrose								
Oenothera elata	X	X	–	–	X	–	X	–

List of some threatened, endangered, and sensitive plant species found in the Sierra Nevada. For more information, you are advised to contact the California Department of Fish and Game, U.S. Fish and Wildlife Service, California Native Plant Society, and the local U.S. Forest Service (California Natural Diversity Database, October 2003).

Family	Species	Designation[1]
Asteraceae	Congdon's Woolly Sunflower (*Eriophyllum congdonii*)	SR
	Layne's Ragwort (*Senecio layneae*)	SR, FT
Berberdiaceae	Truckee Barberry (*Berberis aquifolium* var. *repens*)	SE, FE
Brassicaceae	Tahoe Yellowcress (*Rorippa subumbellata*)	FE
Campanulaceae	Twisselmann's Nemacladus (*Nemacladus twisselmannii*)	SR
Convolvulaceae	Stebbin's Morning-glory (*Calystegia stebbinsii*)	SE, FE
Cyperaceae	Tompkin's Sedge (*Carex tompkinsii*)	SR
Ericaceae	Ione Manzanita (*Arctostaphylos myrtifolia*)	FT
Fabaceae	Long Valley Milkvetch (*Astragalus johannis-howellii*)	SR
	Sodaville Milkvetch (*A. lentiginosus* var. *sesquimetralis*)	SE
	Mono Milkvetch (*A. monoensis* var. *monoensis*)	SR
	Mariposa Lupine (*Lupinus citrinus* var. *deflexus*)	ST
	Father Crowley's Lupine (*Lupinus padre-crowleyi*)	SR
Liliaceae	Yosemite Onion (*Allium yosemitense*)	SR
	Chinese Camp Brodiaea (*Brodiaea pallida*)	SE, FT
	Kaweah Brodiaea (*B. insignis*)	SE
	Striped Adobe Lily (*Fritillaria striata*)	ST
Malvaceae	Owens Valley Checkerbloom (*Sidalcea coville*)	SE
	Keck's Checkermallow (*S. keckii*)	FE
	Scadden Flat Checkerbloom (*S. stipularis*)	SE
Nyctaginaceae	Ramshaw Abronia (*Abronia alpina*)	FC
Onagraceae	Springville Clarkia (*Clarkia springvillensis*)	SE, FT
	Merced Clarkia (*C. linulata*)	SE
Philadelphaceae	Tree-anemone (*Carpenteria californica*)	ST
Polygonaceae	Ione Buckwheat (*Eriogonum apricum* var. *apricum*)	SE, FE
	Irish Hill Buckwheat (*E. apricum* var. *prostratum*	SE, FE
	Twisselmann's Buckwheat (*E. twisselmannii*)	SR
Portulacaceae	Mariposa Pussypaws (*Calyptridium pulchellum*)	FT
	Congdon's Lewisia (*Lewisia congdonii*)	SR
Rhamnaceae	Pine Hill Ceanothus (*Ceanothus roderickii*)	SR, FE
Rutaceae	El Dorado Bedstraw (*Galium californicum* ssp. *sierrae*)	SR, FE
Scrophulariaceae	Succulent Owl's-clover (*Orthocarpus compestris* var. *succulentus*)	SE, FE
	Boggs Lake Hedge-hyssop (*Gratiola heterosepala*)	SE
Staphyleaceae	Pine Hill Flannelbush (*Fremontodendron decumbens*)	SR, FE
Verbenaceae	California Vervain (*Verbena californica*)	ST, FT

1. SE (State-listed Endangered); ST (State-listed Threatened); SR (State-listed Rare); FE (Federally-listed Endangered); FT (Federally-listed Threatened); FC (Federal Candidate)

APPENDIX 3

Scientific and Common Plant Names

Note: Names in all caps indicate scientific names of families.

Abies
Abronia
Acer
ACERACEAE
Achillea
Aconitum
Actaea
Adenocaulon
Adenostoma
Adiantum
Aesculus
Agastache
Ageratina
Agoseris
Agrimonia
Alisma
ALISMATACEAE
Allium
Alnus
AMARANTHACEAE
Amaranthus
Ambrosia
Amelanchier
Amsinckia
ANACARDIACEAE
Anagallis
Anaphalis
Androsace
Anemone
Anemopsis
Angelica
Antennaria
Anthemis
APIACEAE
APOCYNACEAE
Apocynum
Aquilegia
Arabis
Aralia
ARALIACEAE
Arbutus
Arceuthobium
Arctium
Arctostaphylos
Arenaria

Argemone
ARISTOLOCHIACEAE
Arnica
Artemisia
Asarum
ASCLEPIADACEAE
Asclepias
Aspidotis
Asplenium
Aster
ASTERACEAE
Astragalus
Athyrium
Atriplex
Azolla
AZOLLACEAE

Balsamorhiza
Barbarea
BERBERIDACEAE
Berberis
Berula
Betula
BETULACEAE
Bidens
Blechnum
BORAGINACEAE
Boschniakia
Botrychium
Boykinia
Brasenia
Brassica
BRASSICACEAE
Brickellia
Brodiaea

CABOMBACEAE
CACTACEAE
Calandrinia
CALLITRICHACEAE
Callitriche
Calocedrus
Calochortus
Caltha
CALYCANTHACEAE

Calycanthus
Calyptridium
Camassia
Camissonia
Campanula
CAMPANULACEAE
CANNABACEAE
Cannabis
CAPPARACEAE
CAPPARIDACEAE
CAPRIFOLIACEAE
Capsella
Cardamine
Carduus
Carex
CARYOPHYLLACEAE
Cassiope
Castilleja
Castonopsis
Caulanthus
Ceanothus
CELASTRACEAE
Celtis
Centaurium
Cerastium
CERATOPHYLLACEAE
Ceratophyllum
Cercis
Cercocarpus
Chaenactis
Chamaebatia
Chamaesaracha
Chamaesyce
Chamomilla
Cheilanthes
CHENOPODIACEAE
Chenopodium
Chimaphila
Chlorogalium
Chrysolepis
Chrysothamnus
Cichorium
Cicuta
Circaea
Cirsium

CISTACEAE
Clarkia
Claytonia
Clematis
Cleome
Collinsia
Collomia
Comandra
COMPOSITAE
Conium
CONVOLVULACEAE
Convolvulus
Conyza
Corallorrhiza
Cordylanthus
Corethrogyne
CORNACEAE
Cornus
Corydalis
Corylus
CRASSULACEAE
Crepis
CROSSOSOMATACEAE
CRUCIFERAE
Cryptantha
Cryptogramma
CUCURBITACEAE
CUPRESSACEAE
Cupressus
Cuscuta
CUSCUTACEAE
Cymopteris
Cynoglossum
CYPERACEAE
Cyperus
Cypripedium
Cystopteris

Darlingtonia
Datisca
DATISCACEAE
Datura
Daucus
Delphinium
Dendromecon
DENNSTAEDIACEAE
Descurainia
Dianthus
Dicentra
Dichelostemma
DIPSACACEAE
Dipsacus
Disporum
Dodecatheon
Draba
Drosera
DROSERACEAE
DRYOPTERIDACEAE
Dryopteris
Dudleya

Dugaldia
Dulichium

ELAEAGNACEAE
Ephedra
EPHEDRACEAE
Epilobium
Epipactis
EQUISETACEAE
Equisetum
Eremocarpus
Eriastrum
ERICACEAE
Ericameria
Erigeron
Eriodictyon
Eriogonum
Eriophorum
Eriophyllum
Erodium
Eryngium
Erysimum
Eschscholzia
Euphorbia
EUPHORBIACEAE
Eurotia

FABACEAE
FAGACEAE
Floerkea
Fragaria
Frasera
Fraxinus
Fremontodendron
Fritillaria

Gaillardia
Galium
Garrya
GARRYACEAE
Gaultheria
Gayophytum
Gentiana
GENTIANACEAE
Gentianella
Gentianopsis
GERANIACEAE
Geranium
Geum
Gilia
Glossopetalion
Glycyrrhiza
Glyptopleura
Gnaphalium
Goodyera
GRAMINEAE
Grindelia
GROSSULARIACEAE
Gutierrezia
Gypsophila

Habenaria
Hackelia
HALORAGACEAE
Haplopappus
Hastingsia
Hazardia
Helenium
Helianthella
Helianthemum
Helianthus
Heracleum
Hesperolinon
Heteromeles
Heuchera
Hieracium
HIPPOCASTANACEAE
HIPPURIDACEAE
Hippuris
Hoita
Holodiscus
Horkelia
Humulus
HYDROPHYLLACEAE
Hydrophyllum
HYPERICACEAE
Hypericum

Ipomopsis
IRIDACEAE
Iris
ISOETACEAE
Isoetes
Iva
Ivesia

Jamesia
JUNCACEAE
JUNCAGINACEAE
Juncus
Juniperus

Kalmia
Keckellia
Kochia
Krascheninnikovia

Lactuca
LAMIACEAE
Lappula
Lastrea
Lathyrus
LAURACEAE
Layia
Ledum
LEGUMINOSAE
Lemna
LEMNACEAE
LENTIBULARIACEAE
Lepidium
Leptodactylon

Lesquerella
Lessingia
Leucanthemum
Leucothoe
Lewisia
Ligusticum
LILIACEAE
Lilium
LIMNANTHACEAE
Limnanthes
Limosella
LINACEAE
Linanthus
Linnaea
Linum
Lipocarpha
Lithocarpus
Lithophragma
Lithospermum
LOASACEAE
Lomatium
Lonicera
LORANTHACEAE
Lotus
Lupinus
Luzula
Lycium
Lycopus
Lysimachia
LYTHRACEAE
Lythrum

Macharenthera
Madia
Maianthemum
Malacothrix
Malva
MALVACEAE
Malvella
Marah
Marrubium
Marsilea
MARSILIACEAE
Matricaria
Medicago
Melilotus
Mentha
Mentzelia
MENYANTHACEAE
Menyanthes
Mertensia
Microseris
Mimulus
Minuartia
MOLLUGINACEAE
Mollugo
Monardella
Moneses
Monolepsis
Montia

Myrica
MYRICACEAE
Myriophyllum

Nama
Navarretia
Nemophila
Nicotiana
Nuphar
NYCTAGINACEAE
NYMPHAEACEAE
Nymphoides

Oemleria
Oenothera
OLEACEAE
ONAGRACEAE
OPHIOGLOSSACEAE
Opuntia
ORCHIDACEAE
OROBANCHACEAE
Orobanche
Orogenia
Orthilia
Orthocarpus
Oryzopsis
Osmorhiza
OXALIDACEAE
Oxalis
Oxypolis
Oxyria
Oxytropis

Pachistima
Paeonia
PAPAVERACEAE
Parnassia
Pastinaca
Paxistima
Pedicularis
Pellaea
Penstemon
Pentagramma
PEONIACEAE
Pericome
Perideridia
Petrophytum
Phacelia
PHILADELPHACEAE
Philadelphus
Phlox
Pholistoma
Phoradendron
Phragmites
Phyllodoce
Physocarpus
PINACEAE
Pinus
Piperia
Plagiobothrys

PLANTAGINACEAE
Plantago
PLATANACEAE
Platanthera
Platanus
Platystemon
POACEAE
POLEMONIACEAE
Polemonium
Polygala
POLYGALACEAE
POLYGONACEAE
Polygonum
POLYPODIACEAE
Polypodium
Polystichum
Populus
Portulaca
PORTULACACEAE
Potamogeton
POTAMOGETONACEAE
Potentilla
Primula
PRIMULACEAE
Prunella
Prunus
Pseudostellaria
Pseudotsuga
PTERIDACEAE
Pteridium
Pterospora
Purshia
Pycnanthemum
Pyrola

Quercus

RANUNCULACEAE
Ranunculus
Raphanus
RHAMNACEAE
Rhamnus
Rhododendron
Rhus
Ribes
Rorippa
Rosa
ROSACEAE
RUBIACEAE
Rubus
Rudbeckia
Rumex

Sagittaria
SALICACEAE
Salix
Salsola
Salvia
Sambucus
Sanguisorba

Sanicula
SANTALACEAE
Saponaria
Sarcodes
SARRACENIACEAE
SAURURACEAE
Saxifraga
SAXIFRAGACEAE
Scirpus
Sclerolinon
Scrophularia
SCROPHULARIACEAE
Scutellaria
Sedum
Selaginella
SELAGINELLACEAE
Senecio
Sequoiadendron
Shepherdia
Sidalcea
Silene
Silybum
Sisymbrium
Sisyrinchium
Sium
Smilacina
Smilax
SOLANACEAE
Solanum
Solidago
Sonchus
Sorbus
SPARGANIACEAE
Sparganium
Sphaeralcea
Sphaeromeria
Sphenosciadium
Spiraea
Spiranthes

Spirodela
Stachys
Stanleya
Staphylea
STAPHYLEACEAE
Stellaria
Stenotus
Stephanomeria
STERCULIACEAE
Streptopus
STYRACEAE
Styrax
Swertia
Symphoricarpus

Taraxacum
Tauschia
TAXACEAE
TAXODIACEAE
Taxus
Tellima
Thalictrum
Thelypteris
Thermopsis
Thysanocarpus
Torreya
Toxicodendron
Tragopogon
Trautvetteria
Tribulus
Trichostema
Trientalis
Trifolium
Triglochin
Trillium
Triodanis
Triteleia
Tsuga
Turricula

Typha
TYPHACEAE

ULMACEAE
UMBELLIFERAE
Umbellularia
Urtica
URTICACEAE
Utricularia

Vaccinium
Valeriana
VALERIANACEAE
Veratrum
Verbascum
Verbena
VERBENACEAE
Veronica
Viburnum
Vicia
Viola
VIOLACEAE
VISCACEAE
VITACEAE
Vitis

Woodsia
Woodwardia
Wyethia

Xanthium
Xerophyllum

Yucca

Zigadenus
ZYGOPHYLLACEAE

Index of Common Names

Note: Names in all caps indicate common names of families.

Aconite
ADDER'S-TONGUE FAMILY
Agrimony, Common
Alder
Alfalfa
Alkali Mallow
Alpine Mountain Sorrel
Alpine Wintergreen
Alumroot
Amaranth
AMARANTH FAMILY
American Vetch

Amole
Androsace
Anemone
Angelica
Antelope Bush
Arnica
Arrowgrass
ARROWGRASS FAMILY
Arrowhead
ARROWHEAD FAMILY
Arrowleaf Balsamroot
Ash, Oregon

Aspen, Quaking
Aster
Avens
Azalea, Western

Baby Blue Eyes
Baby's-breath
BALD-CYPRESS FAMILY
Baneberry, Red
Barberry
BARBERRY FAMILY
Bastard Toadflax

Beargrass
Beavertail Cactus
Bedstraw
BEECH FAMILY
Bee Plant
Bellflower
Bidens
Big Tree
Birch
BIRCH FAMILY
Bird's-beak
BIRTHWORT FAMILY
Biscuitroot
Bitterbrush
Bittercress
Bitterroot
Blackberry
Bladdernut
BLADDERNUT FAMILY
Bladder-pod, Western
Bladderwort
BLADDERWORT FAMILY
Blazing Star
Bleeding Heart
Bluebell
BLUEBELL FAMILY
Bluecurls
Blue-eyed Grass
Blue-eyed Mary
Bog Kalmia
Boisduvalia
BORAGE FAMILY
Boschniakia
Bouncing-bet
Boxthorn
Boykinia
Bracken Fern
Brickellbush
Brodiaea
Brooklime
Broom-rape
BROOM-RAPE FAMILY
Broom Snakeweed
Buckbean
BUCKBEAN FAMILY
Buckbrush
Buckeye
BUCKEYE FAMILY
Buckthorn
BUCKTHORN FAMILY
BUCKWHEAT FAMILY
Buckwheat, Wild
Buffaloberry
Bugbane, False
Bulrush
Burdock
Burnet
Bur-reed
BUR-REED FAMILY

Bush Poppy
Buttercup
BUTTERCUP FAMILY

CACAO FAMILY
CACTUS FAMILY
California Greenbriar
California Laurel
California Man-root
California Poppy
CALTROP FAMILY
CALYCANTHUS FAMILY
Calyptridium
Camas, Common
Camas, Death
Campion
Canadian Horseweed
Canchalagua
CAPER FAMILY
CARPETWEED FAMILY
Carpetweed, Green
CARROT FAMILY
Cascara
CASHEW FAMILY
Catchfly
Catseye
Cattail
CATTAIL FAMILY
Caulanthus
Chaenactis
Chain Fern, Giant
Chamaesaracha, Dwarf
Chamise
Chamomile
Chamomilla, Western
Checker Mallow
Cheeseweed
Cherry
Chickweed
Chickweed, Mouse-eared
Chicory
Chinese Caps
Chinquapin
Choke Cherry
Cinquefoil
Clarkia
Clematis
Cliffbrake
Cliff Bush
Clover
Cocklebur
Coffeeberry
Collinsia
Collomia
Columbine
Common Burdock
Common Camas
Common Reed
Common Selfheal

COMPOSITE FAMILY
Coneflower
Coon's Tail
Coralroot
Corethrogyne
Corn Lily
Corydalis, Sierra
Cottonwood
Cow-bane
Cow-lily
Cowparsnip
Coyote Thistle
Coyote Tobacco
Cream Cups
CROSSOSOMA FAMILY
Cryptantha
Cucumber, Wild
CUCUMBER FAMILY
Cudweed
Curlleaf Mountain-mahogany
Currant
CURRANT FAMILY
Cymopteris
CYPRESS FAMILY
Cypress, Baker

Dandelion, Common
Dandelion, Mountain
Death Camas
Deer Brush
Deer Fern
Dock
Dodder
DODDER FAMILY
Dogbane
DOGBANE FAMILY
Dogwood
DOGWOOD FAMILY
Douglas-fir
Draba
Duckmeat
Duckweed
DUCKWEED FAMILY
Dulichium
Durango Root
DURANGO ROOT FAMILY

Elderberry
Elk Clover
ELM FAMILY
Ephedra
EPHEDRA FAMILY
Eriastrum, Few-flowered
Eriogonum
Euphorbia
Evening Primrose
EVENING PRIMROSE FAMILY
Everlasting

Fairy Bells
False Bugbane
False Lupine
False Mermaid
FALSE MERMAID FAMILY
False Solomon's-seal
FERN FAMILY
Fiddleneck
Fiesta Flower
Figwort
FIGWORT FAMILY
Filbert
Fireweed
Firs, True
Flannelbush
Flat-sedge
Flax
FLAX FAMILY
Flax, Northwest Yellow
Fleabane
FOUR-O'CLOCK FAMILY
Four-wing Saltbush
Fragile Fern
Fringe-cups
Fringed Gentian
Fritillary

Gentian
Gentian, Fringed
Gentian, Green
Gentian, Little
GENTIAN FAMILY
Geranium
Geranium, Wild
GERANIUM FAMILY
Giant Chain Fern
Giant Helleborine
Giant Hyssop
Giant Sequoia
Gilia
Gilia, Granite
Gilia, Scarlet
Ginger, Wild
GINSENG FAMILY
Globemallow
Glyptopleura
Gnaphalium
Goathead
Goldback Fern
Goldenbush
Goldenrod
Gooseberry
GOOSEBERRY FAMILY
Goosefoot
GOOSEFOOT FAMILY
Granite Gilia
GRAPE FAMILY
Grape-fern
GRASS FAMILY

Grass-of-parnassus
Gray Tansy
Greasewood
Great Duckweed
Greenbriar, California
Green Carpetweed
Gromwell
Groundsel
Groundsmoke
Gumweed

Hackberry, Net-leaved
Hairy Pepperwort
Harebell
Hawksbeard
Hawkweed
Hazardia
Hazelnut
Heather, Red Mountain
Heather, White
HEATH FAMILY
Hedge Mustard
Hedge-nettle
Helianthella
Helleborine
Hemlock, Mountain
Hemlock, Poison
Hemlock, Water
HEMP FAMILY
Hen and Chickens
Holodiscus
Honeysuckle
HONEYSUCKLE FAMILY
Hop, Common
Horehound
Horkelia
HORNWORT FAMILY
Horse-chestnut
Horsetail
HORSETAIL FAMILY
Horseweed, Canadian
Hound's Tongue
Huckleberry
Hyssop, Giant

Incense-cedar
Indian Basket-grass
Indian Blanketflower
Indian's Dream
Iris
Iris, Wild
IRIS FAMILY
Ivesia

Jacob's Ladder
Jimson Weed
Juniper

Kalmia, Bog

Knotweed
Kochia

Labrador Tea, Western
Ladies'-tresses, Hooded
Ladyfern
Lady's Slipper
Lambs's-quarters
Larkspur
Laurel, California
LAUREL FAMILY
Layia
LEGUME FAMILY
Lessingia
Lettuce, Wild
Lewisia
Licorice, Wild
Lilac
Lily
Lily, Corn
Lily, Mariposa
Lily, Sego
LILY FAMILY
Linanthus
Lipfern
Lipocarpha
Little Gentian
Live-forever
LIZARD'S-TAIL FAMILY
Locoweed
Lomatium
Loosestrife, Purple
Loosestrife, Tufted
LOOSESTRIFE FAMILY
Lotus
Lousewort
Lovage, Gray's
Lungwort
Lupine
Lupine, False

MADDER FAMILY
Madrone
Maidenhair Fern
Malacothrix
Mallow, Alkali
Mallow, Checker
MALLOW FAMILY
Man-root
Manzanita
Maple
MAPLE FAMILY
Mare's-tail
MARE'S-TAIL FAMILY
Marijuana
Mariposa Lily
Marsh Marigold
Meadow Foam
Meadow-rue

MENTZELIA FAMILY
Milk Thistle
Milkvetch
Milkweed
MILKWEED FAMILY
Milkwort
MILKWORT FAMILY
Miner's Lettuce
Mint
MINT FAMILY
Mistletoe, Dwarf
Mistletoe, Juniper
MISTLETOE FAMILY
Mock Orange
MOCK ORANGE FAMILY
Monardella
Monkeyflower
Monkshood
Montia
Moonwort
Morning Glory
MORNING GLORY FAMILY
Mosquito Fern
Mountain-ash
Mountain Dandelion
Mountain Hemlock
Mountain Lover
Mountain-mahogany
Mountain-mint
Mountain Misery
Mountain Sorrel
Mouse-ear Chickweed
Mudwort
Mule-ears
Mullein
Mustard
MUSTARD FAMILY

Navarretia
Nemophila
Net-leaved Hackberry
Nettle, Stinging
NETTLE FAMILY
Nightshade, Purple
Nightshade, Enchanters
NIGHTSHADE FAMILY
Ninebark, Pacific
Nodding Silverpuffs
Northern Bugleweed
Nut-grass
Nutmeg, California

Oak
Oak, Tanbark
Oceanspray
OLEASTER FAMILY
OLIVE FAMILY
One-flowered Wintergreen
One-sided Wintergreen
Onion

Orchid, Bog
Orchid, Rein
ORCHID FAMILY
Oregon Ash
Oregon Boxwood
Oregon-grape
Orogenia, California
Oso Berry
Owl's clover
Ox-eye Daisy
Oxytrope

Paintbrush
Parsnip, Wild
PEA FAMILY
Peak Rush-rose
Pearly-everlasting
Pennyroyal
Penstemon
Penstemon, Gaping
Peony
PEONY FAMILY
Peppergrass
Pepperwort, Hairy
Pericome
Phacelia
Phlox
Phlox, Prickly
PHLOX FAMILY
Phlox, Slender
Pigweed
Pincushion
Pineapple Weed
Pine-drops, Woodland
PINE FAMILY
Pines
Pink
PINK FAMILY
Pipsissewa
Pitcher-plant, California
PITCHER-PLANT FAMILY
Plantain
PLANTAIN FAMILY
Plum
Plumeless Thistle
Poison Hemlock
Poison Oak
Ponderosa Pine
Pond-lily, Yellow
Pondweed
PONDWEED FAMILY
Popcorn Flower
Poppy, Bush
Poppy, California
POPPY FAMILY
Poppy, Prickly
Poverty Weed
Prickly Pear Cactus
Prickly Poppy
Prickly Russian Thistle

Primrose, Sierra
PRIMROSE FAMILY
Prince's Plume
Psoralea, Round-leaved
Puccoon
Puncture Vine
Purple Loosetrife
Purple Mat
Purslane
PURSLANE FAMILY
Pussy-paws
Pussy-toes

Quaking Aspen
Quillwort
QUILLWORT FAMILY

Rabbitbrush
Radish, Wild
Ragweed
Ranger's Button
Raspberry
Rattlesnake-plantain
Rattlesnake Weed
Red Baneberry
Redbud
Red Fir
Red Maids
Red Mountain Heather
Red-stem Storksbill
Reed, Common
Rein Orchid
Ribbon Bush
Ricegrass
Rock-brake
Rock Cress
Rockmat
ROCK ROSE FAMILY
Rock-spiraea
Rose
Rose, Wild
ROSE FAMILY
Round-leaved Psoralea
Rush
RUSH FAMILY

Sage
Sagebrush
Salsify
Saltbush
SANDALWOOD FAMILY
Sand Lacepod
Sand Verbena
Sandwort
Saxifrage
SAXIFRAGE FAMILY
Scarlet Gilia
Scouring Rush
Sedge
SEDGE FAMILY

Sego Lily
Selfheal, Common
Sequoia
Serviceberry
Sheperd's Purse
Shield Fern
Shooting Star
Sierra Corydalis
Sierra Laurel
Sierra Sweet-bay
Sierra Water Fern
Silktassel
SILKTASSEL FAMILY
Skullcap
Skunkbush
Slender Cotton-grass
Slender Phlox
Smartweed
Snakeroot
Snakeweed, Broom
SNAPDRAGON FAMILY
Sneezeweed
Snowberry
Snowbush
Snow Plant
Soapberry
Soap Plant
Soapwort
Sorrel, Alpine Mountain
Sorrel, Mountain
Sow Thistle
Spanish Dagger
Speedwell
Spice Bush
Spikemoss
SPIKEMOSS FAMILY
Spiraea
Spleenwort
Spring Beauty
Spurge
Spurge, Thyme-leaved
SPURGE FAMILY
STAFF-TREE FAMILY
Star-flower
Starwort, Sticky
Stenotus
Stickseed
Stinging Nettle
St. John's-wort
ST. JOHN'S-WORT FAMILY
Stonecrop
STONECROP FAMILY
Stone Fruits
Stoneseed
Storksbill, Red-stem
Strawberry
STYRAX FAMILY
Sugar Pine
SUMAC FAMILY
Sun Cup

Sundew
SUNDEW FAMILY
Sunflower
SUNFLOWER FAMILY
Sunflower, Woolly
Sweet-bay, Sierra
Sweet Cicely
Sweetclover
Sweetpea
Sweet Shrub
Sycamore, Western
SYCAMORE FAMILY

Tanbark Oak
Tansy, Gray
Tansy Mustard
Tarweed
Tauschia
Teasel
TEASEL FAMILY
Thimbleberry
Thistle
Thistle, Milk
Thistle, Plumeless
Thistle, Prickly Russian
Thistle, Sow
Toadflax, Bastard
Tobacco, Coyote
Toyon
Trail Plant
Trillium
Triteleia
True Firs
Tule-potato
Tumbleweed
Turkey Mullein
Turricula
Twinflower
Twisted-stalk

Valerian, California
VALERIAN FAMILY
Venus Looking-glass
Verbena
VERBENA FAMILY
Vervain
Vetch, American
Viburnum
Violet
VIOLET FAMILY
Virgin's Bower

Wake-robin
Wallflower
Wapato
WATERCLOVER FAMILY
Watercress
Water Fern, Sierra
WATER FERN FAMILY

Water Fringe
Water Hemlock
Waterleaf
WATERLEAF FAMILY
WATER-LILY FAMILY
Water-milfoil
WATER-MILFOIL FAMILY
Water Parsnip
Water-parsnip, Cut-leaved
Water Plantain
Watershield
WATERSHIELD FAMILY
Water Starwort
WATER STARWORT FAMILY
WAX-MYRTLE FAMILY
Western Azalea
Western Brackenfern
Western Pearly Everlasting
Western Polypody
Western Snakeroot
White Fir
White-flowered Schoenolirion
White Heather
White Layia
Whitlow Grass
Wild Buckwheat
Wild Forget-me-not
Wild Geranium
Wild Ginger
Wild Grape, California
Wild Hyacinth
Wild Iris
Wild Lettuce
Wild Licorice
Wild Parsnip
Wild Radish
Wild Rose
Wild Sarsaparilla
Willow
WILLOW FAMILY
Willow-herb
Windflower
Wintercress
Winterfat
Wintergreen
Wintergreen, Alpine
Wintergreen, One-flowered
Wintergreen, One-sided
Wire-grass
Woodfern
Woodland Star
Woodland Pinedrops
Woodnymph
Wood Rush
Woodsia
Wood-sorrel
WOOD-SORREL FAMILY
Woolly Sunflower
Wormwood
Wyethia

Yampah
Yarrow
Yellowcress
Yellow Pond-lily
Yerba Mansa
Yerba Santa

Yew, Pacific
YEW FAMILY
Yucca

Zygadene

Glossary

Acrid. Sharp, irritating, or biting to the taste.

Alkaloid. A nitrogen-containing, slightly alkaline substance that is often poisonous.

Alterative. A substance that gradually restores the normal functions of the body.

Analgesic. Relieves pain.

Anaphrodisiac. An agent that reduces sexual desire or potency.

Anesthetic. A substance that produces anesthesia.

Anodyne. Helps to quiet or relieve pain.

Antibiotic. Helps to destroy pathogenic action of microbes.

Anti-inflammatory. Reduces or neutralizes inflammation.

Antimicrobial. Something that inhibits the growth or multiplication of microorganisms or kills them.

Antipyretic. Against a fever.

Antiscorbutic. Used against scurvy.

Antiseptic. Prevents infection.

Antispasmodic. Relieves or cures spasms or irregular and painful action of the muscles (for example, epilepsy).

Antisyphilitic. Relieves or cures venereal disease.

Antiviral. An agent used against viruses.

Astringent. Shrinks or binds tissue.

Bitters. Sharp, acrid, or biting medicines, prescribed to stimulate an appetite.

Cardiacs. Have an effect on the heart.

Carminative. Dispels flatulency or gripping pains of the stomach and bowels.

Cathartics. Stimulate the action of the bowels, a purgative.

Decoction. The essence of a plant extracted by boiling it down.

Demulcent. Has a soothing or emollient effect on inflamed surfaces.

Dermititis. Inflammation of the skin.

Diaphoretic. Promotes or increases perspiration.

Diuretic. Increases the flow of urine by acting on the kidneys.

Emetic. Something that causes vomiting.

Emollient. Has a soothing and softening effect on the body tissues.

Expectorant. Causes an increase in expectoration, promoting the excretion of mucous from the chest.

Febrifuge. Helps reduce or control fevers.

Glycocide. Any of the numerous acetal derivatives of sugars that on hydrolysis yield a sugar.

Hydrocarbon. An organic compound containing one hydrogen and one carbon and often occurring in petroleum, natural gas, and coal.

Infusion. A preparation made by soaking a plant in hot water (tea).

Insecticidal. Used against insects.

Laxative. Used to loosen the bowels and relieve constipation.

Liniment. A liquid or semiliquid preparation of an herb to relieve skin irritation and muscle pain.

Mucilaginous. Something that resembles or contains mucilage (slime).

Narcotic. Diminishes the action of the nervous and vascular systems, causing drowsiness, lethargy, stupor, and insensibility.

Nervine. Soothes and calms the nerves, restoring them to a natural state.

Nitrate. A salt or ester of nitric acid.

Nutritive. Nourishes the body, promoting growth or health.

Panacea. A cure-all.

Parch. To toast or scorch with heat.

Photosensitivity. Being sensitive to light.

Poultice. A moist, usually warm or hot mass of plant material applied to the skin, or with a cloth between the skin and plant material, to effect a medicinal action.

Purgative. Evacuates the bowels but more forcefully than a laxative.

Salve. A healing ointment.

Saponin. A glycoside in plants that when shaken with water has a foaming or soapy action.

Sedative. Something used to lessen nervous excitement, irritation, and pain.

Selenium. A nonmetallic element that resembles sulfur chemically.

Stimulant. Something that produces energy.

Styptic. Helps control bleeding by contracting the tissues or blood vessels.

Sudorfic. Produces profuse and visible sweating when taken hot.

Tincture. A diluted alcohol solution of plant parts.

Tonic. Invigorates or stimulates, producing a feeling of well-being or strength.

Topical. Local application.

Vasoconstrictor. An agent that causes the blood vessels to constrict.

Vasodialator. An agent that causes the blood vessels to dilate.

Vermifuge. Expels worms or other parasites from the body.

Volatile. Readily vaporizes at low temperatures.

Vulnerary. Used in the healing of wounds.

BOTANICAL TERMS

Achene. A small, dry, hard, one-celled, one-seeded fruit that is indehiscent.

Acute. Sharp or pointed.

Alpine. Occurring above the tree line in the mountains.

Alternate. Arranged with one structure (for example, leaf, flowers, stem, and so on) per node.

Angiosperm. A plant producing flowers and bearing ovules (seeds) in an ovary (fruit).

Annual. A plant, usually with a slender taproot, completing its life cycle in a single growing season.

Anther. The pollen-bearing portion of the stamen.

Aquatic. Growing in water.

Areole. Small, defined area on a surface of a cactus that bears the spines.

Aromatic. Having a strong, usually agreeable odor.

Asexual. Without sex.

Awn. A slender, bristlelike organ.

Axil. The upper angle formed by a leaf or branch with the stem.

Banner. The upper petal of a papilionaceous flower.

Basal. At the base.

Beak. A prolonged, slender, and tapering projection.

Berry. A fleshy or pulpy fruit developed from a single ovary with more than one seed, such as a grape or blueberry.

Biennial. A plant completing its life cycle in two growing seasons, usually forming a basal rosette the first season and flowering the second.

Borne. Produced or arising from.

Bract. A leaf subtending a flower or flower cluster.

Bulb. An underground organ constituted mostly of fleshy storage leaves and a covered scale (for example, an onion).

Calyx. A flower's sepals considered as a unit.

Campanulate. Bell shaped.

Capillary. Hair- or threadlike.

Capsule. A dry, dehiscent fruit composed of more than one carpel.

Carpel. A modified leaf forming the ovary.

Catkin. In plants such as willows, birches, and alders, the elongated, pendulous, or conelike flower cluster with minute flowers that lacks, or almost lacks, the petals and sepals.

Caudex. The thickened base of some perennial herbs.

Clasping. Partly surrounding the stem.

Clavate. Club shaped.

Claw. The narrow or stalklike base of some petals.

Coma. The tufts of hairs at the ends of some seeds.

Compound Leaf. A leaf that is divided into two or more distinct leaflets.

Cone. A dense cluster of modified, leaflike organs bearing pollen, spores, or seeds, as in horsetails, club mosses, and conifers (for example, a pinecone).

Conifer. A cone-bearing tree.

Coniferous. Having cones or strobili.

Corm. A bulblike underground thickening of the stem.

Corolla. The petals considered as a unit, usually brightly colored.

Corymb. A Convex or flat-topped flower cluster of the racemose type, with the pedicels arising from a different point on the axis.

Cruciform. Cross shaped (for example, the position of petals in the mustard family. Brassicaceae).

Cyathium. The inflorescence in the genus *Euphorbia* (family Euphorbiaceae), that consists of unisexual flowers crowded within a cuplike involucre.

Cyme. A flower cluster, often flat-topped or convex, in which the central or terminal flower blooms the earliest.

Deciduous. Falling off once a year, usually at the end of a growing season.

Dehiscent. Opening to emit the contents.

Dichotomous. Forking regularly by pairs.

Discoid. A flowering head without ray flowers.

Disk. In the Sunflower Family (Asteraceae), the central portion of the head that gives rise to the disk flowers.

Disk flowers. In the Sunflower Family (Asteraceae), the flowers with slender, tubular corollas at the central part of the head.

Divided. Cut or lobed to the base or to a midrib.

Drupe. A fleshy, one-seeded fruit (for example, a cherry).

Emergent. With the lower portion in water and the upper portion extending out.

Endemic. Found only within a limited geographic area.

Entire. With margins not cut, cleft, or otherwise toothed.

Ephemeral. Lasting for only a short time.

Evergreen. Retaining leaves through the winter.

Exotic. Not native but introduced from somewhere else.

Family. A group of related genera.

Fascicle. A small cluster of leaves, flowers, and so on.

Flower. The reproductive portion of the plant consisting of stamens, pistils, or both and including the petals, sepals, or both.

Foliage. The leaves of the plant, collectively.

Follicle. A fruit consisting of a single carpel, dehiscing by the ventral suture.

Forb. A nongrasslike herbaceous plant.

Frond. The leaf of a fern.

Fruit. The mature ovary that includes the attached external structures and enclosed seeds.

Genus. A grouping of related species, the plural of which is *genera*.

Glabrous. Devoid of hairs.

Gland. A secreting cell or group of cells.

Glandular. Having glands, usually hairs.

Glaucous. Covered or whitened with a bloom.

Glochid. A barbed hair or bristle (for example, as in the fine hairs of *Opuntia*).

Gymnosperm. A member of the plant group that is characterized as having ovules not enclosed in an ovary (for example, pines, spruces, firs, and junipers).

Habit. The general appearance or growth form of a plant.

Habitat. The environmental conditions or kind of place in which a plant grows.

Head. A type of inflorescence with mostly sessile flowers densely set on a very short axis or disk, thereby having a round outline; the terminal collection of flowers is surrounded by an involucre, as in the Sunflower Family (Asteraceae).

Herb. A plant with the aerial portion being nonwoody, dying back to the ground at the end of the growing season.

Herbaceous. Not woody, dying back at the end of the growing season.

Host. The plants from which the parasite obtains nutrients (for example, mistletoe).

Immersed. Growing under water.

Indehiscent. Not splitting open.

Inferior. Lower or below.

Inferior ovary. An ovary positioned below the base of other flower parts.

Inflated. Turgid and bladdery.

Inflorescence. The flowering part of plants or the arrangement of flowers.

Inner bark. The cambium layer.

Involucre. A whorl of bracts subtending a flower or flower cluster.

Irregular. A flower where one or more of the organs are unlike the rest.

Latex. The milky sap of certain plants.

Leaflet. One of the divisions of a compound leaf.

Legume. A simple, dry fruit that is dehiscent along both sutures.

Ligulate flower. The same as a ray flower in the Sunflower Family.

Linear. Long and narrow.

Many. For botanical purposes, numbering more than ten.

Meal. Pertaining to flour.

-merous. Parted, having sections—a 5-merous flower has five petals and five sepals.

Montane. Of or pertaining to the mountains.

Naturalized. Plants introduced from somewhere else and now established.

Node. A joint or point of origin for leaves or branches.

Numerous. In botanical terms, more than ten.

Nut. A dry, hard-walled, and indehiscent fruit, usually with one seed.

Nutlet. A small nut.

Opposite. Nodes having two leaves or branches directly across from each other.

Orbicular. Circular in outline.

Pappus. In the Sunflower Family, the highly modified calyx composed of scales, bristles, awns, or short crown at the tip of the achene.

Parasitic. Growing on and deriving nourishment from another living plant.

Parted. Cleft almost to the base.

Pedicel. A stalk of a single flower.

Perennial. Plant with the potential to live more than two years.

Perianth. The corolla and calyx considered collectively.

Petal. A member of the whorl of floral organs, just interior to the sepals and below the stamens.

Petaloid. Brightly colored and petal-like.

Petiole. A leaf stalk.

Pinnate. Having a main central axis with secondary branches or units arranged in two lines on either side of the central axis.

Pinole. The flour from various seeds and grains mixed together.

Pistil. An organ formed from the combination of the stigma, style, and ovary.

Plumose. Feathery and soft.

Pod. Any kind of dry, dehiscent fruit, particularly in the Pea Family (Fabaceae).

Pollen. Dustlike cells produced in the anther.

Pollinium. A mass of waxy or coherent pollen grains (for example, notably in the Asclepiadaceae and Orchidaceae Families).

Potherb. Herb that is boiled and eaten as a vegetable.

Prostrate. Lying flat on the ground.

Pubescent. Covered with hairs.

Raceme. Inflorescence with one main axis and subequal primary branches, each bearing one flower.

Ray flower. In the Sunflower Family (Asteraceae), the straplike flowers attached to the disk.

Receptacle. The end of a flower stalk that bears the floral organs.

Regular. Having members of each part alike in size and shape.

Rhizomatous. Possessing rhizomes.

Rhizome. A creeping underground, usually horizontally oriented stem.

Root. The underground part of a plant.

Rootstalk. An underground, creeping stem.

Rosette. A dense, usually basal cluster of leaves radiating in all directions from the stem.

Sagittate. Arrowhead shaped.

Salverform. A corolla with a long, slender tube, abruptly flaring into a circular limb.

Samara. A winged fruit that does not split at maturity.

Saprophyte. A plant with little or no chlorophyll that obtains nutrients from dead organic matter by a root association with a fungus.

Schizocarp. A dry, indehiscent fruit that splits into separate one-seeded segments at maturity.

Seed. A mature ovule that following germination gives rise to a new plant.

Seed cone. The female seed-producing cone of conifers.

Sepal. One of a whorl of typically green or greenish, leaflike, floral organs originating below the petals.

Septum. A partition that seperates the locules of an ovary.

Serrate. Sharply toothed edges.

Sessile. Lacking a stalk.

Shoot. A young stem or branch.

Shrub. A woody plant, sometimes only at the base, generally with several stems originating from the base.

Silicle. A dry, dehiscent fruit of the Mustard Family, typically less than twice as long as wide.

Silique. A dry, dehiscent fruit of the Mustard Family, typically more than twice as long as wide.

Spatulate. Spatula shaped; having a long, narrow base and a widened, roundish tip.

Spore. A single cell or a small group of undifferentiated cells, each capable of producing a plant.

Spur. A hollow, slender, saclike extension of some part of the flower (for example, the sepal in *Delphinium* or the petal of *Viola*).

Stamen. The male organ of a flower that produces pollen, composed of an anther and filament.

Stellate. Star shaped.

Stem. The main axis of a plant.

Stipules. Appendages on each side of the base of certain leaves.

Subalpine. Growing in the mountains below the alpine zone and above the montane zone.

Succulent. Thick, fleshy, and juicy.

Superior. Above.

Superior ovary. Ovary positioned above the base of the other flower parts.

Talus. Slope of rock rubble, usually at a cliff base.

Taproot. The primary plant root that is considerably larger than any other root-system branches.

Tepal. The perianth part when the perianth is not clearly differentiated into calyx or corolla.

Terrestrial. Growing on ground, not aquatic.

Three ranked. Originating in threes from a common point or level.

Tree. A large woody plant with a single main stem or trunk.

Tuber. A swollen underground stem tip (for example, a potato).

Tubular. In the form of a tube or cylinder.

Umbel. A flower arrangement resembling an "umbrella.."

Urn shaped. Ovoid, with a small opening at the tip.

Vascular plant. A plant having vascular tissue.

Vegetative. The portion of the plant not producing reproductive structures like cones or flowers.

Villous. Sticky.

Weed. An aggressive plant that colonizes disturbed habitats and cultivated lands; a plant out of place.

Whorled. Three or more similar structures (for example, leaves, petals, and bracts) encircling a node.

Woolly. Having soft, curled, or entangled hairs.

References and Additional Reading

Altschul, S. 1973. *Drugs and Food from Little-Known Plants*. Cambridge: Harvard University Press.

Anderson, J. P. 1939. "Plants Used by the Eskimo of the Northern Bering Sea and Arctic Regions of Alaska." *American Journal of Botany* 26 (9): 714–16.

Angier, B. 1966. *Free for the Eating*. Harrisburg, Pa.: Stackpole Books.

——. 1969a. *Feasting Free on Wild Edibles*. Harrisburg, Pa.: Stackpole Books.

——. 1969b. *More Free for the Eating Wild Foods*. Harrisburg, Pa.: Stackpole Books.

——. 1974. *Field Guide to Edible Wild Plants*. Harrisburg, Pa.: Stackpole Books.

Arnason, T., R. J. Hebda, and T. Johns. 1981. "Use of Plants for Food and Medicine by Native Peoples of Eastern Canada." *Canadian Journal of Botany* 59 (11): 2189–325.

Ashton, R. J., and R. D. Walmsley. 1976. "The Aquatic Fern *Azolla* and Its *Anabaena* symbiont." *Endeavour* 35: 39–43.

Avery, A. G., S. Satina, and J. Rietsema. 1959. *Blakeslee: The Genus Datura*. New York: Ronald Press.

Bacon, A. E. 1903. "An Experiment with the Fruit of Red Baneberry." *Rhodora* 5: 77–79.

Bailey, F. L. 1940. "Navajo Foods and Cooking Methods." *American Anthropologist* 42 (2): 270–90.

Baker, M. A. 1981. "The Ethnobotany of the Yurok, Tolowa, and Karok Indians of Northwest California." M.A. thesis, Humboldt State University.

Balls, E. K. 1970. *Early Uses of California Plants*. Berkeley and Los Angeles: University of California Press.

Bank, T. P., II. 1951. "Botanical and Ethnobotanical Studies in the Aleutian Islands: I. Aleutian Vegetation and Aleut Culture." *Botanical and Ethnobotanical Studies Papers: Michigan Academy of Science, Arts, and Letters* 37: 13–30.

Barbour, M., B. Pavlik, F. Drysdale, and S. Lindstrom. 1993. *California's Changing Landscape: Diversity and Conservation of California Vegetation*. Sacramento: California Native Plant Society.

Barbour, M. G., and J. Major, eds. [1977] 1988. *Terrestrial Vegetation of California*. New expanded ed. Sacramento: California Native Plant Society.

Barrett, S. A. 1908. "Pomo Indian Basketry." *University of California Publications in American Archaeology and Ethnology* 7: 134–308.

——. 1917. "The Washoe Indians." *Bulletin of the Public Museum of the City of Milwaukee* 2 (1): 1–52.

——. 1952. "Material Aspects of Pomo Culture." *Bulletin of the Public Museum of the City of Milwaukee*, no. 20.

Barrett, S. A., and E. W. Gifford. 1933. *Miwok Material Culture*. Yosemite National Park: Yosemite Natural History Association.

Barrows, D. P. [1900] 1967. *The Ethno-botany of the Coahuilla Indians of Southern California*. Banning, Calif.: Malki Museum Press.

Bartlett, K. 1943. "Edible Wild Plants of Northern Arizona." *Plateau* (Northern Arizona Society of Science and Art Museum of Northern Arizona, Flagstaff) 16 (1): 11–17.

Baumhoff, M. A. 1963. "Ecological Determinants of Aboriginal California Populations." *University of California Publications in American Archaeology and Ethnology* 49 (2): 155–236.

Bean, L. J., and K. S. Saubel. 1972. *Temalpakh-Cahuilla Indian Knowledge and Usage of Plants*. Morongo Indian Reservation, Calif.: Malki Museum Press.

Bomhard, M. L. 1936. "Leaf Venation as a Means of Distinguishing *Cicuta* from *Angelica.*" *Journal of the Washington Academy of Sciences* 26 (3): 102–7.

Boxer, A. 1974. *Nature's Harvest.* Chicago: Henry Regnery.

Brill, S. 1994. *Identifying and Harvesting Edible and Medicinal Plants in Wild (and Not So Wild) Places.* New York: Hearst Books.

Brown, T. 1985. *Tom Brown's Guide to Wild Edible and Medicinal Plants.* New York: Berkeley Books.

Bryan, N. G., and S. Young. 1940. *Navajo Dyes: Their Preparation and Use.* Chilocco, Okla.: Education Division Publication, U.S. Office of Indian Affairs.

Burgess, R. L. 1966. "Utilization of Desert Plants by Native People." Contribution Committee on Desert and Arid Zone Research. *Bulletin of the American Association for the Advancement of Science* 8: 6–21.

Callegari, J., and K. Durand. 1977. *Wild Edible and Medicinal Plants of California.* El Cerrito, Calif.: by the authors.

Camazine, S., and R. A. Bye. 1980. "A Study of the Medical Ethnobotany of the Zuni Indians of New Mexico." *Journal of Ethnopharmacology* 2: 365–88.

Castetter, E. F. 1935. *Uncultivated Native Plants Used as Sources of Food.* Ethnobiological Studies in the American Southwest. University of New Mexico Bulletin no. 266. Biological Series, vol. 4, no. 1. Albuquerque: University of New Mexico Press.

———. 1944. "The Domain of Ethnobiology." *American Naturalist* 78: 158–70.

Chamberlain, L. S. 1901. "Plants Used by the Indians of Eastern North America." *American Naturalist* 35: 1–10.

Chamberlain, R. V. 1911. "The Ethno-botany of the Gosiute Indians of Utah." *Memoirs of the American Anthropological Association* 2 (5): 331–405.

Chatfield, K. 1997. *Medicine from the Mountains: Medicinal Plants of the Sierra Nevada.* South Lake Tahoe, Calif.: Range of Light Press.

Chestnut, V. K. 1902. "Plants Used by the Indians of Mendocino County, California." *Contributions to the U.S. National Herbarium* 7: 295–408.

Clarke, C. B. 1977. *Edible and Useful Plants of California.* Berkeley and Los Angeles: University of California Press.

Classen, P. W. 1919. "A Possible New Source of Food Supply (Cat-tail Flour)." *Scientific Monthly* 9: 179–85.

Clawson, A. B. 1933. *Alpine Kalmia* (Kalmia microphylla) *as a Stock Poisoning Plant.* USDA Technical Bulletin 391. Washington, D.C.: U.S. Department of Agriculture.

Coffey, T. 1993. *The History and Folklore of North American Wildflowers.* New York: Facts on File.

Coon, N. 1974. *The Dictionary of Useful Plants.* Emmaus, Pa.: Rodale Press.

Cotton, C. M. 1996. *Ethnobotany: Principles and Applications.* New York: John Wiley and Sons.

Coville, F. V. 1897. "Notes on the Plants Used by the Klammath Indians of Oregon." *Contributions to the U.S. National Herbarium* 5 (2).

———. 1904. "Desert Plants as a Source of Drinking Water." In *Smithsonian Institution Annual Report*, 499–505. Washington, D.C.: Smithsonian Institution.

Craighead, J. J., F. C. Craighead, and R. J. Davis. 1963. *A Field Guide to the Rocky Mountain Wildflowers.* Boston: Houghton Mifflin.

Culley, D. D., and E. A. Epps. 1973. "Use of Duckweed for Waste Treatment and Animal Feed." *Journal of Water Pollution Control Federation* 45: 337–47.

Culpeper, N. 1972. *English Physician and Complete Herbal.* Arranged for use as a first aid herbal by C. F. Leyel. North Hollywood: Wilshire Book.

Curtin, L. S. M. 1957. "Some Plants Used by the Yuki Indians . . . : I. Historical Review and Medicinal Plants." *Masterkey* 31: 40–48.

Dall, W. H. 1868. *Useful Indigenous Alaskan Plants.* Report of the Department of Agriculture. Washington, D.C: U.S. Department of Agriculture.

Darlington, W. 1859. *American Weeds and Useful Plants.* New York: A. O. Moore.

Davidson, J. 1919. "Douglas-fir Sugar." *Canadian Field Naturalist* 33 (1): 6–9.

Dawson, R. 1985. *Nature Bound.* Boise: Omnigraphics.

Densmore, F. 1974. *How Indians Use Wild Plants for Food, Medicine, and Crafts.* New York: Dover Publications.

Doebley, J. F. 1984. "'Seeds' of Wild Grasses: A Major Food of Southwestern Indians." *Economic Botany* 38 (1): 52–64.

Douglas, J. S. 1978. *Alternative Foods: A World Guide to Lesser-Known Edible Plants.* London: Pelham Books.

Duke, J. A. 1992a. *Handbook of Medicinal Plants.* Boca Raton, Fla.: CRC Press.

——. 1992b. *Handbook of Phytochemical Constituents of GRAS Herbs and Other Economic Plants.* Boca Raton, Fla.: CRC Press.

Dunmire, W. W., and G. D. Tierney. 1997. *Wild Plants and Native Peoples of the Four Corners.* Santa Fe: Museum of New Mexico Press.

Ebeling, W. 1986. *Handbook of Indian Foods and Fibers of Arid America.* Berkeley and Los Angeles: University of California Press.

Elias, T. S., and P. A. Dykeman. 1982. *A Field Guide to North American Edible Wild Plants.* New York: Outdoor Life Books.

Elliott, D. B. 1976. *Roots: An Underground Botany and Foragers Guide.* Old Greenwich, Conn.: Chatham Press.

——. 1995. *Wild Roots: A Forager's Guide to the Edible and Medicinal Roots, Tubers, Corms, and Rhizomes of North America.* Rochester, Vt.: Healing Arts Press.

Ellis, C. 1941. "Wild Vegetables of the Desert Indians." *Primitive Man,* no. 3: 9–10.

Elmore, F. H. 1944. *Ethnobotany of the Navajo.* Santa Fe, N.M.: School of American Research.

Erichsen-Brown, C. 1979. *Medicinal and Other Uses of North American Plants.* New York: Dover Publications.

Farris, G. 1980. "A Re-assessment of the Nutritional Value of *Pinus monophylla.*" *Journal of California and Great Basin Anthropology* 2 (1): 132–36.

FEB. 1993. "The Foundation for Ethnobotany." Media Launch and Symposium, February 2.

Fernald, M., and A. C. Kinsey. 1958. *Edible Wild Plants of Eastern North America.* Revised by R. C. Rollins. New York: Harper and Row.

Fisher, V. 1965. *Mountain Man: A Novel of Male and Female in the Early American West.* New York: Simon and Schuster.

Ford, R. I. 1978. "Ethnobotany: Historical Diversity and Synthesis." In *The Nature and Status of Ethnobotany,* ed. R. I. Ford, 33–50. Anthropological Papers no. 67. Ann Arbor: University of Michigan Museum of Anthropology.

Foster, S., and J. A. Duke. 1990. *Eastern/Central Medicinal Plants.* Peterson Field Guides. Boston: Houghton Mifflin.

Fowler, C. S. 1989. *Willard Z. Park's Ethnographic Notes on the Northern Paiute of Western Nevada, 1933–1940.* Salt Lake City: University of Utah Press.

Frankton, C., and G. A. Mulligan. 1987. *Weeds of Canada.* Ottawa: N.C. Press Limited and Agricultural.

Frye, T. C. 1934. *Ferns of the Northwest.* Portland, Oreg.: Metropolitan Press.

Gail, F. W. 1916. *Some Poisonous Plants of Idaho.* Bulletin 86. Moscow: University of Idaho Agricultural Experiment Station.

Gibbons, E. 1962. *Stalking the Wild Asparagus.* New York: David McKay.

——. 1966. *Stalking the Healthful Herbs.* New York: David McKay.

——. 1971. *Stalking the Good Life.* New York: David McKay.

Gifford, E. W. 1967. "Ethnographic Notes on the Southwestern Pomo." *Anthropological Records* 25: 10–15.

Gilmore, M. R. 1932. "Importance of Ethnobotanical Investigation." *American Anthropologist* 34: 320–27.

Goodrich, J., and C. Lawson. 1980. *Kashaya Pomo Plants.* Los Angeles: University of California American Indian Studies Center.

Gottesfeld, L. M. J. 1992. "The Importance of Bark Products in the Aboriginal Economies of Northwestern British Columbia, Canada." *Economic Botany* 46 (2): 148–57.

Grant, A. L. 1924. "A Monograph of the Genus *Mimulus.*" Ph.D. diss., Washington University.

Grillos, S. J. 1966. *Ferns and Fern Allies of California.* Berkeley and Los Angeles: University of California Press.

Gunther, E. 1973. *Ethnobotany of Western Washington.* Rev. ed. Seattle: University of Washington Press.

Hall, A. 1976. *The Wild Food Trail Guide.* New York: Holt, Rinehart, and Winston.

Hamel, P. B., and M. U. Chiltoskey. 1975. *Cherokee Plants and Their Uses—a 400 Year History.* Sylva, N.C.: Herald Publishing.

Hardin, J. W., and J. M. Arena. 1974. *Human Poisoning from Native and Cultivated Plants*. Durham: Duke University Press.

Harrington, H. D. 1967. *Edible Native Plants of the Rocky Mountains*. Albuquerque: University of New Mexico Press.

Harshberger, J. W. 1896. "The Purposes of Ethnobotany." *Botanical Gazette* 21: 146–54.

Hart, J. A. 1981. "The Ethnobotany of the Northern Cheyenne Indians of Montana." *Journal of Ethnopharmacology* 4: 1–55.

——. 1996. *Montana—Native Plants and Early Peoples*. Helena: Montana Historical Society.

Harvard, V. 1895. "The Food Plants of North American Indians." *Bulletin of Torrey Botanical Club* 22: 98–123.

Haskin, L. L. 1929. "A Frontier Food, Ipo or Yampa, Sustained the Pioneers." *Nature* 14: 171–72.

Hedges, K. 1986. *Santa Ysabel Ethnobotany*. Ethnic Technology Notes no. 20. San Diego: San Diego Museum of Man.

Hellar, C. A. 1958. *Wild Edible and Poisonous Plants of Alaska*. Bulletin 40. Fairbanks: University of Alaska Extension.

Hellson, J. C. 1974. *Ethnobotany of the Blackfoot Indians*. Ottowa: National Museums of Canada.

Herrick, J. W. 1977. "Iroquois Medical Botany." Ph.D. diss., State University of New York at Albany.

Hickman, J. C., ed. 1993. *The Jepson Manual: Higher Plants of California*. Berkeley and Los Angeles: University of California Press.

Hill, A. F. 1937. *Economic Botany: A Textbook of Useful Plants and Plant Products*. New York: McGraw-Hill.

Hinton, L. 1975. "Notes on La Huerta Diegueno Ethnobotany." *Journal of California Anthropology* 2: 214–22.

Hocking, G. M. 1949. "From Pokeroot to Penicillin." *Rocky Mountain Druggist* (November): 12, 38.

Holloway, P., and G. Alexander. 1990. "Ethnobotany of the Fort Yukon Region, Alaska." *Economic Botany* 44: 214–25.

Holt, C. 1946. "Shasta Ethnography." *Anthropological Records* 3 (4): 308.

Hough, W. 1897. "The Hopi in Relation to Their Plant Environment." *American Anthropologist* 10 (2): 33–44.

Hussey, P. B. 1939. *A Taxonomic List of Some Plants of Economic Importance*. Lancaster, Pa.: Science Press Printing.

Jacobson, C. A. 1915. *Water Hemlock (Cicuta)*. Technical Bulletin 81. Reno: Nevada Agricultural Experiment Station.

Jaynes, R. A. 1997. *Kalmia Mountain Laurel and Related Species*. Portland, Oreg.: Timber Press.

Jencks, Z. 1919. "A Note on the Carbohydrates of the Root of the Cattail (*Typha latifolia*)." *Proceedings of the Society for Experimental Biology and Medicine* 17 (2): 45–46.

Johnston, A. 1970. "Blackfoot Indian Utilization of the Flora of the Northwestern Great Plains." *Economic Botany* 24: 301–24.

Johnston, V. R. 1994. *California Forests and Woodlands: A Natural History*. California Natural History Guides no. 58. Berkeley and Los Angeles: University of California Press.

Kartesz, J. 1994. *A Synonymized Checklist of the Vascular Flora of the United States, Canada, and Greenland*. Vols. 1–2. Portland, Oreg.: Timber Press.

Kavash, B. 1979. *Native Harvests: Recipes and Botanicals of the American Indian*. New York: Vintage Books.

Kelly, I. 1932. "Ethnography of the Surprise Valley Paiute." *University of California Publications in American Archaeology and Ethnology* 31 (3): 67–210.

Kephart, H. 1909 and later editions. "Edible Plants of the Wilderness." Chap. 17 of *The Book of Camping and Woodcraft*. New York: Century.

Kindscher, K. 1987. *Edible Wild Plants of the Prairie*. Lawrence: University Press of Kansas.

Kinghorn, D. 1979. *Toxic Plants*. New York: Columbia University Press.

Kingsbury, J. M. 1964. *Poisonous Plants of the United States and Canada*. Englewood Cliffs, N.J.: Prentice-Hall.

——. 1965. *Deadly Harvest: A Guide to Common Poisonous Plants.* New York: Holt, Rinehart, and Winston.

Kirk, D. R. 1975. *Wild Edible Plants of the Western United States.* Heraldsburg, Calif.: Naturegraph.

Knap, A. H. 1975. *Wild Harvest: An Outdoorsman's Guide to Edible Wild Plants in North America.* Toronto: Pagurian Press.

Krochmal, A., and C. Krochmal. 1973. *A Guide to the Medicinal Plants of the United States.* New York: Quadrangle.

Krochmal, A., S. Paur, and P. Duisberg. 1951. "Useful Native Plants in the American Deserts." *Economic Botany* 8 (1): 3–20.

Lands, M. 1959. "Folk Medicine and Hygiene." *Anthropological Papers of the University of Alaska* 8: 1–75.

Lee, D. 1989. *Exploring Nature's Uncultivated Garden.* Tacoma Park, Md.: Havelin Communications.

Le Strange, R. 1977. *A History of Herbal Plants.* New York: Arco Publishing.

Lewis, W. H., and M. Elvin-Lewis. 1977. *Medical Botany.* New York: John Wiley and Sons.

Life-Support Technology. 1963. *Foods in the Wilderness.* N.p.: Life-Support Technology.

Linn, J. G., E. J. Sraba, R. D. Goodrich, J. C. Meiske, and D. E. Otterby. 1975. "Nutritive Value of Dried and Ensiled Aquatic Plants: I. Chemical Composition." *Journal of Animal Science* 41: 601–9.

Lust, J. B. 1987. *The Herb Book.* 20th ed. New York: Bantam Books.

Mabey, R. 1977. *Plantcraft: A Guide to Everyday Use of Wild Plants.* New York: Universe Books.

Mahar, J. M. 1953. "Ethnobotany of the Oregon Paiutes of the Warm Springs Indian Reservation." Bachelor's thesis, Reed College.

Martin, G. J. 1995. *Ethnobotany: A Conservation Manual.* London: Chapman and Hall.

Martin, L. C. 1984. *Wildflower Folklore.* Chester, Conn.: Globe Pequot Press.

McHarg, I. L. 1969. *Design with Nature.* Garden City, N.Y.: Doubleday.

Medsger, O. P. 1974. *Edible Wild Plants.* New York: Collier-Macmillan Publishers.

Merriam, C. H. 1918. "The Acorn, a Possibly Neglected Source of Food." *National Geographic* 34 (2): 129–37.

——. 1966. *Ethnographic Notes on California Indian Tribes.* Berkeley: University of California Archaeological Research Facility.

Merrill, R. E. 1923. "Plants Used in Basketry by the California Indians." UC-PAAE 20: 215–42.

Meuninck, J. 1988. *The Basic Essentials of Edible Wild Plants and Useful Herbs.* Merrillville, Ind.: ICS Books.

Miller, J. A. 1973. "Naturally Occurring Substances That Can Induce Tumors." In *Toxicants Occurring Naturally in Foods.* Washington, D.C.: National Academy of Sciences.

Millspaugh, C. F. 1974. *American Medicinal Plants: An Illustrated and Descriptive Guide to Plants Indigenous to and Naturalized in the United States Which Are Used in Medicine.* New York: Dover Publications.

Mitchell, R. S., and J. K. Dean. 1982. "Ranunculaceae (Crowfoot Family) of New York State." *New York State Museum Bulletin,* no. 446: 1–100.

Moerman, D. E. 1977. *American Medical Ethnobotany: A Reference Dictionary.* New York: Garland Publishing.

——. 1986. *Medicinal Plants of the Native Americans.* Technical Report no. 19. Ann Arbor: University of Michigan Museum of Anthropology.

Moore, M. 1979. *Medicinal Plants of the Mountain West.* Santa Fe: Museum of New Mexico Press.

——. 1989. *Medicinal Plants of the Desert and Canyon West.* Santa Fe: Museum of New Mexico Press.

Morton, J. 1963. "Principal Wild Food Plants of the United States, Excluding Alaska and Hawaii." *Economic Botany* 17: 319–30.

——. 1975. "Cattails (*Typha* spp.)—Weed Problem or Potential Crop?" *Econominc Botany* 29 (1): 7–29.

Moulton, G. E. 1983–1999. *The Journals of the Lewis and Clark Expedition.* Vols. 1–12. Lincoln: University of Nebraska Press.

Muenscher, W. C. 1962. *Poisonous Plants of the United States.* New York: Macmillan.

Muir, J. 1894. *The Mountains of California.* New York: Century.

———. 1901. *Our National Parks.* Boston: Houghton Mifflin.

———. 1911. *My First Summer in the Sierra.* Boston: Houghton Mifflin.

———. 1912. *The Yosemite.* New York: Century.

———. 1918. *Steep Trails.* Boston: Houghton Mifflin.

———. 1938 [1979]. *John of the Mountains.* Madison: University of Wisconsin Press.

———. 1950. *Studies in the Sierra.* San Francisco: Sierra Club Books.

Munz, P. A. 2003. *Introduction to California Mountain Wildflowers.* Rev. ed. California Natural History Guides no. 68. Berkeley and Los Angeles: University of California Press.

Munz, P. A., and D. D. Keck. 1949. "California Plant Communities." *El Aliso* 2(1): 87-105.

———. 1973. *A California Flora and Supplement.* Berkeley and Los Angeles: University of California Press.

Murphey, E. V. A. 1990. *Indian Uses of Native Plants.* Glenwood, Ill.: Meyerbooks.

National Research Council. 1985. *Amaranth: Modern Prospects for an Ancient Crop.* Emmaus, Pa.: Rodale Press.

Nelson, R. A. 1992. *Handbook of Rocky Mountain Plants.* 4th ed. Niwot, Colo.: Roberts Rinehart Publishers.

Nequakewa, E. 1943. "Some Hopi Recipes for the Preparation of Wild Plant Foods." *Plateau,* no. 18: 18–20.

Newberry, J. S. 1887. "Food and Fiber Plants of the North American Indians." *Popular Scientific Monthly* 32: 31–46.

Niehaus, T. F. 1974. *Sierra Wildflowers: Mt. Lassen to Kern Canyon.* Berkeley and Los Angeles: University of California Press.

Norton, C. 1942. "Would You Starve?" *Nature* 35 (6): 295–97.

Norton, H. 1981. "The Association between Anthropogenic Prairies and Important Food Plants in Western Washington." *Northwest Anthropological Research Notes* 13: 175–200.

Ohiyesa [Charles Alexander Eastman]. 1911. *The Soul of an Indian: An Interpretation.* N.p.

Olsen, L. D. 1990. *Outdoor Survival Skills.* Provo, Utah: Brigham Young University Press.

Oswalt, W. H. 1957. "A Western Eskimo Ethnobotany." *Anthropological Papers of the University of Alaska* 6: 17–36.

Palmer, E. 1878. "Plants Used by the Indians of the United States." Parts 1 and 2. *American Naturalist* 12 (September): 593–606; (October): 646–55.

Parsons, M. E. 1966. *The Wild Flowers of California.* 3d ed. New York: Dover Publications.

Perry, E. 1952. "Ethno-botany of the Indians in the Interior of British Columbia." *Museum and Art Notes* 2 (2): 36–43.

Peterson, L. A. 1978. *A Field Guide to Edible Wild Plants of Eastern and Central North America.* Peterson Field Guides. Boston: Houghton Mifflin.

Pfeiffer, N. E. 1922. "Monograph of the Isoetaceae." *Annuals Missouri Botanical Gardens* 9: 79–232.

Phillips, J. 1979. *Wild Edibles of Missouri.* Jefferson City, Mo.: Missouri Department of Conservation.

Plotkin, M. 1988. "The Outlook for New Agricultural and Industrial Products from the Tropics." In *Biodiversity,* ed. E. O. Wilson, 106–16. Washington, D.C.: National Academy Press.

Pojar, J., and A. MacKinnon, eds. 1994. *Plants of the Pacific Northwest Coast.* Washington, D.C.: Lone Pine Publishing.

Porsild, A. E. "Edible Plants of the Arctic." *Arctic* 6 (1): 15–34.

Powers, S. 1874. "Aboriginal Botany." *Proceedings of the California Academy of Science* 5: 373–79.

Price, L. W. 1981. *Mountains and Man.* Berkeley and Los Angeles: University of California Press.

Reagan, A. B. 1929. "Plants Used by the White Mountain Apache Indians of Arizona." *Wisconsin Archeologist* 8: 143–61.

———. 1934. "Various Uses of Plants by West Coast Indians." *Washington Historical Quarterly* 25: 133–37.

Risk, P. 1983. *Outdoor Safety and Survival.* New York: John Wiley.

Ritchie, G. A., ed. 1979. *New Agricultural Crops.* Boulder: Westview Press.

Romero, J. B. 1954. *The Botanical Lore of the California Indians.* New York: Vantage Press.

Saunders, C. F. 1976. *Edible and Useful Wild Plants of the United States and Canada.* New York: Dover Publications.

Sawyer, J. O., and T. Keeler-Wolf. 1995. *A Manual of California Vegetation.* Sacramento: California Native Plant Society.

Schery, R. W. 1972. *Plants for Man.* 2d ed. Englewood Cliffs, N.J.: Prentice-Hall.

Schoenherr, A. A. 1992. *A Natural History of California.* California Natural History Guide no. 56. Berkeley and Los Angeles: University of California Press.

Schofield, J. J. 1989. *Discovering Wild Plants: Alaska, Western Canada, the Northwest.* Seattle: Alaska Northwest Books.

Scully, V. 1970. *A Treasury of American Indian Herbs: Their Lore and Their Use for Food, Drugs, and Medicine.* New York: Crown.

Smith, C. E. 1973. *Man and His Foods: Studies in the Ethnobotany and Nutrition— Contemporary, Primitive, and Prehistoric Non-European Diets.* Tuscaloosa: University of Alabama Press.

Smith, G., ed. 2000. *Sierra East: Edge of the Great Basin.* California Natural History Guide no. 60. Berkeley and Los Angeles: University of California Press.

Smith, H. H. [1923] 1978. "Ethnobotany of the Menomini Indians." *Bulletin of the Public Museum of the City of Milwaukee* 4 (1).

Snow, C. R. 1935. "Vegetables of the Alaska Wilderness." *Alaska Sportsman* 1 (4): 6–8.

Sparkman, P. S. 1908. "The Culture of the Luiseiio Indians." *University of California Publications in American Archaeology and Ethnology* 8 (4): 187– 234.

Spellenberg, R. 1979. *The Audubon Society Field Guide to North American Wildflowers.* New York: Alfred A. Knopf.

Spier, L. 1930. "Klamath Ethnography." *University of California Publications in American Archaeology and Ethnology* 30: 1–338.

Steward, J. H. 1933. "Ethnography of the Owens Valley Paiute." *University of California Publications in American Archaeology and Ethnology* 33 (3): 233– 350.

Stewart, K. M. 1965. "Mohave Indian Gathering of Wild Plants." *Kiva* 31 (1): 46–53.

Storer, T. I., and R. L. Ursinger. 1963. *Sierra Nevada Natural History: An Illustrated Handbook.* Berkeley and Los Angeles: University of California Press.

Strike, S. S. 1994. *Ethnobotany of the California Indians.* Vol. 2, *Aboriginal Uses of California's Indigenous Plants.* Champaign, Ill.: Koeltz Scientific Books.

Stuart, J. D., and J. O. Sawyer. 2001. *Trees and Shrubs of California.* California Natural History Guides no. 62. Berkeley and Los Angeles: University of California Press.

Swartz, B. K., Jr. 1958. "A Study of Material Aspects of Northeastern Maidu Basketry." *Kroeber Anthropological Society Publications* 19: 67–84.

Sweet, M. 1976. *Common Edible and Useful Plants of the West.* Heraldsburg, Calif.: Naturegraph Publishers.

Taylor, S. J. Elk. 1994. *Eat the Weeds at Your Feet: An Edible Plant Guide of Sonoma County.* Citrus Heights, Calif.: Rose of Sharon Press.

Teit, J. A. 1930. *Ethnobotany of the Thompson Indians of British Columbia.* Washington, D.C.: U.S. Government Printing Office.

Terrell, E. E. 1977. *A Checklist of Names for 3,000 Vascular Plants of Economic Importance.* Handbook no. 505. Washington, D.C.: U.S. Department of Agriculture.

Thompson, S., and M. Thompson. 1972. *Wild Plant Foods of the Sierra.* Berkeley: Dragtooth Press.

———. 1977. *Huckleberry Country: Wild Food Plants of the Pacific Northwest.* Berkeley: Wilderness Press.

Tilford, G. L. 1993. *The Ecoherbalist's Fieldbook: Wildcrafting in the Mountain West.* Conner, Mont.: Mountain Weed Publishing.

———. 1997. *Edible and Medicinal Plants of the West.* Missoula, Mont.: Mountain Press Publishing.

Train, P., J. R. Henriches, and W. A. Archer. 1957. *Medicinal Uses of Plants by Indian Tribes of Nevada.* Lawrence, Mass.: Quarterman.

Truax, R. E., D. D. Culley, M. Griffith, W. A. Johnson, and J. P. Wood. 1972. "Duckweed for Chick Feed." *Louisiana Agriculture* 16 (1): 8–9.

Tull, D. 1987. *A Practical Guide to Edible and Useful Plants.* Austin: Texas Monthly Press.

Turner, N., and H. V. Kuhnlein. 1991. *Traditional Plant Foods of Canadian Indigenous Peoples.* Philadelphia: Gordon and Breach Science Publishers.

Turner, N. J., and A. F. Szczawinski. 1991. *Common Poisonous Plants and Mushrooms of North America.* Portland, Oreg.: Timber Press.

Turney-High, H. 1933. "Cooking Camass and Bitterroot." *Scientific Monthly* 36: 262–63.

Tyler, V. E. 1987. *The Honest Herbal.* Philadelphia: George F. Stickley.

Underhill, J. E. 1974. *Wild Berries of the Pacific Northwest.* Seattle: Superior Publishing.

Uphof, J. C. T. 1959. *Dictionary of Economic Plants.* N.p.: H. R. Engleman.

Usher, G. 1976. *A Dictionary of Plants Used by Man.* London: Constable.

Van Etten, C. H., R. W. Miller, I. A. Wolff, and Q. Jones. 1963. "Amino Acid Composition of Seeds from 200 Angiosperm Plant Species." *Journal of Agriculture and Food Chemistry* 11 (5): 399–410.

Vestal, P. A. 1952. *Ethnobotany of the Ramah Navaho.* Papers of the Peabody Museum of American Archaeology and Ethology. Vol. 40, no. 4. Cambridge: Harvard University.

Vizgirdas, R. 1999a. "Courting the Conifers: The Pines." *Wilderness Way* 4 (1).

———. 1999b. "The Fallacy of Plant Edibility Tests." *Wilderness Way* 5 (2).

———. 1999c. "Fireweed." *Wilderness Way* 4 (4).

———. 2000a. "Butterflies and Edible Plants." *Wilderness Way* 5 (3).

———. 2000b. "The Mustard Family." *Wilderness Way* 6 (1).

———. 2003a. *Useful Plants of Idaho.* Pocatello: Idaho State University Press.

———. 2003b. "Useful Plants of the Southern California Mountains." *San Bernardino County Museum Association Quarterly* 50 (2).

Von Schiller, J. C. F. [1803] 1962. *The Bride of Messina.* New York: F. Ungar.

Walker, M. 1984. *Harvesting the Northern Wild: A Guide to Traditional and Contemporary Uses of Edible Forest Plants of the Northwest Territories.* Yellowknife, Northwest Territories: n.p.

Weber, W. A. 1987. *Colorado Flora, Western Slope.* Boulder: University Press of Colorado.

———. 1990. *Colorado Flora, Eastern Slope.* Boulder: University Press of Colorado.

Webster, J. 1980. *Fungi.* 2d ed. Cambridge: Cambridge University Press.

Weedon, N. F. 1996. *A Sierra Nevada Flora.* 4th ed. Berkeley: Wilderness Press.

Weiner, M. A. 1972. *Earth Medicine–Earth Food: Plant Remedies, Drugs, and Natural Foods of the North American Indians.* New York: Macmillan.

Wherry, E. T. 1942. "Go Slow on Eating Fern Fiddleheads." *American Fern Journal* 32 (3): 108–9.

Whiting, A. F. 1939. *Ethnobotany of the Hopi.* Museum of Northern Arizona Bulletin no. 15. Flagstaff: Northern Arizona Society of Science and Art.

Whitney, S. 1979. *A Sierra Club Naturalists' Guide to the Sierra Nevada.* San Francisco: Sierra Club Books.

Whittlesey, R. 1985. *Familiar Friends: Northwest Plants.* Portland, Oreg.: Rose Press.

Wickens, G. E. 1990. "What Is Economic Botany?" *Economic Botany* 44: 12–28.

Wilford, W. R., J. P. Harrington, and B. Freire-Marreco. 1916. *Ethnobotany of the Tewa Indians.* Bureau of American Ethnology Bulletin 55. Washington D.C.: U.S. Government Printing Office.

Willard, T. 1992. *Edible and Medicinal Plants of the Rocky Mountains and Neighboring Territories.* Alberta, Canada: Wild Rose College of Natural Healing.

Wilson, E. O. 1989. *Biodiversity.* Washington, D.C.: National Academy Press.

Wyman, L. C., and S. K. Harris. 1941. *Navajo Indian Medical Ethnobotany.* Albuquerque: University of New Mexico Press.

Zigmond, M. L. 1981. *Kawaiisu Ethnobotany.* Salt Lake City: University of Utah Press.

Zwinger, A. H., and B. E. Willard. 1972. *Land above the Trees: A Guide to American Alpine Tundra.* New York: Harper and Row.

Index